MODERN HIGHER ALGEBRA

MODERN HIGHER ALGEBRA

By

A. ADRIAN ALBERT
Associate Professor of Mathematics
The University of Chicago

CAMBRIDGE
UNIVERSITY PRESS
1938

CAMBRIDGE
UNIVERSITY PRESS

University Printing House, Cambridge CB2 8BS, United Kingdom

Cambridge University Press is part of the University of Cambridge.

It furthers the University's mission by disseminating knowledge in the pursuit of education, learning and research at the highest international levels of excellence.

www.cambridge.org
Information on this title: www.cambridge.org/9781107544628

Published December 1937
Second impression August 1938
First paperback edition 2015

A catalogue record for this publication is available from the British Library

ISBN 978-1-107-54462-8 Paperback

PREFACE

During the present century modern abstract algebra has become more and more important as a tool for research not only in other branches of mathematics but even in other sciences. Many discoveries in abstract algebra itself have been made during the past ten years and the spirit of algebraic research has definitely tended toward more abstraction and rigor so as to obtain a theory of greatest possible generality. In particular the concepts of group, ring, integral domain, and field have been emphasized.

The notion of an abstract group is fundamental in all science, and it is certainly proper to begin our subject with this concept. Commutative additive groups are made into rings by assuming closure with respect to a second operation having some of the properties of ordinary multiplication. Integral domains and fields are rings restricted in special ways and may be thought of as respective generalizations of ordinary integers and rational numbers.

These fundamental concepts and their more elementary properties are the basis for modern algebra. They are certainly abstract notions but their ultimate absorption by the reader of modern algebra is absolutely necessary and the best place for them is at the beginning. This mode of presentation has not been used in the present textbooks on algebra in the English language but is the customary presentation in all of the more recent texts in foreign languages. We treat the concepts in our first two chapters and present what is basic not only for what follows in the text but for all modern algebra and algebraic number theory.

Our exposition continues in Chapters III, IV, and V with the theory of matrices with elements in a completely general field. Recent trends in algebraic investigation have made it important to know the extent of the validity of the classical theorems on matrices. It is no more difficult to carry out the proofs, where they are valid, for general fields instead of the classical case of subfields of the field of all complex numbers. But it is true that the proofs and results of the classical theory are not always valid. This is brought out clearly in Chapter V, where it is necessary to restrict the types of fields considered.

Essential clarifications in the arguments used in proving matrix theorems are obtained here by extensive use of elementary transformations. These transformations are familiar, except in name, to any reader who has had college training in the theory of determinants and make our proofs of a non-computational character and easy to understand. The author has also attempted to give as much as possible of the algebraic manipulative technique which he has used in his recent investigations in algebras and on Riemann matrices.

The final chapter on matrices presents a rather novel complete generalization of the theory of symmetric matrices which arises naturally in the algebraic geometric study of Riemann matrices. It is here that even the elementary theorems on symmetric matrices are not valid unless the fundamental assumption is made that the so-called characteristic of the field is not two.

The Galois theory is of great importance in algebra and the theory of algebraic numbers. The older treatments defined the Galois group of an equation with distinct roots as a certain group of permutations on these roots. This treatment is not very simple and the final theorems not in a very good form for certain algebraic applications. One may say that the essential trouble is that the Galois group is defined as a subgroup of the group of all permutations. The more modern treatment is that of the theory of the Galois group of a normal field. This is the set of all automorphisms of the given field and the fact that we take *all* automorphisms makes our proofs quite simple. We present this treatment in Chapters VI–IX. The first of these chapters gives all of the results needed from the theory of finite groups. Chapter VII gives the essentials of the theory of algebraic extensions of a given field, and in Chapter VIII the Galois theory of fields is given and applied to obtain as a consequence the Galois theory of equations. The final chapter of this set is an application of the theory to obtain structure theorems for the simplest type of algebraic extension, the cyclic field.

The theory of linear associative algebras is a fundamental, if quite advanced, branch of modern algebra. It is natural, however, to introduce this subject from the matrix point of view and we do so in Chapter X. Many quite abstract notions are made concrete by such a treatment, and a quite adequate introduction to the theory is made in this way without going at all deeply into the abstract structure theorems on algebras.

Our exposition closes with an introduction to the theory of p-adic numbers. This subject is best studied by considering the general theory of fields with a valuation, and we do here. The author has collected his material from a large number of sources and hopes that the present exposition is an adequate foundation of the theory. It is certainly true that no progress can be made in reading modern papers on algebraic numbers and their applications to the theory of algebras without a knowledge of p-adic number theory. It is equally true that it has heretofore been necessary to read a forbidding number of articles in order to get even a meager acquaintance with the theory.

The author has written the present text as a foundation for future exposition of the modern theory of algebraic numbers and class fields, and of the theory of linear associative algebras. He wishes to acknowledge as sources for much of the material L. E. Dickson's *Modern Algebraic Theories* and *Algebren und ihre Zahlentheorie*, A. Speiser's *Theorie der Gruppen von endlicher Ordnung*, B. L. van der Waerden's *Moderne Algebra*, J. H. M. Wedderburn's *Lectures on Matrices*, as well as a number of papers by C. Chevalley, H. Hasse, J. Kürschák, and A. Ostrowski. These latter were used as sources, particularly in the last two chapters. Final thanks are due to Mr. Sam Perlis, a graduate student at the University of Chicago, who has materially assisted in the preparation of the manuscript; to Drs. Daniel Dribin and Nathan Jacobson, who have read it critically; and to his colleagues of the Department of Mathematics of the University of Chicago, who have given very valuable advice on its preparation.

A. A. ALBERT

UNIVERSITY OF CHICAGO
October 1, 1936

TABLE OF CONTENTS

CHAPTER I

GROUPS AND RINGS

1. Introduction. The usual textbook on college algebra begins with a *review of fundamental operations.* This review is really a rather incomplete formulation of the postulates for the mathematical system studied in elementary algebra and is really not even a good description of the system.

Modern algebra has many applications and requires a consideration of certain general, rigorously defined mathematical systems. We shall define these systems abstractly and most of the resulting properties will be the usual ones of elementary algebra. Any bizarre properties obtained will be due either to the generality of the systems studied, to the fact that *less* is assumed about our systems than is assumed in elementary algebra, or to the fact that we later specialize our general systems so as to give number systems different from those ordinarily used in algebra.

We shall usually replace the systems of real or complex numbers used as coefficients in elementary algebra by more general systems called "fields." There will generally be a fundamental or basic field whose elements will act as coefficients in our discussions, and we shall usually designate this field by the letter $\mathfrak{F}$. Whenever we talk about sets of elements we shall use Gothic letters to designate these sets.

The theorems and equations occurring in the text will be used in subsequent sections, and references will be made to them. We shall number the equations and theorems in each chapter separately and, after the first chapter, refer to them by a number consisting of the chapter number and the number of the theorem or equation. Thus, for example, we shall refer to Theorem 5 of Chapter X as Theorem 10.5 and to equation (22) of Chapter VII by (7.22).

2. Sets. Our first abstract notion is the elementary one of a *set* or aggregate $\mathfrak{S}$ of undefined entities $a, b, \ldots$ called the *elements* or *quantities* of $\mathfrak{S}$. We shall sometimes specify these elements, but they will usually be any abstract entities. For example, $\mathfrak{S}$ may consist of all even integers, or all real numbers, or all functions of x, or all rotations of a line in a plane about a point on the line. However, they will simply be unspecified. This is desirable as we thereby obtain mathematical systems of great generality and results applicable to many special theories.

When the elements of a set $\mathfrak{H}$ are all elements of a set $\mathfrak{G}$ we call $\mathfrak{H}$ a *subset* of $\mathfrak{G}$, say that $\mathfrak{H}$ *is contained in* $\mathfrak{G}$, and write

$$\mathfrak{H} \leqq \mathfrak{G} .$$

For example, $\mathfrak{G}$ may be the set of all rational numbers, $\mathfrak{H}$ the set of all integers. We also say that $\mathfrak{G}$ *contains* $\mathfrak{H}$, and indicate this by writing

$$\mathfrak{G} \geqq \mathfrak{H} .$$

If $\mathfrak{G}$ contains $\mathfrak{H}$ and $\mathfrak{H}$ contains $\mathfrak{G}$, then $\mathfrak{G}$ and $\mathfrak{H}$ are identical, that is, *equal* sets, and we write

$$\mathfrak{G} = \mathfrak{H} .$$

But when $\mathfrak{G}$ contains $\mathfrak{H}$ and also elements not in $\mathfrak{H}$, we say that $\mathfrak{G}$ contains $\mathfrak{H}$ *properly* and write

$$\mathfrak{G} > \mathfrak{H} ,$$

or that $\mathfrak{H}$ is a proper subset of $\mathfrak{G}$,

$$\mathfrak{H} < \mathfrak{G} .$$

Let $\mathfrak{G}$ and $\mathfrak{K}$ be two sets and $\mathfrak{H}$ consist of all elements which are in common to $\mathfrak{G}$ and $\mathfrak{K}$. Then the set $\mathfrak{H}$ is called the *intersection* (or cross-cut) of $\mathfrak{G}$ and $\mathfrak{K}$. This concept will occur frequently.

A set with no elements in it is called an *empty* set. When $\mathfrak{G}$ has at least one element we call $\mathfrak{G}$ a *non-empty* set. A set consisting of one element and the element are logically distinct concepts, but we may identify them. Then we indicate that g is an element of $\mathfrak{G}$ by writing

$$g \leqq \mathfrak{G} .$$

3. Correspondences. Let $\mathfrak{G}$ and $\mathfrak{G}'$ be any two sets and assume that to every element g of $\mathfrak{G}$ there corresponds a unique element g' of $\mathfrak{G}'$. We write

$$S: \qquad g \to g'$$

(read g goes to g') and call S a *correspondence from* $\mathfrak{G}$ *to* $\mathfrak{G}'$. It is clear that S need not be a correspondence from $\mathfrak{G}'$ to $\mathfrak{G}$.

For example, $\mathfrak{G}$ may be the set of all football games in a season and $\mathfrak{G}'$ the set of all possible scores. Every game corresponds to a unique pair of integers called a score. But the set of scores so obtained will not exhaust $\mathfrak{G}'$. Also the same score may be obtained for more than one game. A more abstract example is that given by the set $\mathfrak{G}$ of all ordinary integers and the set $\mathfrak{G}'$ of their squares. The correspondence $g \to g' = g^2$ from $\mathfrak{G}$ to $\mathfrak{G}'$ is evidently not a correspondence from $\mathfrak{G}'$ to $\mathfrak{G}$ since $-g \to g^2$. Another example is that in which both $\mathfrak{G}$ and $\mathfrak{G}'$ consist of all rational functions of x with real coefficients. We let S be the correspondence $g(x) \to g'(x)$ where $g'(x)$ is the derivative of $g(x)$ and is uniquely determined by $g(x)$. However, it is well known to the reader that not every rational function is the derivative of a rational function so that the correspondents $g'(x)$ do not exhaust the set $\mathfrak{G}'$. Finally, let $\mathfrak{G}$ be the set of all integers and $\mathfrak{G}'$ the set of all integral multiples of three. The correspondence $g \to 3g$ from $\mathfrak{G}$ to $\mathfrak{G}'$ is now also a correspondence $3g \to g$ from $\mathfrak{G}'$ to $\mathfrak{G}$.

Let S be a correspondence $g \to g'$ from $\mathfrak{G}$ to $\mathfrak{G}'$ and g range over all elements of $\mathfrak{G}$. Then g' ranges over all elements of a subset $\mathfrak{G}'_S$ of $\mathfrak{G}'$ and when S is simultaneously a correspondence from $\mathfrak{G}'$ to $\mathfrak{G}$ we must have $\mathfrak{G}'_S = \mathfrak{G}'$. If g and h in $\mathfrak{G}$ go to g' and h' respectively in $\mathfrak{G}'$ under a correspondence S from $\mathfrak{G}$ to $\mathfrak{G}'$ and $g' = h'$ then we must have $g = h$ if S is a correspondence from $\mathfrak{G}'$ to $\mathfrak{G}$. These necessary conditions are obviously sufficient and S from $\mathfrak{G}$ to $\mathfrak{G}'$ is a correspondence from $\mathfrak{G}'$ to $\mathfrak{G}$ if and only if $\mathfrak{G}'_S = \mathfrak{G}'$, and $g' = h'$ if and only if $g = h$. Correspondences of this type are quite important, and we call them one-to-one correspondences

$$S: \qquad g \longleftrightarrow g'$$

(read g corresponds to g') *between* $\mathfrak{G}$ and $\mathfrak{G}'$. We shall often write (1–1) instead of one-to-one.

A *transformation* S of a set $\mathfrak{G}$ is a (1–1) correspondence between $\mathfrak{G}$ and itself indicated by

$$S: \qquad g \longleftrightarrow g^S .$$

We say that g goes to g^S under the transformation S and notice that both g and g^S range over all elements of $\mathfrak{G}$. Every transformation S has what we will call an *inverse* defined by

$$S^{-1}: \qquad g^S \longleftrightarrow g ,$$

and the *identical transformation* is the particular transformation

$$I: \qquad g \longleftrightarrow g$$

carrying every g of $\mathfrak{G}$ into itself.

In some environments it is more natural to use the word *function* instead of correspondence. The concepts are identical but the notation and terminology are sometimes changed as follows.

A correspondence f from a set $\mathfrak{G}$ to a set $\mathfrak{K}$ may be called a *function*

$$f: \qquad g \to k = f(g) ,$$

on $\mathfrak{G}$ to $\mathfrak{K}$. As usual the elements of $\mathfrak{G}$ may be any entities whatever and in particular may be systems $(g_1, \ldots, g_r)$ where g_i is an element of a set $\mathfrak{G}_i$ $(i = 1, \ldots, r)$. The above correspondence may now be written

$$f: \qquad (g_1, \ldots, g_r) \to f(g_1, \ldots, g_r) ,$$

and we say that f is a *function on* $\mathfrak{G}_1\mathfrak{G}_2 \ldots \mathfrak{G}_r$ *to* $\mathfrak{K}$. The sets $\mathfrak{G}_i$ need not be distinct. If they are all equal to a set $\mathfrak{G}$, f is on $\mathfrak{G}\mathfrak{G} \ldots \mathfrak{G}$ to $\mathfrak{K}$.

A particularly important example of a function on sets is suggested by the operations of elementary algebra. For example, let $\mathfrak{G}$ be the set of all non-zero integers and $\mathfrak{K}$ the set of rational numbers. Then division is a function on $\mathfrak{G}\mathfrak{G}$ to $\mathfrak{K}$. Considering arbitrary sets $\mathfrak{G}$, $\mathfrak{H}$, $\mathfrak{K}$ any function

$$\mathrm{O} \text{ on } \mathfrak{G}\mathfrak{H} \text{ to } \mathfrak{K}$$

will be called an *operation*. For operations* every a of $\mathfrak{G}$ and b of $\mathfrak{H}$ define a unique element $\mathrm{O}(a, b)$, in $\mathfrak{K}$ and we shall prefer to write

$$a \,\mathrm{O}\, b$$

instead of $\mathrm{O}(a, b)$.

When $\mathfrak{H} = \mathfrak{K} = \mathfrak{G}$ and therefore every a and b of $\mathfrak{G}$ define a unique $a\mathrm{O}b$ in $\mathfrak{G}$ we have O on $\mathfrak{G}\mathfrak{G}$ to $\mathfrak{G}$. We then say that $\mathfrak{G}$ *is closed* with respect to the operation O. This will be our most important type of operation.

The equality, that is, actual identity of elements of $\mathfrak{G}$ is a relation among its elements. The reader has of course met many other relations in elementary mathematics and sees that they are all functions R on $\mathfrak{G}\mathfrak{G}$ to the set $\mathfrak{K}$ consisting of two elements, *true, false*. We are accustomed to writing either $a = b$ or $a \neq b$. Thus for relations we write

$$a \, R \, b$$

if $R(a, b) = \textit{true}$, and

$$a \, \not{R} \, b$$

if $R(a, b) = \textit{false}$. Our notation for relations is seen to be different from that used for operations.

* An example where $\mathfrak{G}$, $\mathfrak{H}$, $\mathfrak{K}$ are all distinct may be given as follows. Let $\mathfrak{G}$ be the set of all real numbers, $\mathfrak{H}$ consist of the numbers 1, $i = \sqrt{-1}$. Then multiplication is an operation on $\mathfrak{G}\mathfrak{H}$ to the set $\mathfrak{K}$ of all real or pure imaginary numbers.

The example above of a set $\mathfrak{K}$ whose elements are the two concepts true, false indicates again that we are considering sets whose elements are absolutely arbitrary. Our notations are familiar to the reader who has written repeatedly $a = b$ and $a \neq b$, the latter of course being read a not equal to b. Similarly one writes $a > b$ and $a \not> b$, or $a \geqq b$ and $a \ngeqq b$. The first of these well-known relations, that of equality, is an example of a type of relation which will arise very frequently in our further work and which we shall wish to recognize when it arises. We write $\cong$ instead of R for this relation and make the

DEFINITION. *A relation* $\cong$ *among the elements of a set* $\mathfrak{G}$ *is called an equivalence relation if*

I. *For every* a *of* $\mathfrak{G}$ *it is true that* a $\cong$ a;

II. *If* a $\cong$ b *then* b $\cong$ a;

III. *If* a $\cong$ b *and* b $\cong$ c *then* a $\cong$ c.

The reader should verify whether or not the relations $=$, $>$, $\geqq$ in the set $\mathfrak{G}$ of all real numbers satisfy these postulates. He should also do this for other elementary relations—for example, the inclusion relation in sets.

Every equivalence relation enables us to classify the elements of $\mathfrak{G}$ into subsets called *classes of equivalent elements.* We put into a class $\{a\}$ (read class a) all the elements of $\mathfrak{G}$ equivalent to a, and our above postulates show that a is always in $\{a\}$, and that $\{a\} = \{b\}$ if and only if $a \cong b$. We shall call the element a appearing in $\{a\}$ a *representative* of the class. Then any b equivalent to a will serve equally well as a representative of the same class.

ORAL EXERCISES

1. Show that addition is an operation on $\mathfrak{G}\mathfrak{G}$ to $\mathfrak{G}$ where $\mathfrak{G}$ is the set of all even integers.

2. Let $\mathfrak{G}$ be the set of all non-zero even integers and ordinary division be an operation on $\mathfrak{G}\mathfrak{G}$ to $\mathfrak{K}$. Find a $\mathfrak{K}$.

3. Let $\mathfrak{G}$ be the set of all integers, $\mathfrak{H}$ consist of the integer 2. Describe the operation of ordinary multiplication as a function on $\mathfrak{G}\mathfrak{H}$ to a set $\mathfrak{K}$ to be determined.

4. Call two integers equivalent if they are both odd or both even, and otherwise inequivalent. Show that the relation so defined in the set of all integers is an equivalence relation and find the corresponding two classes of integers and representatives thereof.

5. Let a and b be any integers and write $a\ R\ b$ or $a\ \not R\ b$ according as the relation $|a - b| = 3$ is or is not true. Is this relation an equivalence relation? Does it become an equivalence relation if we replace the above equality by the statement $a\ R\ b$ if and only if $a - b$ is divisible by 3?

4. Integers. The ordinary integers

$$0, \pm 1, \pm 2, \ldots$$

occur frequently in all mathematics. This is not only true when they are elements of selected number systems but also when they are not. For they are sometimes used as exponents and subscripts on elements of arbitrary number systems. We shall discuss some of their elementary properties.

One of the most important properties of the set $\mathfrak{P}_P$ of all positive integers is called the *principle of complete induction.* Consider a subset $\mathfrak{G}$ of $\mathfrak{P}_P$ such that $\mathfrak{G}$ contains 1, and $a + 1$ for every a of $\mathfrak{G}$. Then the principle states that $\mathfrak{G} = \mathfrak{P}_P$. Many of our proofs will be inductive proofs and will thus depend on this principle. An illustration is given by the proof of the theorem on the **Division Algorithm** in the set $\mathfrak{P}$ of all integers.

Theorem 1. *Let* f *and* g $\neq$ 0 *be integers and define* $|\mathrm{g}| = \mathrm{g}$ *or* $-\mathrm{g}$ *according as* $\mathrm{g} > 0$, $\mathrm{g} < 0$. *Then there exist unique integers* q, r *such that*

$$\text{(1)} \qquad \mathrm{f} = \mathrm{qg} + \mathrm{r}, \qquad 0 \leqq \mathrm{r} < |\mathrm{g}| .$$

We first take $f > 0$. If $f = 1$ then $q = 0$, $r = 1$ when $|g| > 1$ and $r = 0$, $q = \pm 1$ when $|g| = 1$ so this case is complete. We make an induction on f and assume that $f = qg + r$. Then if $r < |g| - 1$ we have $f + 1 = qg + (r + 1)$, while if $r = |g| - 1$ then $f + 1 = (q \pm 1)g$ according as $g > 0$ or $g < 0$. This completes our induction on f and proves the existence of q and r when $f > 0$. If $f = 0$ we have $q = r = 0$ while if $f < 0$ we let $f_0 = -f = q_0 g + r_0$ by proof. Then $f = -q_0 g - r_0 = qg + r$, where $r = |g| - r_0$ and $q = -(q_0 \pm 1)$ according as $|g| = \pm g$. It remains to prove q, r unique. If then $f = sg + t = qg + r$ we have $(s - q)g = r - t$ where $|r - t| < |g|$. But g cannot divide an integer $r - t$ with $|r - t| < |g|$ unless $r - t = 0$. Hence $r = t$, $(s - q)g = 0$, $s = q$.

The principle of complete induction was used in the above proof as we have said. We shall of course use it later in many other situations.

Two integers a and b with the same remainder $r = 0, 1, \ldots, |g| - 1$ on division by g are said to be *congruent modulo* g, and it is customary in the elementary theory of numbers to write

$$a \equiv b \pmod{g} ,$$

or the simpler form

$$a \equiv b \ (g) .$$

Evidently $a \equiv b \ (g)$ if and only if $a - b$ is divisible by g. We shall sometimes use this congruence notation.

The Division Algorithm may be used to prove

Theorem 2. *Let* f *and* g *be integers not both zero. Then there exist a unique positive integer divisor* d *of* f *and* g *and integers* a *and* b *such that*

$$\text{(2)} \qquad \mathrm{d} = \mathrm{af} + \mathrm{bg} .$$

For let $\mathfrak{L}$ be the set of all positive integers $xf + yg$ for integers x, y. Since $\mathfrak{L}$ contains one of f, g, $-f$, $-g$, it is not an empty set. Thus there is a least positive integer $d = af + bg$ in $\mathfrak{L}$. We may write $f = qd + r$ with $0 \leqq r < d$ and obtain $(1 - aq)f + (-bq)g = r$ which is in $\mathfrak{L}$ or is zero. But the definition of d and $0 \leqq r < d$ imply that r is not in $\mathfrak{L}$, $r = 0$, d divides f. Similarly d divides g. If also d_1 divides f and g it must divide $d = af + bg$. When $d_1 = a_1f + b_1g$ we have d a divisor of d_1, $d = d_1$. This proves that d is unique.

We call d the *greatest common divisor* (abbreviated g.c.d.) of f and g. It is evidently the largest positive integral divisor of f and g.

Two integers f and g are called *relatively prime* if their g.c.d. is unity. We also say that f is *prime* to g or g is prime to f. When this occurs there always exist integers a, b such that $af + bg = 1$. This case of Theorem 2 is applied in

Theorem 3. *Let* f *divide* gh *and let* f *be prime to* g. *Then* f *divides* h.

For $af + bg = 1$, $gh = qf$, $afh + bgh = (ah + bq)f = h$ is divisible by f.

An integer $p \neq \pm 1, 0$ is called a *prime* if the only divisors of p are ± 1, $\pm p$. Theorem 3 then gives

Theorem 4. *Every integer* f *not zero or* ± 1 *is expressible uniquely apart from the order of the factors as a product*

$$f = \pm p_1^{e_1} p_2^{e_2} \dots p_r^{e_r}$$

where the p_i *are positive primes.*

We leave the proof of Theorem 4 to the reader.

EXERCISES

1. Let m be a positive integer and call two integers *equivalent* (or *congruent*) if they have the same positive remainder r on division by m as in Theorem 1. Prove that the relation so defined is an equivalence relation.

2. The equivalence relation of Exercise 1 defines classes $\{a\}$ of elements a of $\mathfrak{B}$. Define $\{a_1\} + \{a_2\} = \{a_1 + a_2\}$, $\{a_1\}\{a_2\} = \{a_1a_2\}$ and prove that the classes $\{a_1 + a_2\}$, $\{a_1a_2\}$ are independent of the particular a_1, a_2 used.

5. Groups. The notion of a group is fundamental in our subject. We shall define groups and obtain some of their elementary properties.

DEFINITION. *A non-empty set* $\mathfrak{G}$ *of elements* a, b, . . . *is said to form a group with respect to an operation* O *if:*

I. $\mathfrak{G}$ *is closed with respect to* O;

II. *The associative law holds in* $\mathfrak{G}$, *that is,*

$$a \; O \; (b \; O \; c) = (a \; O \; b) \; O \; c$$

for every a, b, c *of* $\mathfrak{G}$;

III. *For every* a *and* b *of* 𝔊 *there exist solutions* x *and* y *in* 𝔊 *of the equations*

$$\text{(3)} \qquad a \mathrm{O} x = b, \qquad y \mathrm{O} a = b\,.$$

A group is thus a *system* consisting of a set of elements 𝔊 and an operation O with respect to which 𝔊 forms a group. We shall generally designate the entire system by the set 𝔊 of its elements and shall call 𝔊 a group. The notation used for the operation is generally unimportant and may be taken in as convenient a way as possible. When 𝔊 is an abstract set of elements and multiplication is not already defined for these elements we may designate any given O as multiplication and write ab instead of $a\mathrm{O}b$. We designate this by calling 𝔊 a *multiplicative* group. We may similarly write $a + b$ and call 𝔊 an *additive* group. However, we shall generally not use the addition symbol except when $a \mathrm{O} b = b \mathrm{O} a$.

DEFINITION. *A group* 𝔊 *is called commutative or abelian if*

$$a \mathrm{O} b = b \mathrm{O} a$$

for every a *and* b *of* 𝔊.

An elementary physical example of an abelian group is a certain rotation group. We let 𝔊 consist of the rotations of the spoke of a wheel through multiples of 90° and $a\mathrm{O}b$ be the result of the rotation a followed by the rotation b. The reader will easily verify that 𝔊 forms a group with respect to O and that $a\mathrm{O}b = b\mathrm{O}a$. Slightly more complicated examples will be found in the exercises at the end of this section.

There is no loss of generality when we restrict our attention to multiplicative groups, that is, write ab instead of $a\mathrm{O}b$. *We shall do this in our proofs* and shall obtain some elementary properties of groups. Let a be in 𝔊 so that Postulate III implies the existence of elements e, f in 𝔊 such that

$$ea = af = a\,.$$

By the same postulate every b of 𝔊 has the form

$$b = ac = da\,,$$

and thus $eb = e(ac) = (ea)c = ac = b$ by the associative law. Similarly $bf = b$. But then, taking $b = f$, $b = e$ in turn, we get $ef = f = e$ and have proved that there exists an element e in 𝔊 such that

$$\text{(4)} \qquad eb = be = b$$

for every b of 𝔊. If also either $e_0a_0 = a_0$ or $a_0e_0 = a_0$ for some a_0 our proof shows that $e_0b = be_0 = b$ for every b of 𝔊. Then $e_0e = ee_0 = e = e_0$ so that

e is a unique element of $\mathfrak{G}$. We call e the *identity element* of $\mathfrak{G}$. This is a very important concept.

Let e satisfy (4) so that Postulate III implies that for every a of $\mathfrak{G}$ there exists an element a^{-1} of $\mathfrak{G}$ such that $aa^{-1} = e$. Then $a^{-1}(aa^{-1}) = a^{-1}e = a^{-1} = (a^{-1}a)a^{-1}$ and $a^{-1}(a^{-1})^{-1} = e = [(a^{-1}a)a^{-1}](a^{-1})^{-1} = (a^{-1}a)[a^{-1}(a^{-1})^{-1}] = (a^{-1}a)e = a^{-1}a$. We have proved that $aa^{-1} = a^{-1}a = e$. If also $ab = e$ then $a^{-1}(ab) = a^{-1}e = a^{-1} = (a^{-1}a)b = eb = b$ so that a^{-1} is unique. Moreover, $ax = b$ implies that $x = a^{-1}b$, $ya = b$ implies that $y = ba^{-1}$. We state the properties above in

Theorem 5. *There exists a unique identity element* e *of any group* $\mathfrak{G}$ *such that*

$$a \bigcirc e = e \bigcirc a = a$$

for every a *of* $\mathfrak{G}$. *Every* a *of* $\mathfrak{G}$ *has a unique* ***inverse*** a^{-1} *for which*

$$a \bigcirc a^{-1} = a^{-1} \bigcirc a = e . \tag{5}$$

Moreover (3) *have the unique solutions*

$$x = a^{-1} \bigcirc b, \qquad y = b \bigcirc a^{-1} . \tag{6}$$

The uniqueness in (6) gives immediately

COROLLARY. *Let* $\mathfrak{G}$ *be a group and let* a, f *be in* $\mathfrak{G}$ *such that either* $a \bigcirc f = a$ *or* $f \bigcirc a = a$. *Then* $f = e$ *is the identity element of* $\mathfrak{G}$.

If $\mathfrak{G}$ is an additive group we call e the *zero element* of $\mathfrak{G}$ and write 0 for e. We also write $-a$ for the inverse of a with respect to addition and write $x = -a + b$, $y = b - a$ in (6). An additive abelian group is frequently called a *modul.*

Let $\mathfrak{G}$ be a set of elements, O be an operation on $\mathfrak{G}\mathfrak{G}$ to $\mathfrak{G}$ such that the associative law, Postulate II, holds. Suppose that $\mathfrak{G}$ contains an element e such that $a \bigcirc e = e \bigcirc a = a$ for every a of $\mathfrak{G}$, and that for this e and every a there exists an element a^{-1} such that $a \bigcirc a^{-1} = a^{-1} \bigcirc a = e$. Then $\mathfrak{G}$ is a group. For clearly Postulate III is satisfied by (6). The criterion that $\mathfrak{G}$ be a group thus obtained is in a sense a converse of Theorem 5, and is often simpler to apply than our definition of a group.

EXERCISES

1. Verify that the set $\mathfrak{B}$ of all integers is an additive abelian group. Prove the same result for the set $\mathfrak{E}$ of all even integers.
2. Prove that no subset of $\mathfrak{B}$ with more than two elements is a multiplicative group.
3. Determine whether the elements of $\mathfrak{E}$, $\mathfrak{B}$ form a group with respect to the operations defined by $O(a, b) = 2(a + b)$, $2a + b$, $a - b$.

4. Show that the set $\mathfrak{A}$ of all classes $\{a\}$ defined by division by m as in Ex. 1 of Section 4 is an additive abelian group with $0 = \{0\}$.

5. Prove that if m is a prime p in Exercise 4 then the set $\mathfrak{A}$ with $\{0\}$ omitted is a multiplicative abelian group.

6. Let $m = pq$ with $p > 1$, $q > 1$ in Exercise 4. Prove that then the set $\mathfrak{A}$ with $\{0\}$ omitted does not form a multiplicative group.

7. Show that if a and b are elements of a multiplicative group $\mathfrak{G}$, then there exist elements x and y in $\mathfrak{G}$ such that $abx = ba$, $yab = ba$. We call x and y the *right* and *left commutators*, respectively, of the pair a, b. How are the commutators of a, b related to those of b, a? Of a^{-1}, b^{-1} and of b^{-1}, a^{-1}?

6. Equivalence, subgroups. In any study of mathematical systems the concept of equivalence of systems of the same kind always arises. Equivalent systems are logically distinct but we usually can replace any one by any other in a mathematical discussion with no loss of generality. For groups this notion is given by the

DEFINITION. *Let $\mathfrak{G}$ and $\mathfrak{G}'$ be groups with respective operations* O, O′ *and let there be a* (1–1) *correspondence*

$$S: \qquad a \longleftrightarrow a' \qquad (a \text{ in } \mathfrak{G},\ a' \text{ in } \mathfrak{G}')$$

between $\mathfrak{G}$ and $\mathfrak{G}'$ such that

$$(a \mathrm{O} b)' = a' \mathrm{O}' b'$$

for all a, b *of $\mathfrak{G}$. Then we call $\mathfrak{G}$ and $\mathfrak{G}'$ equivalent* (*or simply-isomorphic*) *groups.*

The relation of equivalence is an equivalence relation in the technical sense in the set of all groups. We again emphasize that while equivalent groups may be logically distinct they have identical properties.

The groups $\mathfrak{G}$, $\mathfrak{G}'$ of the above definition need not be distinct of course, and O′ may be O. When this is the case the *self-equivalence* S of $\mathfrak{G}$ is called an *automorphism* of $\mathfrak{G}$. A particular automorphism is the identity automorphism

$$I: \qquad a \longleftrightarrow a\,,$$

of $\mathfrak{G}$, but other automorphisms may also exist. We notice though that in every automorphism the identity element is self-corresponding, $e \longleftrightarrow e$. In fact, if $\mathfrak{G}$ and $\mathfrak{G}'$ are equivalent groups with respective identity elements e, e', then $e \longleftrightarrow e'$ under any correspondence S defining the equivalence of $\mathfrak{G}, \mathfrak{G}'$. For $(a\mathrm{O}e) = a$, $(a\mathrm{O}e)' = a' \mathrm{O}' e'$ where $e \longleftrightarrow e'$ in $\mathfrak{G}'$. Thus e' is a solution of $a' \mathrm{O}' e' = a'$. But $\mathfrak{G}'$ is a group and e' is the identity element of $\mathfrak{G}'$ by Theorem 5.

Another important concept is that of a subsystem. For groups we have the

DEFINITION. *A subset $\mathfrak{H}$ of a group $\mathfrak{G}$ is called a subgroup of $\mathfrak{G}$ if $\mathfrak{H}$ forms a group with respect to the defining operation* O *of $\mathfrak{G}$.*

The associative law holds always in any subset $\mathfrak{H}$ of a group $\mathfrak{G}$ so that $\mathfrak{H} \leqq \mathfrak{G}$ is a subgroup of $\mathfrak{G}$ if $\mathfrak{H}$ is closed with respect to O and (3) have solutions in $\mathfrak{H}$ for every a and b in $\mathfrak{H}$. By the last paragraph in Section 5 we also have

Theorem 6. *Let $\mathfrak{G}$ be a group with operation* O *and $\mathfrak{H} \leqq \mathfrak{G}$. Then $\mathfrak{H}$ is a subgroup of $\mathfrak{G}$ if and only if $\mathfrak{H}$ is closed with respect to* O, *$\mathfrak{H}$ contains the identity element* e *of $\mathfrak{G}$, and $\mathfrak{H}$ contains the inverse in $\mathfrak{G}$ of every* a *of $\mathfrak{H}$.*

We notice that a most important part of Theorem 6 is the statement that *the identity element of a subgroup $\mathfrak{H}$ of a group $\mathfrak{G}$ is the identity element of $\mathfrak{G}$ and the inverse in $\mathfrak{H}$ of any* a *of $\mathfrak{H}$ is the same as its inverse in $\mathfrak{G}$.*

The *order* of a group $\mathfrak{G}$ is the number of elements in $\mathfrak{G}$, whether the number is finite or infinity. We call $\mathfrak{G}$ a *finite group* if it has finite order and otherwise an *infinite group.*

If a is in $\mathfrak{G}$ so is the subgroup

$$[a] \tag{7}$$

of all powers a^i of a under the operation O of $\mathfrak{G}$. We again pass to multiplicative groups and then define $a^0 = e$, $a^{-m} = (a^{-1})^m$ for every positive integer m. The groups (7) are called the *cyclic subgroups* of $\mathfrak{G}$, and we say that *a generates* $[a]$. An infinite group $\mathfrak{G}$ may have finite subgroups $[a]$.

The order of $[a]$ is the number of distinct powers of a and is called the *order of* a. *Either all powers of* a *are distinct and* [a] *is an infinite group,* or $a^s = a^t$ for $t - s > 0$. Then $e = a^0 = a^{t-s}$. We let r be *the least positive integer such that* $a^r = e$. Then $e, a, a^2, \ldots, a^{r-1}$ are distinct since otherwise $a^g = a^h$ for $r > g > h \geqq 0$, $a^{g-h} = e$, $g - h < r$, a contradiction. Every integer n has the form

$$n = rk + q \qquad (0 \leqq q < r)\,, \tag{8}$$

and $a^n = (a^r)^k \cdot a^q = a^q$. But then [a] *has order* r *and is a finite group.*

If $\mathfrak{G}$ is a finite group the finite integer r exists for every a of $\mathfrak{G}$, and every subset $\mathfrak{H}$ of $\mathfrak{G}$ which is closed with respect to the operation O of $\mathfrak{G}$ contains $[a]$ for every a of $\mathfrak{H}$. But then $\mathfrak{H}$ contains e and every a^{-1}. By Theorem 6 we have

Theorem 7. *A subset $\mathfrak{H}$ of a finite group $\mathfrak{G}$ is a subgroup of $\mathfrak{G}$ if and only if $\mathfrak{H}$ is closed with respect to the operation* O *of $\mathfrak{G}$.*

EXERCISES

1. Show that the set of all integral multiples of an integer m forms a subgroup of the additive group of all integers.

2. Let n be a positive integer, $\mathfrak{C}$ a circle of fixed center Q, radius QP. Rotate QP counterclockwise through *all integral* multiples of $2\pi n^{-1}$ radians and obtain a set $\mathfrak{G}_n$ of points on $\mathfrak{C}$. Prove that $\mathfrak{G}_n$ is a cyclic group of order n with respect to addition of points defined as the result of the corresponding two rotations performed successively.

3. Replace n in Ex. 2 by any positive non-rational real number. Show that the resulting points form an infinite abelian group with respect to the operation of Ex. 2 and that all such groups are subgroups of the group of all points on $\mathfrak{C}$.

4. Let $\mathfrak{G}$ be a multiplicative group and S be the correspondence $a \longleftrightarrow a^{-1}$ for every a of $\mathfrak{G}$. Show that S is a one-to-one correspondence between $\mathfrak{G}$ and itself and determine a necessary and sufficient condition that S be an automorphism of $\mathfrak{G}$.

5. Show that if n is any positive integer the correspondence $a \longleftrightarrow na$ is an automorphism of the additive group of all rational numbers.

6. Give the notation for the elements of a cyclic additive group $[a]$. What is the inverse of any element?

7. What are all automorphisms of the additive group of all integers? Hint: Find the elements which correspond to unity.

8. Two elements a and b of a multiplicative group $\mathfrak{G}$ are called *commutative* if $ab = ba$. Prove that the subset $\mathfrak{C}$ of all quantities of $\mathfrak{G}$ commutative with every element of $\mathfrak{G}$ forms a subgroup of $\mathfrak{G}$. This is called the *centrum* of $\mathfrak{G}$.

7. Transformation groups. One of the most important types of groups is the transformation group. We consider an abstract set $\mathfrak{M}$ and have already defined what we mean by a transformation

$$S: \qquad a \longleftrightarrow a^S$$

in Section 3. Any other transformation

$$T: \qquad a \longleftrightarrow a^T$$

on $\mathfrak{M}$ carries every a^S into an element of $\mathfrak{M}$ which we naturally call $(a^S)^T$. But a^S ranges over all elements of $\mathfrak{M}$ and T is actually the same transformation as

$$T: \qquad a^S \longleftrightarrow (a^S)^T .$$

We now define the *product* of S and T to be

$$ST: \qquad a \longleftrightarrow a^{ST} = (a^S)^T .$$

The multiplication we have defined is associative. For $(ST)U$ is the transformation

$$a \longleftrightarrow (a^{ST})^U = [(a^S)^T]^U ,$$

and $S(TU)$ is

$$a \longleftrightarrow (a^S)^{TU} = [(a^S)^T]^U ,$$

that is, the associative law holds. Notice that we have actually shown that no matter how we group the symbols $S_1, \ldots, S_n$ in any product of transformations, we find the product by applying first S_1, then S_2, then S_3, etc., to a.

The set $\mathfrak{G}$ of all transformations on $\mathfrak{M}$ forms a group since $\mathfrak{G}$ is closed with respect to multiplication, the associative law holds in $\mathfrak{G}$, I is the identity element of $\mathfrak{G}$, S^{-1} is in $\mathfrak{G}$ for every S of $\mathfrak{G}$. Moreover, any set $\mathfrak{H}$ of transformations on $\mathfrak{M}$ is a subgroup of $\mathfrak{G}$ (called a *transformation group*) if and only if I, S^{-1}, ST are in $\mathfrak{H}$ for every S and T of $\mathfrak{H}$.

EXERCISES

1. Let $1, 2, \ldots, n$ be integers and $r, r+1, \ldots, n, 1, 2, \ldots, r-1$ be any cyclic permutation of $1, \ldots, n$. Show that the set of all such permutations S_r is a transformation group on the set $1, 2, \ldots, n$ and that $S_r = S_2^{r-1}$; that is, the group is cyclic.

2. There are six permutations of three letters a, b, c. Show that these permutations form the group of all transformations on the set consisting of a, b, c and that the group is not abelian.

8. Rings. The mathematical systems of elementary mathematics are special cases of what we call rings. The so-called *axioms of algebra* are a part of the postulates defining these systems and may be extracted as the postulates defining the more general systems called rings. A great part of modern algebra is concerned with these general systems and questions of their structure. Since they are of such importance we shall define them here and study some of their simpler properties. Notice in particular that we are making an abelian group into a ring by the introduction of a second operation.

DEFINITION. *A ring is an additive abelian group* $\mathfrak{A}$ *such that*

I. *The set* $\mathfrak{A}$ *is closed with respect to a second operation designated by multiplication; that is, every* a *and* b *of* $\mathfrak{A}$ *define a unique element* ab *of* $\mathfrak{A}$;

II. *Multiplication is associative; that is,*

$$\mathrm{a(bc)} = \mathrm{(ab)c}$$

for every a, b, c, *of* $\mathfrak{A}$;

III. *The distributive laws*

$$a(b + c) = ab + ac, \qquad (b + c)a = ba + ca$$

hold for every a, b, c *of* $\mathfrak{A}$.

The concepts of equivalence, subsystems again arise. We shall write

$$\mathfrak{A} \cong \mathfrak{A}'$$

to mean that $\mathfrak{A}$ and $\mathfrak{A}'$ are equivalent rings and define this concept in the

DEFINITION. *Two rings* $\mathfrak{A}$, $\mathfrak{A}'$ *are called equivalent if there is a* (1–1) *correspondence* $a \longleftrightarrow a'$ *between them such that*

$$(a + b)' = a' + b', \qquad (ab)' = a'b'$$

for every a, b *of* $\mathfrak{A}$ *and corresponding* a′, b′ *of* $\mathfrak{A}'$.

If $\mathfrak{A} \cong \mathfrak{A}'$ the zero elements of $\mathfrak{A}$, $\mathfrak{A}'$ correspond. This is an immediate consequence of our argument of Section 6 and the fact that $\mathfrak{A}$ and $\mathfrak{A}'$ are equivalent additive groups.

The notion of a subring is given by

DEFINITION. *A subset* $\mathfrak{B}$ *of a ring* $\mathfrak{A}$ *is called a subring of* $\mathfrak{A}$ *if* $\mathfrak{B}$ *is a ring with respect to the operations of addition and multiplication in* $\mathfrak{A}$.

The above definition leads to

Theorem 8. *A subset* $\mathfrak{B}$ *of a ring* $\mathfrak{A}$ *is a subring of* $\mathfrak{A}$ *if and only if* $\mathfrak{B}$ *is closed with respect to addition, subtraction, and multiplication.*

The proof of the above theorem consists in a verification of the fact that the above closure properties and $\mathfrak{B} \leq \mathfrak{A}$ imply that the postulates for a ring are satisfied by $\mathfrak{B}$. We leave this verification as an exercise for the reader.

A ring $\mathfrak{A}$ is called an *extension* of $\mathfrak{B}$ if $\mathfrak{B}$ is a subring of $\mathfrak{A}$. A very important principle in existence theorems about extensions is given by

Theorem 9. *Let* $\mathfrak{A}'$ *and* $\mathfrak{B}$ *be rings with no elements in common and let* $\mathfrak{A}'$ *contain a ring* $\mathfrak{B}'$ *equivalent to* $\mathfrak{B}$. *Then there exists a ring* $\mathfrak{A}$ *equivalent to* $\mathfrak{A}'$ *and such that* $\mathfrak{A}$ *contains* $\mathfrak{B}$.

For let $\mathfrak{S}$ be the set of all elements in $\mathfrak{A}'$ which are not in $\mathfrak{B}'$ and $\mathfrak{A}$ be the set consisting of all the elements of $\mathfrak{S}$ and all the elements of $\mathfrak{B}$. In any equations $a' + b' = c'$ or $a'\,b' = c'$ in ring $\mathfrak{A}'$ we replace those of a', b', c' which are in $\mathfrak{B}'$ by the corresponding a, b, c of $\mathfrak{B}$ and leave unaltered the a', b', or c' in $\mathfrak{S}$. This is possible when $\mathfrak{B} \cong \mathfrak{B}'$ and defines the operations of addition and multiplication for $\mathfrak{A}$. Evidently $\mathfrak{A} \cong \mathfrak{A}'$ and $\mathfrak{A}$ is the desired ring.

Theorem 9 is not of passing importance but is a very useful tool for existence theorems. We shall use it very frequently and shall use not merely the formal statement of the theorem but also the construction of $\mathfrak{A}$. Notice particularly that *the elements of* $\mathfrak{A}$ *are the elements of* $\mathfrak{B}$ *taken together with those elements of* $\mathfrak{A}'$ *which are not in* $\mathfrak{B}'$.

9. Some properties of rings. Every ring $\mathfrak{A}$ certainly contains a zero element 0, the identity element of the additive group $\mathfrak{A}$, and the negatives of elements of $\mathfrak{A}$. We shall investigate some simple properties of zero, $-a$, with respect to multiplication. These properties are consequences of our definitions of a ring and are well known in the elementary systems.

Theorem 10. *The equations*

$$0a = a0 = 0, \qquad (-a)b = a(-b) = -(ab) \tag{9}$$

are valid for every a *and* b *of* $\mathfrak{A}$.

For $\mathfrak{A}$ is an additive group and 0 is the unique solution of $a0 + x = a0$. But $0 + 0 = 0$, $a(0 + 0) = a0 + a0 = a0$, $0 = a0$. Similarly $0a = 0$. We use this property and obtain $0b = 0 = [a + (-a)]b = ab + (-a)b$, $(-a)b = -(ab)$. Similarly, $a(-b) = -(ab)$.

The element $-a$ is defined as the unique solution of $a + x = 0$. Hence $-(-a) = a$, $(-a)(-b) = -[a(-b)] = -[-(ab)] = ab$.

Corollary. *In any ring* $(-a)(-b) = ab$.

A ring $\mathfrak{A}$ *is said to have a* unity element *if there exists an* e *in* $\mathfrak{A}$ *such that*

$$ae = ea = a$$

for every a *of* $\mathfrak{A}$. The element e is unique since if also $fa = af = a$ then $ef = e = f$. We usually designate this element by 1. When $\mathfrak{A} \cong \mathfrak{A}'$ and $e \longleftrightarrow e'$ the element e' is a unity element of $\mathfrak{A}'$ since $ae = ea = a$ implies that $a'e' = e'a' = a'$. For rings with unity elements we now have

Theorem 11. *Equivalent rings have corresponding zero and unity elements.*

In a ring we have $a0 = 0$ but there may also exist quantities $a \neq 0$, $b \neq 0$ in $\mathfrak{A}$ such that $ab = 0$. Such elements are called *divisors of zero.* They are rather bizarre but occur in a really elementary example of a ring. We take the set $\mathfrak{A}$ of all classes $\{a\}$ of Exercise 4, Section 5. The reader may verify that $\mathfrak{A}$ is a ring of m elements with $0 = \{0\}$. Moreover, if $m = pq$ with $p > 1$, $q > 1$ then $\{p\} \neq 0$, $\{q\} \neq 0$, $\{p\}\{q\} = 0$.

Elements which have a property essentially opposite to that of divisors of zero are given by those satisfying the

Definition. *Let* $\mathfrak{A}$ *be a ring with unity element* 1 *and let* a *in* $\mathfrak{A}$ *have the property*

$$ab = ba = 1$$

for some b *of* $\mathfrak{A}$. *Then* a *is called a regular element of* $\mathfrak{A}$.

EXERCISES

1. Verify that $(a + b)(c + d) = ac + ad + bc + bd$ in a ring $\mathfrak{A}$.

2. The ring $\mathfrak{A}$ of the classes $\{a\}$ of integers is one with a unity element $\{1\}$. Verify that $\{r\}$ is a regular element of $\mathfrak{A}$ if r is prime to m.

3. Show that every non-zero element of the above $\mathfrak{A}$ is regular if m is a prime.

4. Prove that a regular element of any $\mathfrak{A}$ is not a divisor of zero.

5. Show that $\mathfrak{A}$ has no divisors of zero if and only if $\mathfrak{A}$ has the following property: Let $a \neq 0$, b, b' be in $\mathfrak{A}$. Then $ab = ab'$ if and only if $b = b'$.

6. Show that if every non-zero element of $\mathfrak{A}$ is regular there are no divisors of zero in $\mathfrak{A}$. This is true for rational numbers, real numbers.

10. Linear sets over a ring. The notion of a *linear set* is a fundamental one in all mathematics. The coefficient domain is usually the set of all real or all complex numbers but this would be insufficient for modern algebra. We shall define linear sets over any ring $\mathfrak{A}$.

Consider an additive abelian group $\mathfrak{L}$ of elements $\alpha, \beta, \ldots$ and an operation S_R on $\mathfrak{L}\mathfrak{A}$ to $\mathfrak{L}$. We designate the *right scalar product* $\alpha S_R a$ for α in $\mathfrak{L}$ and a in $\mathfrak{A}$ by αa and our closure property states that $\alpha a = \pi$ is in $\mathfrak{L}$ for every a of $\mathfrak{A}$ and α of $\mathfrak{L}$. Then $\mathfrak{L}$ is a *right linear set* (vector space, linear space, $\mathfrak{A}$-modul) if

$$(10) \qquad (\alpha b)a = \alpha(ba), \quad \alpha(a + b) = \alpha a + \alpha b, \quad (\alpha + \beta)a = \alpha a + \beta a$$

for every a, b of $\mathfrak{A}$ and α, β of $\mathfrak{L}$.

Left scalar products $aS_L\alpha$ and left linear sets are defined similarly. It is sometimes customary to refer to right linear sets as linear sets. We shall not do this in the present text but shall make the

DEFINITION. *An additive abelian group $\mathfrak{L}$ is called a linear set over $\mathfrak{A}$ if $\mathfrak{L}$ is both a right and a left linear set over $\mathfrak{A}$.*

Right linear subsets, left linear subsets, and linear subsets may now be defined. Similarly, we may define the notion of equivalence in linear sets. The definitions are entirely analogous to the similar ones made for rings and are left as exercises for the reader.

11. Sequences. General linear sets are not of as great importance as a certain type of linear set defined as follows. We consider the set $\mathfrak{B}$ of all sequences

$$(11) \qquad \alpha = (a_0, a_1, \ldots)$$

with a_i in a ring $\mathfrak{A}$. If also

$$(12) \qquad \beta = (b_0, b_1, \ldots)$$

is in $\mathfrak{B}$ we define

(13) $$\alpha + \beta = (a_0 + b_0, a_1 + b_1, \ldots),$$

(14) $$a\alpha = (aa_0, aa_1, \ldots), \qquad \alpha a = (a_0 a, a_1 a, \ldots),$$

for every α and β of $\mathfrak{B}$, a of $\mathfrak{A}$. The operations (14) are left and right scalar multiplication in $\mathfrak{B}$ and it is easily verified that $\mathfrak{B}$ is a linear set over $\mathfrak{A}$. We notice in particular that $\mathfrak{B}$ is an additive abelian group with

$$0 = (0, 0, \ldots), \qquad -\alpha = (-a_0, -a_1, \ldots).$$

The linear set $\mathfrak{B}$ is not an absolutely general linear set. One of its important special properties is that $a\alpha = \alpha a$ if a is commutative with every a_i of α. In particular, consider the case where $\mathfrak{A}$ is a ring with a unity element 1 and let μ_i be the sequence with 1 in the ith place and zero elsewhere. Then every α of $\mathfrak{L}$ is the formal infinite sum

$$\alpha = \sum_{i=1}^{\infty} a_{i-1}\mu_i = \sum_{i=1}^{\infty} \mu_i a_{i-1}$$

and $a\mu_i = \mu_i a$ for every a of $\mathfrak{A}$. We shall not consider this case further but shall make an additional specialization of $\mathfrak{B}$.

12. Polynomials over $\mathfrak{A}$. Any linear set $\mathfrak{L}$ may be made into a ring when multiplication is properly defined in $\mathfrak{L}$. For this is possible in any additive abelian group.* Such rings are called *algebras*. The algebras obtained from a given $\mathfrak{L}$ have various properties dependent on the definition of multiplication and the particular set $\mathfrak{L}$. An important case is that where $\mathfrak{L}$ is the set $\mathfrak{P}$ of all sequences of $\mathfrak{B}$ with a finite number of non-zero elements.

Write

$$\pi = (p_0, p_1, \ldots, p_n, 0, 0, \ldots), \quad \rho = (q_0, q_1, \ldots, q_m, 0, 0, \ldots)$$

and define

(15) $$\pi\rho = (\sigma_0, \sigma_1, \ldots, \sigma_{m+n}, 0, 0, \ldots)$$

in $\mathfrak{P}$ by

(16) $$\sigma_j = \sum_{i+k=j} p_i q_k .$$

* For example, define every $ab = 0$.

The set $\mathfrak{P}$ is actually the set of all formal polynomials

$$p_0 + p_1x + \dots + p_nx^n \tag{17}$$

in x with coefficients p_i in $\mathfrak{A}$. But the products p_0x^0, p_jx^j and the sums (17) are not ordinary sums and products in a ring but are merely formal expressions (17) which are really sequences. However, they have most of the usual properties of polynomials in elementary algebra except that care must be taken in forming products to preserve the order of the p_i and the q_k since in $\mathfrak{A}$ we may have $p_iq_k \neq q_kp_i$.

The constant polynomials are the sequences $(p_0, 0, \dots, 0)$. They form a subring $\mathfrak{A}_0$ of $\mathfrak{P}$ equivalent to $\mathfrak{A}$. By Theorem 1.9 there is no loss of generality if we assume that $\mathfrak{A}_0 = \mathfrak{A}$, that is identify the constant polynomials with the constants, that is elements of $\mathfrak{A}$. When we do this we may assert that $\mathfrak{P} > \mathfrak{A}$.

Every ring $\mathfrak{A}$ is contained in a ring with a unity element (as we shall show in Chap. II). There is then little loss of generality if we assume that $\mathfrak{A}$ has a unity element. Do this. Then $\mathfrak{P}$ contains x and the properties of $\mathfrak{P}$ may be obtained in the following important way.

Let $\mathfrak{Q}$ be a ring and let k in $\mathfrak{Q}$ have the property that $kq = qk$ for every q of $\mathfrak{Q}$. Then k is called a *scalar* of $\mathfrak{Q}$. We now assume that $\mathfrak{A}$ is a subring of $\mathfrak{Q}$ and define

$$\mathfrak{A}[k]$$

to be the set of all polynomials in k with coefficients in $\mathfrak{A}$. Here addition and multiplication are the ordinary operations in $\mathfrak{Q}$. We now make the

DEFINITION. *A scalar* x *of a ring* $\mathfrak{Q}$ *containing a ring* $\mathfrak{A}$ *is called an indeterminate over* $\mathfrak{A}$ *or an algebraic quantity over* $\mathfrak{A}$ *according as there do not or do exist quantities* $p_0, p_1, \dots, p_n$ *not all zero and in* $\mathfrak{A}$ *such that*

$$p_0 + p_1x + \dots + p_nx^n = 0 .$$

It is now clear that if x is an indeterminate over $\mathfrak{A}$ the ring $\mathfrak{A}[x]$ is equivalent to $\mathfrak{P}$. Moreover, when $\mathfrak{A}$ has a unity element our definition of $\mathfrak{P}$ implies that if we take $\mathfrak{P} = \mathfrak{Q}$ above then $\mathfrak{P} = \mathfrak{A}[x]$. Hence we have the following existence theorem. *Let* $\mathfrak{A}$ *be a ring with a unity element. Then there exists a ring* $\mathfrak{A}[x]$ *where* x *is an indeterminate over* $\mathfrak{A}$.

Our concept of $\mathfrak{A}[x]$ may be readily generalized to polynomial domains $\mathfrak{A}[x_1, \dots, x_n]$. For $\mathfrak{A}[x] = \mathfrak{A}_1$ is itself a ring with a unity element and we may define $\mathfrak{A}_1[y] = \mathfrak{A}[x, y]$. We see that $\mathfrak{A}[x, y] = \mathfrak{A}[y, x]$ and an easy induction gives us the ring $\mathfrak{Q}_m = \mathfrak{A}[x_1, \dots, x_m]$. Here $\mathfrak{Q}_m$ is a ring containing the scalars $x_1, \dots, x_m$ and $\mathfrak{A}$, and we make the

DEFINITION. *The indeterminates* $x_1, \dots, x_m$ *over* $\mathfrak{A}$ *of* $\mathfrak{Q} > \mathfrak{A}$ *are called* *independent indeterminates over $\mathfrak{A}$ if no polynomial in* $x_1, \ldots, x_m$ *with coefficients in $\mathfrak{A}$ is zero unless these coefficients themselves are all zero.*

The quantities $x_1, \ldots, x_m$ of $\mathfrak{Q}_m$ are thus independent indeterminates and we see that $\mathfrak{Q}_m$ is the algebra of all polynomials in $x_1, \ldots, x_m$ with coefficients in $\mathfrak{A}$. We notice that if $y_1, \ldots, y_m$ are any other independent indeterminates over $\mathfrak{A}$ in $\mathfrak{Q}_m$ the ring $\mathfrak{A}[y_1, \ldots, y_m]$ of all polynomials in $y_1, \ldots, y_m$ with coefficients in $\mathfrak{A}$ is equivalent to $\mathfrak{Q}_m$ under the correspondence

$$f(x_1, \ldots, x_m) \longleftrightarrow f(y_1, \ldots, y_m)$$

for any f. Hence $\mathfrak{Q}_m$ can actually be equivalent to a proper subring of itself.

We introduced the notion of scalars of a ring $\mathfrak{A}$ above. The set $\mathfrak{C}$ of all scalars of $\mathfrak{A}$ is called the *centrum* of $\mathfrak{A}$. It is easily proved that $\mathfrak{C}$ is a subring of $\mathfrak{A}$ and we leave the proof for the reader.

Definition. *A ring $\mathfrak{A}$ is called a commutative ring if* $ab = ba$ *for every* a *and* b *of $\mathfrak{A}$.*

The centrum of any ring $\mathfrak{A}$ is a commutative ring and $\mathfrak{A}$ is commutative if and only if $\mathfrak{A}$ is its own centrum. The centrum of $\mathfrak{A}[x_1, \ldots, x_m]$ is $\mathfrak{C}[x_1, \ldots, x_m]$ where $\mathfrak{C}$ is the centrum of $\mathfrak{A}$. For it is sufficient to prove this for $\mathfrak{A}[x]$ and $\mathfrak{C}[x]$ since we may then make an induction on m. If $k = p_0 + p_1x + \ldots + p_nx^n$ is in the centrum of $\mathfrak{A}[x]$ and p_j is not in $\mathfrak{C}$ then $p_ja \neq ap_j$ for some a in $\mathfrak{A}$. But then $ka - ak = \sum_i (p_ia - ap_i)x^i \neq 0$, a contradiction. Hence $\mathfrak{C}[x_1, \ldots, x_m]$ consists of all scalars of $\mathfrak{A}[x_1, \ldots, x_m]$. *We call these elements the scalar polynomials.*

We have given a proof above of the existence of a ring $\mathfrak{A}[x_1, \ldots, x_n]$ of all polynomials in n independent symbols $x_1, \ldots, x_n$ with coefficients in $\mathfrak{A}$ with a unity element. This ring is seen to be obtained as follows. The elements of $\mathfrak{A}[x_1]$ are the usual polynomials in x_1 with coefficients in the ring $\mathfrak{A}$. But these elements are actually nothing more than sequences of elements of $\mathfrak{A}$, the coefficients of the polynomials; and we have the usual laws of combination in algebra holding. To form $\mathfrak{A}[x_1, x_2]$ we take sequences whose elements are themselves sequences. But of course any polynomial in x_1 and x_2 is a sequence of coefficients which are polynomials in x_2 and these coefficients are themselves sequences with elements in $\mathfrak{A}$. Consequently our definition is natural. Similarly we pass to $\mathfrak{A}[x_1, \ldots, x_n]$.

We now have a rigorous notion of what is meant by the elementary concept of forming a polynomial in a symbol x. Of course it is necessary to know this since for example the sum of a quantity of a ring and the letter x is not defined by itself but must be formulated somehow. We have done this and now shall allow ourselves to form such polynomials in an indeterminate without further comment. It is important to know that by the notation $\mathfrak{A}[x]$, where $\mathfrak{A}$ is a ring with a unity element we mean a ring containing x and a subring $\mathfrak{A}$ and which consists of all polynomials in x with coefficients in $\mathfrak{A}$. When x is an indeterminate over $\mathfrak{A}$ we shall say so. We shall later consider a case where $\mathfrak{A}$ is a commutative ring but x is merely a scalar. The polynomial ring $\mathfrak{A}[x]$ is commutative if it exists, and we shall have to prove an existence theorem in this case. It always exists when $\mathfrak{A}$ and x are in the same ring.

The fundamental concepts of this, our introductory chapter, are those of groups, rings, linear sets, polynomials, equivalence, and subsystems. It is important to remember the definition of our fundamental concept, a group, as a set closed with respect to an operation, such that the associative law holds and two-sided division is always possible. But a group always has the properties that division is unique, there is a unique identity element, and every element has a unique inverse.

Rings are our number systems. They are commutative, additive groups with closure with respect to an associative operation called multiplication, and such that the distributive laws hold. The element zero has the usual properties that $0 \cdot a = a \cdot 0 = 0$ for every a, and this gives rise to the laws of signs; that is, $(-a)(-b) = (ab)$, $-(-a) = a$, $(-a)(b) = (a)(-b) = -(ab)$. But a ring does not have all properties of ordinary number systems since it may contain divisors of zero, and not all elements are regular. This latter property is not at all surprising since it is a property of the ring of all ordinary integers.

The concepts of equivalence and subsystems are quite natural and need no comment. Polynomials have been fully discussed above and we have done little more with linear sets than to define them and to define polynomials by their use. We shall discuss them further in the next chapter. We note finally that linear sets over a ring $\mathfrak{A}$ may be made into rings called *algebras over* $\mathfrak{A}$.

CHAPTER II

RINGS WITH A UNITY ELEMENT

1. The ring $\mathfrak{E}$. The rings which will be used as our coefficient domains will be certain types of rings with a unity element. Before defining these special rings (called fields and integral domains) we shall classify general rings according to a certain subring $\mathfrak{E}$.

A ring $\mathfrak{A}$ is an additive abelian group and contains the cyclic subgroups $[a]$ generated by its elements a. When we considered multiplicative groups we wrote a^n for the general element of $[a]$. In our present additive groups we write $n \cdot a$ where n ranges over all ordinary integers. The quantity $n \cdot a$ is not a product but is defined for all integers n by

$$(1) \qquad m \cdot a = a + a + \ldots + a$$

with $m > 0$ summands, $0 \cdot a = 0$ the identity element of the additive group $\mathfrak{A}$, $-m \cdot a = -(m \cdot a)$ the inverse (under addition) in $\mathfrak{A}$ of $m \cdot a$.

The group $\mathfrak{A}$ is closed under the second operation, multiplication. The distributive law implies that

$$(2) \qquad (n \cdot a)(q \cdot a) = nq \cdot a^2 .$$

Since a^2 need not be in $[a]$ the cyclic group $[a]$ need not be a ring. But if $\mathfrak{A}$ has a unity element e the group $[e]$ is a ring. For $e^2 = e$. We call this ring $\mathfrak{E}$.

The correspondence

$$S: \qquad n \rightarrow n \cdot e$$

from the set $\mathfrak{B}$ of all ordinary integers to $\mathfrak{E}$ is preserved under addition and multiplication since

$$(3) \qquad n \cdot e + q \cdot e = (n + q) \cdot e, \qquad (n \cdot e)(q \cdot e) = (nq) \cdot e .$$

But we saw in Section 1.6* that a cyclic additive group $\mathfrak{E}$ is an infinite group if and only if the condition $n \cdot e = q \cdot e$ implies that $n = q$. Thus $\mathfrak{E}$ has infinite order only when S is a (1–1) correspondence between $\mathfrak{B}$ and $\mathfrak{E}$. This proves

* By this notation we of course mean Sec. 6 of Chap. I.

Theorem 1. *Let $\mathfrak{E}$ be an infinite additive group. Then $\mathfrak{E}$ is a ring equivalent to the ring of all ordinary integers.*

Suppose that $\mathfrak{E}$ is a finite group. We have shown in Section 1.6 that then $\mathfrak{E}$ consists of m distinct elements

$$0, e, 2 \cdot e, \ldots, (m-1) \cdot e, \qquad m > 0,$$

and that $m \cdot e = 0$. This gives

Theorem 2. *Let $\mathfrak{E}$ be a finite additive group. Then $\mathfrak{E}$ is equivalent to the ring of all residue classes modulo a positive integer* m.

The ring of residue classes modulo $m > 0$ is the ring considered in Exercise 4 of Section 1.5. When $\mathfrak{E}$ is a subring of a given ring $\mathfrak{A}$ we have $(m \cdot e)a = m \cdot (ea) = m \cdot a = 0$ since $m \cdot e = 0$. This proves that every cyclic additive subgroup $[a]$ of a ring $\mathfrak{A}$ with $\mathfrak{E}$ finite is a finite additive subgroup of $\mathfrak{A}$.

If $m = qr$ where $q > 1$ and $r > 1$ are integers we use (2) and obtain $m \cdot e = (q \cdot e)(r \cdot e) = 0$. But $\mathfrak{E}$ has order m and $q < m$, $r < m$. Hence $q \cdot e \neq 0$, $r \cdot e \neq 0$, that is $\mathfrak{A}$ has divisors of zero. We state this result as

Theorem 3. *Let $\mathfrak{A}$ contain no divisors of zero and the subring $\mathfrak{E}$ of $\mathfrak{A}$ have* m *elements. Then* m *is a prime* p.

We now proceed to classify our rings according to the properties of $\mathfrak{E}$.

2. The characteristic of a ring. A ring $\mathfrak{A}$ is said to have *finite characteristic* m > 0 if $m \cdot a = 0$ for every a of $\mathfrak{A}$ and no smaller positive integer has this property. Then every cyclic subgroup $[a]$ of $\mathfrak{A}$ has finite order r_a. By Section 1.6 we see that m is divisible by all the r_a, m is divisible by the least common multiple d of the r_a. A trivial argument shows that $m = d$.

If $\mathfrak{A}$ does not have finite characteristic m it either possesses an infinite additive subgroup $[a]$ or there are infinitely many subgroups $[a]$ of $\mathfrak{A}$ with distinct r_a. In either case we say that $\mathfrak{A}$ *has infinite characteristic.*

The characteristic of a ring $\mathfrak{A}$ with a unity element is the same as the characteristic of its subring $\mathfrak{E}$. For we have already shown that $m \cdot a = 0$ for every a of $\mathfrak{A}$ and m is the least integer for which $m \cdot e = 0$. We also have

Theorem 4. *The characteristic* $m_{\mathfrak{A}}$ *of a subring $\mathfrak{A}$ of a ring $\mathfrak{B}$ with a unity element is less than or equal to the characteristic of $\mathfrak{B}$.*

We may assume that $\mathfrak{B}$ has finite characteristic m. Then $m \cdot a = 0$ for every a of $\mathfrak{A}$. Hence $m \geqq m_{\mathfrak{A}}$.

We now prove the important

Theorem 5. *Every ring $\mathfrak{A}$ is contained in a ring $\mathfrak{B}$ with a unity element and characteristic that of $\mathfrak{A}$.*

For let $\mathfrak{C}$ be the set of all pairs of elements (α, a) where α ranges over all integers and a over all elements of $\mathfrak{A}$. Define

$$(\alpha, a) + (\beta, b) = (\alpha + \beta, a + b), \qquad (\alpha, a)(\beta, b) = (\alpha\beta, \alpha b + \beta a + ab) .$$

The element (α, a) has all the formal properties of an abstract sum $\alpha + a$ and we easily see that $\mathfrak{C}$ is a ring with unity element $1 = (1, 0)$ and infinite characteristic. Moreover, $\mathfrak{C}$ contains the subring $(0, a)$ equivalent to $\mathfrak{A}$ and Theorem 1.9 states that there exists a ring $\mathfrak{B}$ equivalent to $\mathfrak{C}$ and with $\mathfrak{A}$ as a subring. When $\mathfrak{A}$ has infinite characteristic so has $\mathfrak{B}$ and this case is complete.

Next let $\mathfrak{A}$ have finite characteristic m. Then we let $\mathfrak{C}$ be defined as above and call two pairs (α, a), (β, b) of $\mathfrak{C}$ equivalent if $a = b$ and $\alpha - \beta$ is divisible by m. This defines an equivalence relation and we may use it to classify $\mathfrak{C}$ into classes $[\alpha, a]$ of equivalent pairs. Define addition and multiplication in the set $\mathfrak{B}_0$ of all classes of $\mathfrak{C}$ by

$$[\alpha, a] + [\beta, b] = [\alpha + \beta, a + b] , \quad [\alpha, a] [\beta, b] = [\alpha\beta, \alpha b + \beta a + ab] .$$

This definition is unique since if $\alpha' = \alpha + \lambda m$ and $\beta' = \beta + \mu m$ then $\alpha' + \beta' = \alpha + \beta + (\lambda + \mu)m$. Also $\alpha'\beta' = \alpha\beta + (\lambda\beta + \alpha\mu + \lambda\mu m)m$, $\alpha'b + \beta'a = \alpha b + \beta a + \lambda mb + \mu ma = \alpha b + \beta a$ since $mb = ma = 0$ in $\mathfrak{A}$. As before the set $\mathfrak{B}_0$ is a ring with $(1, 0)$ as unity element. But now $(m, 0) = 0$ and $\mathfrak{B}_0$ has characteristic m. The equivalent ring $\mathfrak{B}$ of Theorem 1.9 which contains $\mathfrak{A}$ has the same characteristic as $\mathfrak{A}$ and contains $\mathfrak{A}$ as a subring.

The rings with a unity element are the most important and interesting rings. The above theorem indicates that there will be no loss of generality if we always discuss rings with a unity element and their subrings. We shall do so and shall discuss some of their properties. A most important type of such a ring is that given by the

DEFINITION. *A ring $\mathfrak{A}$ with a unity element is called a division ring (or quasi-field) if every non-zero element of $\mathfrak{A}$ is a regular element.*

This definition is equivalent to the alternative

DEFINITION. *A division ring $\mathfrak{A}$ is an additive abelian group whose non-zero elements form a multiplicative group such that* $a(b + c) = ab + ac$, $(b + c)a = ba + ca$, $a0 = 0a = 0$ *for every* a, b, c *of* $\mathfrak{A}$.

The equivalence of these two definitions follows from the definition of a ring and the property that a set $\mathfrak{G}$ closed with respect to an associative operation (multiplication) is a group if and only if $\mathfrak{G}$ has an identity element e for which $ea = ae = a$, and every element of $\mathfrak{G}$ is regular. Division rings are extremely important in modern algebra. The real and complex number systems as well as the number systems we shall use and define presently are such rings. There are many, more complicated, division rings. We shall not discuss them here but shall think of our present discussion as preparatory for a more advanced text on the subject of rings.

EXERCISES

1. Prove that when the characteristic of a division ring is finite it is a prime p.

2. Let p be the characteristic of a division ring $\mathfrak{A}$ and $a \neq 0$ be in $\mathfrak{A}$. Prove that the sum $n \cdot a = 0$ if and only if n is divisible by p.

3. Let a and b be commutative elements of a ring of characteristic two. Show that $(a + b)^2 = (a - b)^2 = a^2 + b^2 = a^2 - b^2$.

3. The ring $\mathfrak{A}[x]$. A number of the properties of polynomials in elementary algebra hold when the coefficients are in an arbitrary ring $\mathfrak{A}$. We let $\mathfrak{A}$ be a ring with a unity element 1, x be an indeterminate over $\mathfrak{A}$, and

$$f = f(x) = a_0 + a_1x + \dots + a_nx^n \qquad (a_i \text{ in } \mathfrak{A}) \tag{4}$$

be any element of the polynomial set $\mathfrak{A}[x]$. When $f = a_0$ is in $\mathfrak{A}$ we call f a *constant polynomial* and say that *f has degree zero.* Every f not in $\mathfrak{A}$ has the form (4) with $a_n \neq 0$ and we call n the degree of f. We have therefore defined the degree of any polynomial $f(x)$. Polynomials of degree one are called *linear,* of degree two *quadratic,* and so forth. The coefficient a_n is called the *leading coefficient* of f and we shall call f a *monic** polynomial when $a_n = 1$.

Write $g = g(x) = b_0 + b_1x + \dots + b_mx^m$. Then the degree of $f + g$ is at most the maximum of m and n. The degree of

$$fg = a_0b_0 + (a_1b_0 + a_0b_1)x + \dots + a_nb_mx^{n+m} \tag{5}$$

is $n + m$ unless $a_nb_m = 0$. But if $f \neq 0$ has degree n and $g \neq 0$ has degree m the degree of fg is $n + m$ unless a_n and b_m are divisors of zero whose product is zero. Thus the degree of fg is $n + m$ if either a or b is not a divisor of zero.

A fundamental theorem in $\mathfrak{A}[x]$ is that of the *Division Algorithm.*

Theorem 6. *Let* $f \neq 0$ *and* g *be given as above such that* b_m *is a regular element of* $\mathfrak{A}$. *There exist unique polynomials* s, s_1, r, r_1 *in* $\mathfrak{A}[x]$ *such that*

$$f = sg + r = gs_1 + r_1 , \tag{6}$$

where s *and* s_1 *are either zero or have degree* $n - m \geqq 0$ *and the degrees of* r *and* r_1 *are less than* m.

We prove the above theorem by an induction. If $n < m$ we take $s = s_1 = 0$, $r = r_1 = f$. Assume therefore that $n \geqq m$ and, as the basis of an

* This notion occurs so frequently in algebra that the author believes it would be wise to adopt the above term. We shall do so in the present text.

induction on $n - m$, that the theorem is true for all polynomials f_1 of degree at most $n - 1$. Then b_m has an inverse b_m^{-1} in $\mathfrak{A}$,

$$f_1 = f - a_n b_m^{-1} x^{n-m} g$$

has degree less than n, $f_1 = f_2 g + r$. The degree of r is at most $m - 1$ and that of f_2 is at most $n - 1 - m$. But then $f = sg + r$, where

$$s = f_2 + a_n b_m^{-1} x^{n-m}$$

has degree $n - m$ since its leading coefficient is $a_n b_m^{-1}$. The existence of s_1 and r_1 is proved similarly.

The uniqueness of s, s_1, r, r_1 is proved by the following consideration of degrees. Assume that $f = sg + r = tg + u$ and therefore that $(s - t)g = u - r$. The leading coefficient of g is not a divisor of zero and the degree of $(s - t)g$ is at least m unless $s - t = 0$. But $u - r$ has degree at most $m - 1$. Hence $s - t = 0 = u - r$. This proves s and r unique. The uniqueness of s_1, r_1 follows similarly.

The element s is called the *right-hand quotient*, r the *right remainder* in the division of f by g on the right, and we say that f has g as a *right-hand factor*, g is a *right divisor* of f if $r = 0$, that is $f = sg$. Similarly, s_1 and r_1 are called the left-hand quotient and remainder respectively and we define left-hand factors, left divisibility.

The polynomial $g(x) = x - c$, c in $\mathfrak{A}$, has regular leading coefficient and we apply Theorem 2.6 to prove what is usually called the *Remainder Theorem of Algebra.*

Theorem 7. *The right- and left-hand remainders on the division of* f(x) *of* (4) *by* x − c, c *in* $\mathfrak{A}$, *are respectively*

$$f_R(c) = a_0 + a_1 c + a_2 c^2 + \dots + a_n c^n, \tag{7}$$

$$f_L(c) = a_0 + c a_1 + c^2 a_2 + \dots + c^n a_n. \tag{8}$$

For $x^n - c^n = (x - c)Q_n(x, c) = Q_n(x, c)(x - c)$ where

$$Q_n(x, c) = Q_n(c, x) = x^{n-1} + cx^{n-2} + \dots + c^{n-2}x + c^{n-1}. \tag{9}$$

Then $f(x) - f_R(c) = \sum_{i=0}^{n} a_i(x^i - c^i) = \left[\sum_{i=0}^{n} a_i Q_i(x, c)\right](x - c)$. By the uniqueness of r in Theorem 2.6 we have $r = f_R(c)$ and in fact have $s = \sum_{i=0}^{n} a_i Q_i(x, c)$. Similarly, $r_1 = f_L(c)$, $s_1 = \sum_{i=0}^{n} Q_i(x, c) a_i$.

In elementary college algebra the *Factor Theorem of Algebra* follows as a corollary of the Remainder Theorem. This is also true here and in fact follows immediately from Theorem 2.7 and the uniqueness of r, r_1, s, s_1 in Theorem 2.6.

Theorem 8. *The polynomial* $f(x)$ *has* $x - c$ *as a right (left) hand factor if and only if* $f_R(c) = 0 (f_L(c) = 0)$.

We have obtained all of the general properties of rings needed for our further discussions and shall leave the subject. Rings will now be specialized and will become the domains used as the coefficient domains in our further discussions.

EXERCISES

1. Let $\mathfrak{A}$ be the commutative ring consisting of the residue classes modulo 6. Find the quotients and remainders on division of $ax^2 + bx + c$ with a, b, c in $\mathfrak{A}$ by $5x + 1$; by $x + 4$.

2. If $\mathfrak{A}$ is as in Ex. 1 there is no linear polynomial $ax + b$ such that $x^2 + 5x + 1 = (ax + b)(3x + 1) + c$ where c is in $\mathfrak{A}$ and $ax + b$ is then the quotient on division by $3x + 1$. Why is this true?

4. Integral domains and fields. The number systems used as coefficient domains are integral domains and fields. *An integral domain is defined as a commutative ring with a unity element and no divisors of zero.* The set $\mathfrak{E}$ of all ordinary integers is an integral domain. Integral domains are to be thought of as generalizations of $\mathfrak{E}$ and much of our discussion of integral domains will be a study of the properties of integral domains which are analogous to those of integers.

One of the principal properties of an integral domain $\mathfrak{J}$ is that given by the statement that

$$ab = ab'$$

for $a \neq 0$, b, b' in $\mathfrak{J}$ if and only if $b = b'$. This follows since $a(b - b') = 0$ in $\mathfrak{J}$, $a \neq 0$, $\mathfrak{J}$ has no divisors of zero. The above property is in fact equivalent to the property that $\mathfrak{J}$ has no divisors of zero since if $ab = 0$ in a ring $\mathfrak{J}$ with the above uniqueness property and $a \neq 0$ then $ab = a \cdot 0$ and $b = 0$, $\mathfrak{J}$ has no divisors of zero.

An element α of an integral domain $\mathfrak{J}$ is called *a unit* of $\mathfrak{J}$ if there exists an element β in $\mathfrak{J}$ such that $\alpha\beta = 1$. This is evidently our earlier definition of a regular element α of $\mathfrak{J}$ but we shall prefer this new name when $\mathfrak{J}$ is an integral domain. Every $\mathfrak{J}$ has the units $1, -1$ and possibly other units. When α is a unit so is α^{-1}, since $(\alpha\beta)^{-1} = \beta^{-1}\alpha^{-1}$ is in $\mathfrak{J}$ when α^{-1} and β^{-1} are in $\mathfrak{J}$, and the product of two units is a unit.

An element a' of $\mathfrak{J}$ is said to be *associated* with a of $\mathfrak{J}$ if $a' = \alpha a$ for a unit α of $\mathfrak{J}$. Evidently $a = 1 \cdot a$ is associated with a, $a = \alpha^{-1}a'$ is associated with a', $a'' = \gamma a' = (\gamma\alpha)a$ is associated with a. *The relation of association is an equivalence relation.*

The most important type of integral domain is the field. A field $\mathfrak{F}$ is an integral domain whose non-zero elements form a multiplicative group. A definition equivalent to this but depending on fewer concepts is given by

DEFINITION. *A field $\mathfrak{F}$ is an additive abelian group whose non-zero elements form a multiplicative abelian group such that* $\mathrm{a}(\mathrm{b} + \mathrm{c}) = \mathrm{ab} + \mathrm{ac}$, $\mathrm{a}0 = 0\mathrm{a} = 0$ *for every* a, b, c *of* $\mathfrak{F}$.

Thus *any commutative division ring is a field.* The notion of a field is of fundamental importance in algebra. By far the greater half of the present text will be devoted to a discussion of the properties of fields. Their elements will be used as coefficients in most of our algebraic work and in many other ways.

The set of all rational numbers is a field and the notion of a field is to be thought of as a generalization of that of the rational number system. The set of all integers is a subring of the set of all rational numbers and this is proved as a special case of the generalization

Theorem 9. *Every integral domain $\mathfrak{J}$ is contained as a subring in a field $\mathfrak{F}$.*

Consider the set $\mathfrak{S}$ of all ordered pairs

$$\frac{a}{b} \qquad (a, b \neq 0, \text{ in } \mathfrak{J}) .$$

A pair a'/b' is said to be equivalent to a/b,

$$\frac{a'}{b'} \cong \frac{a}{b} , \tag{10}$$

if

$$ab' = a'b .$$

Evidently $a/b \cong a/b$ and $a/b \cong a'/b'$ when $a'/b' \cong a/b$. Moreover, if also $a''/b'' \cong a'/b'$ then $ab' = ba'$, $a'b'' = b'a''$, $(ab'')b' = (ab')b'' = ba'b'' = bb'a''$. But $b' \neq 0$ and $\mathfrak{J}$ is an integral domain so that $ab'' = ba''$, $a''/b'' \cong a/b$. *This proves that the relation* (10) *is an equivalence relation.*

We may now classify the elements of $\mathfrak{S}$ into classes

$$\alpha = \left\{\frac{a}{b}\right\}$$

of equivalent elements. The set $\mathfrak{F}_0$ of all these classes will be equivalent to the desired field $\mathfrak{F}$.

Define the operations in $\mathfrak{F}_0$ for α, $\beta = \{c/d\}$ by

$$\alpha + \beta = \left\{\frac{ad + bc}{bd}\right\}, \qquad \alpha\beta = \left\{\frac{ac}{bd}\right\}. \tag{11}$$

The right-hand members of the above equations are elements of $\mathfrak{F}_0$ since $bd \neq 0$ in an integral domain $\mathfrak{J}$. A trivial computation also shows that, while the definitions of $\alpha + \beta$, $\alpha\beta$ depend on the particular representatives of a/b, c/d, respectively, the resulting classes depend only on α, β.

The set $\mathfrak{F}_0$ is an additive abelian group. For $+$ is on $\mathfrak{F}_0\mathfrak{F}_0$ to $\mathfrak{F}_0$, $\alpha + \beta = \beta + \alpha$, $\alpha + (\beta + \gamma) = (\alpha + \beta) + \gamma$ since $\mathfrak{J}$ is commutative and associative with respect to both addition and multiplication, $\alpha + x = \beta$ has the unique solution

$$x = \left\{\frac{cb - da}{bd}\right\}.$$

The identity of the additive group $\mathfrak{F}_0$ is

$$0 = \left\{\frac{0}{1}\right\}$$

and $\mathfrak{F}_0$, with 0 omitted, forms a multiplicative abelian group with

$$1 = \left\{\frac{1}{1}\right\}, \qquad \alpha^{-1} = \left\{\frac{b}{a}\right\}.$$

Finally, if $\gamma = \{g/h\}$ then

$$\alpha(\beta + \gamma) = \left\{\frac{a}{b} \cdot \frac{ch + dg}{dh}\right\} = \left\{\frac{(ac)(bh) + (bd)(ag)}{(bd)(bh)}\right\} = \alpha\beta + \alpha\gamma,$$

and $\mathfrak{F}_0$ is a field.

The field $\mathfrak{F}_0$ contains as a subset $\mathfrak{J}_0$ the classes

$$\left\{\frac{a}{1}\right\}.$$

Evidently $\mathfrak{J}_0$ is an integral domain equivalent to $\mathfrak{J}$. Moreover if $\mathfrak{J}$ is a field then $\mathfrak{J}_0 = \mathfrak{F}_0$ is equivalent to $\mathfrak{J}$. By Theorem 1.9 there exists a field $\mathfrak{F} \geqq \mathfrak{J}$ and equivalent to $\mathfrak{F}_0$.

The field $\mathfrak{F}$ of Theorem 2.9 is called the *quotient field* of $\mathfrak{J}$. It is the smallest field containing $\mathfrak{J}$ in that if $\mathfrak{K}$ is a field containing $\mathfrak{J}$ then $\mathfrak{K}$ has a subfield equivalent to $\mathfrak{F}$. For $\mathfrak{K}$ is a field and contains ab^{-1} for every $a, b \neq 0$ in $\mathfrak{J}$. Then the correspondence

$$\alpha = \begin{Bmatrix} a \\ b \end{Bmatrix} \longleftrightarrow ab^{-1}$$

defines a field of all ab^{-1} equivalent to $\mathfrak{F}$.

The non-zero elements of an integral domain form a set closed under multiplication and such that the associative and commutative laws hold. Then these elements form a multiplicative abelian group if and only if every non-zero a of $\mathfrak{J}$ has an inverse in $\mathfrak{J}$. Evidently every field is an integral domain. But the above argument shows that *an integral domain $\mathfrak{J}$ is a field if and only if every non-zero element of $\mathfrak{J}$ is a unit of $\mathfrak{J}$.*

EXERCISES

1. Verify the last statement above. Give complete sets of postulates for a field, an integral domain, without using the group concept.

2. The field $\mathfrak{F} = \mathfrak{R}(\sqrt{2})$ consists of all rational functions with rational coefficients of $\sqrt{2}$. Show that every quantity of $\mathfrak{F}$ may be expressed in the form $a + b\sqrt{2}$ with rational a, b.

3. We define the field $\mathfrak{C}$ of all complex numbers as $\mathfrak{R}(i)$ where $\mathfrak{R}$ is the field of all real numbers, $i^2 = -1$, and the notation is as in Ex. 2. Prove that every element of $\mathfrak{C}$ has the form $a + bi$ with real a, b.

4. Prove that if $\mathfrak{J}$ is an integral domain the sum and product in (11) are independent of the representatives a/b, c/d.

5. Let m be a prime p and $\mathfrak{P}$ the set of all classes $\{a\}$ of Ex. 4 of Section 1.5. Prove that $\mathfrak{P}$ is a field consisting of p elements.

6. Show that our definitions imply that the characteristic of any integral domain is either infinity or a prime p. Prove that in the latter case every integral domain contains a subfield, the field $\mathfrak{P}$ of Ex. 5.

7. Let $\mathfrak{J}$ be an integral domain and x be an indeterminate over $\mathfrak{J}$. Show that $\mathfrak{J}[x]$ is an integral domain whose units are those of $\mathfrak{J}$. Find the quotient field.

8. Use an induction on n to generalize the result of Ex. 7 to $\mathfrak{J}[x_1, \ldots, x_n]$ where $x_1, \ldots, x_n$ are independent indeterminates over $\mathfrak{J}$.

5. Equivalence, subsystems. Any integral domain or field is a ring and we have already defined equivalence for rings. This definition thus holds with the words fields, integral domains replacing rings. We cannot do this, however, for the definitions of subsystems.

A field may have subrings which are not fields. An elementary example is furnished by the field of all rational numbers and its subring, the integral domain of all integers. We therefore make the

DEFINITION. *A subset $\mathfrak{K}$ of a field (integral domain) $\mathfrak{F}$ is called a subfield* *(integral subdomain) of* $\mathfrak{F}$ *if* $\mathfrak{K}$ *is a field (integral domain) with respect to the defining operations of* $\mathfrak{F}$.

The above definition evidently implies that a subset $\mathfrak{K}$ of a field $\mathfrak{F}$ is a subfield of $\mathfrak{F}$ if and only if $\mathfrak{K}$ is closed with respect to addition, subtraction, multiplication, and division by non-zero elements of $\mathfrak{K}$. Integral subdomains have similar properties.

The set $\mathfrak{P}$ of all elements in common to all subfields of a field $\mathfrak{F}$ is a subfield of $\mathfrak{F}$. For $\mathfrak{P}$ evidently has the above closure properties. The field $\mathfrak{P}$ is contained in every subfield of $\mathfrak{F}$ and is a subfield of $\mathfrak{F}$. Thus $\mathfrak{P}$ is the smallest subfield of $\mathfrak{F}$ and is called the *prime subfield* of $\mathfrak{F}$. We seek the structure of $\mathfrak{P}$.

The field $\mathfrak{P}$ contains the unity element e of $\mathfrak{F}$. Then $\mathfrak{P}$ contains the ring $\mathfrak{E}$ of Section 2.1. Moreover, $\mathfrak{E}$ is an integral domain and $\mathfrak{P}$ is the quotient field of $\mathfrak{E}$. If $\mathfrak{E}$ is equivalent to the domain of all integers then $\mathfrak{P}$ is equivalent to the field $\mathfrak{R}$ of all rational numbers. In this case we call $\mathfrak{F}$ a *non-modular field.* The field $\mathfrak{F}$ then has *infinite characteristic.**

The other case is that where $\mathfrak{E}$ is equivalent to the ring $\mathfrak{B}_p$ of Theorem 2.2. But $\mathfrak{B}_p$ is a field and $\mathfrak{P} = \mathfrak{E}$ is equivalent to the field of all residue classes modulo p, a prime. We call $\mathfrak{F}$ a modular field of characteristic p. Such fields are rather bizarre but occur frequently and are of great importance. Notice that any sum with p equal summands $a + a + \ldots + a = p \cdot a = 0$ in $\mathfrak{F}$. The binomial theorem then implies that $(a + b)^p = a^p + b^p$ in $\mathfrak{F}$.

EXERCISE

Let $\mathfrak{F}$ be a field of characteristic 2 and $x^2 + ax + b = 0$ be a quadratic equation with coefficients $a \neq 0$ and b in $\mathfrak{F}$. Show that a linear transformation $x = cy + d$ with coefficients in $\mathfrak{F}$ cannot carry our quadratic into a reduced quadratic in y. But this can be done when $\mathfrak{F}$ does not have characteristic two.

6. Divisibility. The questions about divisility form one of the most important subjects for investigation in the integral domain of all natural integers. Such questions also arise in general integral domains and we shall discuss them.

A quantity a of an integral domain $\mathfrak{J}$ *is said to be divisible* by $b \neq 0$ in $\mathfrak{J}$ if

$$a = bc$$

for some c of $\mathfrak{J}$. If α is any unit of $\mathfrak{J}$ we write $a = (\alpha b)(\alpha^{-1}c)$ and see that a is also divisible by αb for every unit α of $\mathfrak{J}$.

* In the literature this is called a field of characteristic zero. That name is a misnomer since the characteristic should be thought of as the order of the additive abelian group $\mathfrak{E}$ and is infinity when $\mathfrak{F}$ is non-modular.

A unit a of $\mathfrak{J}$ is divisible by b if and only if b is a unit of $\mathfrak{J}$. For if $a = bc$ then a^{-1} is in $\mathfrak{J}$ and $1 = b(a^{-1}c)$. Thus $b^{-1} = a^{-1}c$ is in $\mathfrak{J}$ and b is a unit. The converse is trivial.

Let $a \neq 0$ and $b \neq 0$ be in $\mathfrak{J}$. *Then* a *divides* b *and* b *divides* a *if and only if* a *and* b *are associated.* For $a = bc$, $b = ad$ imply that $a = acd$ and $cd = 1$ in an integral domain $\mathfrak{J}$. But d is then a unit of $\mathfrak{J}$ and $b = ad$ is associated with a. Conversely if $b = ad$ with d a unit we have $a = d^{-1}b$ divisible by b.

Every non-zero quantity a *of* $\mathfrak{J}$ *is divisible by its associates and by units of* $\mathfrak{J}$. When a is not a unit and these are its only divisors we call it a *prime* or *irreducible* element of $\mathfrak{J}$. Thus a is irreducible *if and only if every associate of* a *is irreducible.*

DEFINITION. *Two elements* a *and* b *of* $\mathfrak{J}$ *which are not both zero are said to have a greatest common divisor (abbreviated g.c.d.) in* $\mathfrak{J}$ *if* $\mathfrak{J}$ *contains a quantity*

$$d = (a, b) \neq 0 \tag{12}$$

dividing both a *and* b, *and such that every divisor of both* a *and* b *divides* d.

Two elements a, b may not have a g.c.d. But when they do have a g.c.d., say d, any other g.c.d. d_0 is an associate of d. For the above definition implies that d and d_0 divide each other.

A pair of elements a, b of $\mathfrak{J}$ may have a g.c.d. in $\mathfrak{J}$ but may not have a g.c.d. in an integral domain $\mathfrak{J}_0 > \mathfrak{J}$. However, if it happens that there exist elements x and y in $\mathfrak{J}$ for which

$$d = (a, b) = ax + by$$

then d is a g.c.d. of a and b in any integral domain $\mathfrak{J}_0 \geqq \mathfrak{J}$. For d divides a and b and if g is a divisor of a, b in $\mathfrak{J}_0$ we have $a = ga_0$, $b = gb_0$ with a_0, b_0 in $\mathfrak{J}_0$. Thus $d = g(a_0x + b_0y)$ is divisible by g and d is a g.c.d. of a and b in $\mathfrak{J}_0$.

The above notion of a g.c.d. may evidently be extended to a set $a_1, \ldots, a_r$ of elements of $\mathfrak{J}$, and we write

$$d = (a_1, \ldots, a_r) .$$

Thus d divides all the a_i, and every common divisor of the a_i divides d. If every two elements not zero of $\mathfrak{J}$ have a greatest common divisor then this is true of every r elements of $\mathfrak{J}$. For let it be true of every $r - 1$ elements of $\mathfrak{J}$ where we are taking $r > 2$. Then every common divisor of $a_1, \ldots, a_r$ divides a_r, $a_1, \ldots, a_{r-1}$ and hence divides $d_{r-1} = (a_1, \ldots, a_{r-1})$ and a_r. Thus every common divisor divides $d = (d_{r-1}, a_r)$. Conversely d divides d_{r-1}, a_r and divides $a_1, \ldots, a_r$ since d_{r-1} divides $a_1, \ldots, a_{r-1}$, and $d = (a_1, \ldots, a_r)$ as desired.

DEFINITION. *The elements* $a_1, \ldots, a_r$ *of* $\mathfrak{J}$ *are said to be relatively prime or prime to each other if their only common divisors are units of* $\mathfrak{J}$.

This definition implies that when $a_1, \ldots, a_r$ are relatively prime and have a g.c.d. their g.c.d. is unity or any unit of $\mathfrak{J}$. We now prove the elementary

Theorem 10. *An irreducible element* p *is either a divisor of any* a *of* $\mathfrak{J}$ *or is prime to* a.

For if a is not prime to p there is a quantity b of $\mathfrak{J}$ which divides both a and p and is not a unit of $\mathfrak{J}$. But our definition of p implies that its divisor b is an associate αp and $a = bc$ implies that $a = p(\alpha c)$ is divisible by p.

7. The integral domain $\mathfrak{F}[x]$. The most important and interesting case of polynomial domains $\mathfrak{A}[x]$ is the case where $\mathfrak{A}$ is a field $\mathfrak{F}$. This is the domain considered in a great portion of high school algebra, and many of our properties were observed there by the reader.

The set $\mathfrak{J} = \mathfrak{F}[x]$ is an integral domain. Its quotient field is the field of all rational functions of x with coefficients in $\mathfrak{F}$. Similarly, if $x_1, \ldots, x_n$ are independent indeterminates over $\mathfrak{F}$ the set $\mathfrak{F}(x_1, \ldots, x_n)$ is the field of all rational functions of $x_1, \ldots, x_n$ and is the quotient field of $\mathfrak{F}[x_1, \ldots, x_n]$. We call such fields *function fields* over $\mathfrak{F}$. The units of $\mathfrak{F}[x]$ are the quantities $a \neq 0$ of $\mathfrak{F}$. We have called these quantities constant polynomials and, if f in $\mathfrak{J}$ has degree $n > 0$, we call f a *non-constant* polynomial. Every f of $\mathfrak{J}$ is associated with a monic polynomial of $\mathfrak{J}$. If f divides g and g divides f then $f \neq 0$, $g \neq 0$, $f = ag = abf$ so that $ab = 1$, a and b are units of $\mathfrak{J}$. Hence f and g divide each other if and only if they are associated polynomials, that is, differ only by a factor in $\mathfrak{F}$. When f and g are also monic polynomials they must be identical. We now prove

Theorem 11. *Let* f *and* g *be any two quantities of* $\mathfrak{J} = \mathfrak{F}[x]$ *not both zero. Then there exists a unique monic divisor* d *of* f *and* g *such that*

$$d = af + bg \tag{13}$$

for a *and* b *in* $\mathfrak{J}$.

For consider the set $\mathfrak{L}$ of all non-zero quantities $Af + Bg$ with A and B in $\mathfrak{J}$. It is not an empty set since it contains one of f and g. This implies that there exist a_0, b_0 in $\mathfrak{J}$ such that $d_0 = a_0f + b_0g$ in $\mathfrak{L}$ has the minimum degree n of all quantities of $\mathfrak{L}$. Then $d = d_0\alpha$ is a monic polynomial of degree n associated with d_0 and $d = af + bg$ is in $\mathfrak{L}$ for $a = a_0\alpha$, $b = b_0\alpha$ in $\mathfrak{J}$. By Theorem 2.6 we may write $f = sd + r$ where r has degree at most $n - 1$. Then $r = (1 - sa)f + (-sb)g$ cannot be in $\mathfrak{L}$ and must be zero. Thus d divides f and similarly d divides g. If d_1 is any second polynomial with the same properties as d then d_1 divides f and g and thus $af + bg = d = \alpha d_1$.

Similarly, $d_1 = \beta d$, $d = \alpha\beta d$, $\alpha\beta = 1$, α is a unit. But d and d_1 are monic polynomials and $\alpha = 1$, $d = d_1$.

The quantity (13) is the g.c.d. of f and g. For every divisor of f and g divides d of (13) and d divides f and g. We may therefore state that d is independent of the field $\mathfrak{F}$ and therefore depends only on the *smallest* subfield of $\mathfrak{F}$ containing the coefficients of f and g. For we saw in Section 2.6 that if $\mathfrak{K}$ is a field containing $\mathfrak{F}$, and x is an indeterminate over $\mathfrak{K}$ as well as $\mathfrak{F}$ then all the g.c.d. in $\mathfrak{K}[x]$ of f and g have the form αd where α is a unit and hence is in $\mathfrak{K}$. When αd is a monic polynomial we have $\alpha = 1$.

For polynomial domains $\mathfrak{F}[x]$ we define a unique g.c.d. It is the polynomial (13) and we again think of the very important theorem that states that d is unchanged when we extend the coefficient field $\mathfrak{F}$. This also implies that f and g are relatively prime polynomials of $\mathfrak{F}[x]$ if and only if they have no non-constant polynomial factor in common in any $\mathfrak{K}[x]$ containing them.

An *irreducible* (or *prime*) polynomial of $\mathfrak{F}[x]$ need not remain irreducible in $\mathfrak{K}[x]$, $\mathfrak{K} > \mathfrak{F}$. We shall designate this dependence on $\mathfrak{F}$ by sometimes saying that f is irreducible in $\mathfrak{F}$. We now prove

LEMMA 1. *Let* p *and* q *be in* $\mathfrak{F}[\mathrm{x}]$ *and* p *be irreducible in* $\mathfrak{F}$. *Then either* p *divides* q *or is prime to* q, *that is* $\mathrm{ap} + \mathrm{bq} = 1$ *for* a, b *in* $\mathfrak{F}[\mathrm{x}]$.

For the g.c.d. of p and q is $d = ap + bq$. If $d \neq 1$ then d is associated with p and p divides q. Otherwise $1 = ap + bq$.

LEMMA 2. *Let an irreducible* p *divide* gh. *Then* p *divides* g *or* h.

For if p does not divide g we have $1 = ap + bg$ by Lemma 1, $gh = pr$ by hypothesis, $h = aph + bgh = (ah + br)p$ as desired.

The assumption that a non-constant polynomial f is reducible implies the existence of factors r, s of degrees less than that of $f = rs$. We shall not discuss the complicated question of finding such a factorization but it must of course exist. If r were not irreducible it would have a factorization. This process lowers the degrees of the factors and ultimately gives $f = \alpha p_1 \ldots p_u$ where the p_i are non-constant monic irreducible factors of f and α is clearly the leading coefficient of f. If also $f = \alpha q_1 \ldots q_v$ we have $p_1 \ldots p_u = q_1 \ldots q_v$ and p_1 divides $q_1 \ldots q_v$. By Lemma 2 p_1 divides one q_i and must be a q_i. We may then assume that $p_1 = q_1$, $p_2 \ldots p_u = q_2 \ldots q_v$. Continuing in this fashion we ultimately prove that the q_i may be rearranged so that $u = v$, $q_i = p_i$. Equal p_i are multiplied to yield $p_i^{e_i}$ and we have proved

Theorem 12. *Every non-constant polynomial* f *of* $\mathfrak{F}[\mathrm{x}]$ *is uniquely factorizable apart from the order of the factors as a product*

$$\alpha \mathrm{p}_1^{\mathrm{e}_1} \ldots \mathrm{p}_t^{\mathrm{e}_t},$$

where the p_i *are distinct monic irreducible polynomials of* $\mathfrak{F}[\mathrm{x}]$ *and* α *in* $\mathfrak{F}$ *is the leading coefficient of* f.

EXERCISES

1. Show that an irreducible polynomial $f(x)$ is prime to every $g(x)$ whose degree is less than that of f.

2. State the analogue of Ex. 1 for the integral domain of all ordinary integers.

3. Let f and f_1 be both not zero and in $\mathfrak{J} = \mathfrak{F}[x]$. Divide f by f_1 and get a remainder f_2. Divide f_1 by f_2 and obtain a remainder f_3, and so forth until ultimately f_{r-1} is divisible by f_r, $f_{r+1} = 0$. Show that $f_r = a_0 f + b_0 f_1$ divides f and f_1 and is associated with the g.c.d. of f and f_1.

4. State and prove the analogue of Ex. 3 for the domain of all ordinary integers.

5. The process of Ex. 3, 4 is called Euclid's process. Apply it to find the g.c.d. of $f = x^4 - 7x - 2$ and $f_1 = x^3 - x^2 - x - 2$.

6. Prove that $f = x^3 + 2x^2 - x + 1$ and $g = x^2 + x + 2$ are relatively prime.

7. Let $af + bg = 1$. Show that if also $a_0 f + b_0 g = 1$ then $a_0 = a + gc$, $b_0 = b - fc$ for polynomials a, b, c, a_0, b_0, f, g of $\mathfrak{F}[x]$.

8. Unique factorization domains. The set of all natural integers possesses the unique factorization property of Theorem 1.4. We have also obtained an analogous property for the integral domain $\mathfrak{F}[x]$. In the present section and in Section 2.9 we shall study other domains with such a property. The notions and results obtained are not needed in the text except for a single application in Chapter X. Their omission will not disturb the continuity of the text, and so, although their natural position in the logical development of our subject is the present one, the reading of these two sections will usually be postponed until after the study of Chapter IX. Thus the reader may pass on to the summary in Section 2.10.

Consider an integral domain $\mathfrak{J}$. The elements of $\mathfrak{J}$ which are not prime elements, zero, or units have a factorization $a = bc$ with neither b nor c a unit of $\mathfrak{J}$. Then also $a = b_0 c_0$ for b_0 any associate ba of b and $c_0 = a^{-1}c$ in $\mathfrak{J}$. Since every integral domain $\mathfrak{J}$ has at least two units -1, 1 we cannot expect a factorization of the elements of $\mathfrak{J}$ into unique factors. Also the order of the factors is of course not unique.

It is well known to the reader that rational *positive* integers are uniquely factorizable into *positive* prime factors apart from their order. We shall wish to consider integral domains with the analogous property and shall make the

DEFINITION. *An integral domain $\mathfrak{J}$ is called a unique factorization domain (abbreviated u.f. domain) if*

1. *Every element of $\mathfrak{J}$ which is neither zero nor a unit is expressible in the form*

$$a = p_1 p_2 \dots p_r , \tag{14}$$

where the p_i are primes of $\mathfrak{J}$.

2. *Let also* a $=$ $q_1 \dots q_s$ *for primes* q_i *of* $\mathfrak{J}$. *Then* r $=$ s *and there exists a permutation* $i_1, \dots, i_r$ *of* 1, 2, . . . , r *such that* q_{i_j} *is associated with* p_j.

In a u.f. domain we may write every a of $\mathfrak{J}$ in the form

$$a = \alpha p_1^{e_1} \dots p_t^{e_t}, \tag{15}$$

where α is a unit of $\mathfrak{J}$ and p_i is a prime not associated with p_j for $i \neq j$. Then the only other factorizations of a of this type are those obtained by replacing the p_i by associates with the corresponding change in α and by permuting the factors. Thus the exponents e_i are unique and the primes p_i are unique apart from unit factors. If $a = bc$ and we factor b and c into prime factors q_j we see that the q_j must be associates of the p_i. Thus every divisor of a has the form

$$b = \beta p_1^{f_1} \dots p_t^{f_t}, \tag{16}$$

where β is a unit of $\mathfrak{J}$ and the f_j are integers such that

$$0 \leqq f_j \leqq e_j .$$

If g is in $\mathfrak{J}$ and b divides both a and g the prime factors of a occurring in b with non-zero exponents must be prime factors of g with exponents at least as large as those of b. Hence we may write

$$g = \gamma p^{g_1}{}_1 \dots p_t^{g_t} p_{t+1}^{g_{t+1}} \dots p_s^{g_s}, \tag{17}$$

where the $g_i \geqq 0$ and every common divisor of a and g is a common divisor of a and $g_0 = \gamma p_1^{g_1} \dots p_t^{g_t}$. We let k_i be the smaller of g_i and e_i and write

$$d = p_1^{k_1} \dots p_t^{k_t} .$$

Then if b divides both a and g we have $f_i \leqq g_i$, $f_i \leqq e_i$ so that b divides d. Evidently d divides a and g and we have proved

Theorem 13. *Any two elements not both zero of a u.f. domain* $\mathfrak{J}$ *have a g.c.d. in* $\mathfrak{J}$.

Two evident corollaries of our above factorization are stated in

Theorem 14. *Let* a, b, c *be in a u.f. domain* $\mathfrak{J}$ *and let* c *be prime to* a *and divide* ab. *Then* c *divides* b.

Corollary. *Let a prime quantity* p *divide* ab *and not* a. *Then* p *divides* b.

In an arbitrary integral domain it is possible that some element a which is neither zero, a prime, nor a unit has a factorization $a = bc$, $b = b_1c_1$, and so forth, and that we never arrive at a factorization (14) of a. We assume that $\mathfrak{J}$ is not such a domain and derive the criterion

Theorem 15. *Let $\mathfrak{J}$ be an integral domain with the property that every* $\mathfrak{a} \neq 0$ *of $\mathfrak{J}$ has a factorization*

$$\mathfrak{a} = a\mathfrak{p}_1 \ldots \mathfrak{p}_r$$

with a a unit and the $\mathfrak{p}_i$ *primes of $\mathfrak{J}$. Suppose also that a prime* $\mathfrak{p}$ *of $\mathfrak{J}$ divides a product* $\mathfrak{ab}$ *with* $\mathfrak{a}$, $\mathfrak{b}$ *in $\mathfrak{J}$ if and only if one of* $\mathfrak{a}$ *and* $\mathfrak{b}$ *is divisible by* $\mathfrak{p}$. *Then $\mathfrak{J}$ is a unique factorization domain.*

The proof of the above is trivial. It is left as an exercise for the reader.

EXERCISES

1. Show that the field $\mathfrak{P}$ of all residue classes modulo a prime p is a u.f. domain.

2. Prove that every a and b of $\mathfrak{P}$ have a g.c.d. of the form $ag + bh$.

3. Prove that the polynomials $f = xy + y + 1$, $g = xy - x$ in indeterminates x, y over the rational number field are relatively prime. Prove also that there exist no polynomials $a(x, y)$, $b(x, y)$ such that $af + bg = 1$. Hint: Find integral values of x and y which make $f = g = 0$.

4. Verify that Theorem 2.12 is equivalent to the statement that $\mathfrak{F}[x]$ is a u.f. domain.

9. The domain $\mathfrak{J}[x]$. The non-zero elements of a field $\mathfrak{F}$ are all units and $\mathfrak{F}$ is vacuously a u.f. domain. The extension $\mathfrak{F}[x]$ has been shown to be a u.f. domain and we shall generalize this result and shall show that if $\mathfrak{J}$ is any u.f. domain, and x an indeterminate over $\mathfrak{J}$, then $\mathfrak{J}[x]$ is a u.f. domain.

Let $\mathfrak{J}$ be a u.f. domain and x be an indeterminate over the quotient field $\mathfrak{F}$ of $\mathfrak{J}$. The elements of $\mathfrak{J}[x]$ have the form

$$f = f(x) = a_0 + a_1x + \ldots + a_nx^n \qquad (a_i \text{ in } \mathfrak{J}) . \tag{18}$$

We call f a *primitive element* of $\mathfrak{J}[x]$ if the a_i are relatively prime. Evidently every f of $\mathfrak{J}[x]$ has the form $f = af_0$ where f_0 is primitive and $a = (a_0, \ldots, a_n)$ is in $\mathfrak{J}$. For $\mathfrak{J}$ is a u.f. domain.

A quantity b of $\mathfrak{J}$ divides f if and only if b divides all the a_i. For if $f = bf_0$ where f_0 is in $\mathfrak{J}[x]$ we have $f_0 = b_0 + b_1x + \ldots + b_mx^m$, b_i in $\mathfrak{J}$, and obtain $n = m$, $b_ib = a_i$ as desired.

When f is primitive it is divisible by b in $\mathfrak{J}$ if and only if b is a unit. For $a_i = b_ib$ and the a_i are relatively prime.

LEMMA 1. *The product of two primitive polynomials of $\mathfrak{J}[x]$ is primitive.*

For let f and g be primitive and fg be not primitive. Then $fg = bh$ where b is in $\mathfrak{J}$. Since b is in a u.f. domain there exists a prime factor of b in $\mathfrak{J}$ which divides fg. Hence we may assume that b is prime. Write

$$f = f_0 + f_1b, \qquad g = g_0 + g_1b ,$$

where no coefficient of f_0 or g_0 is divisible by b and f_0, f_1, g_0, g_1 are in $\mathfrak{J}[x]$. Then $fg = f_0g_0 + b(f_0g_1 + f_1g_0 + f_1g_1b)$ is divisible by b and so is f_0g_0. The leading coefficient of f_0g_0 is the product of the leading coefficients c of f_0 and d of g_0. Neither c nor d is divisible by b but cd is divisible by b. This is impossible in a u.f. domain $\mathfrak{J}$. Thus $f = f_1b$, $g = g_1b$ which is contrary to hypothesis.

LEMMA 2. *Let* c *in* $\mathfrak{J}$ *divide* fg *and be prime to* f. *Then* c *divides* g.

For $f = af_0$, $g = bg_0$, $fg = abh$ where $h = f_0g_0$ is primitive. Then c divides all the coefficients of fg and must divide their g.c.d. which is evidently ab. Since c is prime to f it is prime to a and must divide b. But then c divides g.

LEMMA 3. *Let* $\mathfrak{F}$ *be the quotient field of* $\mathfrak{J}$, f *be a primitive element of* $\mathfrak{J}$[x], *and*

$$\mathrm{f} = \mathrm{gh} \qquad (\mathrm{g, h} \textit{ in } \mathfrak{F}[\mathrm{x}]),$$

be reducible in $\mathfrak{F}$. *Then*

$$\mathrm{g} = \mathrm{ag_0}, \qquad \mathrm{h} = \mathrm{bh_0} \qquad (\mathrm{a, b} \textit{ in } \mathfrak{F}), \tag{19}$$

where $\mathrm{g_0}$ *and* $\mathrm{h_0}$ *are primitive elements of* $\mathfrak{J}$[x], *and for every factorization* (19) *of* g *and* h *the product* ab *is a unit of* $\mathfrak{J}$.

For the coefficients of g are quotients of elements of $\mathfrak{J}$ and we write these with a common denominator, factor out the g.c.d. of the numerators and obtain (19). Let (19) hold and write $ab = cd^{-1}$ where c and d are relatively prime and we may take $d = 1$ if d is a unit. But Lemma 1 states that $f_0 = g_0h_0$ is a primitive polynomial, $f = cf_0d^{-1}$, and Lemma 2 states that d divides f_0. Hence d is a unit and c is in $\mathfrak{J}$. Both f_0 and $cf_0 = f$ are primitive so that c is a unit.

As a corollary of Lemma 3 we may state what is usually called the *Gauss Lemma.*

Theorem 16. *Let* f(x) *be a monic polynomial with coefficients in a u.f. domain* $\mathfrak{J}$ *and*

$$\mathrm{f(x)} = \mathrm{g(x)} \cdot \mathrm{h(x)}$$

where g(x) *and* h(x) *are monic polynomials with coefficients in the quotient field* $\mathfrak{F}$ *of* $\mathfrak{J}$. *Then these coefficients are all in* $\mathfrak{J}$.

For one coefficient of $f = f(x)$ is unity and f is a primitive element of $\mathfrak{J}[x]$. By Lemma 3 we have $g = ag_0$, $h = bh_0$ where g_0 and h_0 are primitive and a, b are in $\mathfrak{F}$, ab is a unit of $\mathfrak{J}$. Write

$$a = cd^{-1}, \qquad b = qr^{-1} \qquad (c, d, q, r \text{ in } \mathfrak{J}),$$

where c and d are relatively prime, q and r are relatively prime. Then cq is divisible by dr and the quotient is a unit ab of $\mathfrak{J}$. By Theorem 2.14 we have

$$c = rc_0, \qquad q = dq_0 \qquad ab = c_0q_0$$

for c_0, q_0 in $\mathfrak{J}$. Then c_0 and q_0 are units of $\mathfrak{J}$ and c_0g_0, q_0h_0 are primitive. We replace g_0 by c_0g_0, h_0 by q_0h_0 and may thus assume with no loss of generality that $c_0 = q_0 = 1$,

$$a = rd^{-1}, \qquad b = dr^{-1} = a^{-1}, \qquad f = g_0h_0 .$$

The leading coefficient of $g(x) = ag_0(x)$ is $rd^{-1}\gamma_0$ where γ_0 is the leading coefficient of g_0. Thus

$$rd^{-1}\gamma_0 = 1, \qquad dr^{-1}\delta_0 = 1 \qquad\qquad (\gamma_0, \delta_0 \text{ in } \mathfrak{J}),$$

where δ_0 is the leading coefficient of h_0. Now $r\gamma_0 = d$, $d\delta_0 = r$ so that r and d divide each other. Hence a and b are in $\mathfrak{J}$ and $g = ag_0$, $h = bh_0$ have coefficients in $\mathfrak{J}$.

Another consequence of our lemma is given by

Theorem 17. *Let $\mathfrak{J}$ be a u.f. domain and* x *be an indeterminate over $\mathfrak{J}$. Then $\mathfrak{J}$[x] is a u.f. domain.*

For let $f(x)$ be in $\mathfrak{J}[x]$. Then $f = a_0f_0$ where f_0 is primitive and a_0 is in $\mathfrak{J}$. We let $\mathfrak{F}$ be the quotient field of $\mathfrak{F}$ and factor f in $\mathfrak{F}[x]$ into irreducible factors as in Theorem 2.12. Applying Lemma 3 we may replace each irreducible factor of f_0 by a primitive polynomial of $\mathfrak{J}[x]$ and have $f_0 = \alpha f_1 \ldots f_r$ where the f_i are primitive polynomials of $\mathfrak{J}[x]$ and are irreducible in $\mathfrak{F}[x]$, α is a unit of $\mathfrak{J}$.

The polynomials $f_i(x)$ are prime elements of $\mathfrak{J}[x]$. For otherwise $f_i(x) = g_i(x) \cdot h_i(x)$ with neither factor a unit of $\mathfrak{J}[x]$ and, since $f_i(x)$ is an irreducible polynomial of $\mathfrak{F}[x]$, one of $h_i(x)$ and $g_i(x)$ is in $\mathfrak{F}$. We let this be $h_i(x)$ and see that $h_i = h_i(x)$ is in $\mathfrak{J}[x]$ and in $\mathfrak{F}$ and hence is in $\mathfrak{J}$. But $f_i = g_i(x)h_i$ is primitive so that h_i is a unit of $\mathfrak{J}$, h_i is a unit of $\mathfrak{J}[x]$, a contradiction. We thus have $a = \alpha a_0$

$$f = af_1 \ldots f_r \qquad\qquad (a \text{ in } \mathfrak{J}).$$

The factorization of $a = a_1 \ldots a_s$ into prime elements of $\mathfrak{J}$ provides a factorization of f,

$$f = a_1 \ldots a_sf_1 \ldots f_r ,$$

into prime elements of $\mathfrak{J}[x]$ since every prime element of $\mathfrak{J}$ is evidently prime in $\mathfrak{J}[x]$.

Let now $f = b_1 \ldots b_tg_1 \ldots g_u$ with $g_i(x)$ primes of $\mathfrak{J}[x]$ not in $\mathfrak{J}$ and b_j primes of $\mathfrak{J}$. Then Theorem 2.12 implies that $u = r$ and the f_i differ from the g_i by factors h_i in $\mathfrak{F}$. But if g_i is a prime of $\mathfrak{J}[x]$ it is primitive and

$f_i = h_i g_i$. Thus h_i is a unit of $\mathfrak{J}$ and $b_1 \ldots b_t$ is associated with $a_1 \ldots a_s$. Since $\mathfrak{J}$ is a u.f. domain the factorization of a is unique apart from unit factors and the b_i are associated with the a_i in some order. This proves our theorem.

An immediate induction gives

Theorem 18. *Let $\mathfrak{J}$ be a u.f. domain and* $x_1, \ldots, x_n$ *be independent indeterminates over $\mathfrak{J}$. Then* $\mathfrak{J}[x_1, \ldots, x_n]$ *is a u.f. domain.*

The above result is usually applied to the special case of

Theorem 19. *Let* $x_1, \ldots, x_n$ *be independent indeterminates over a field $\mathfrak{F}$. Then* $\mathfrak{F}[x_1, \ldots, x_n]$ *is a u.f. domain.*

We also apply Theorem 2.16 and obtain

Theorem 20. *Let* $x, x_1, \ldots, x_n$ *be independent indeterminates over $\mathfrak{F}$ and* $f(x)$ *have leading coefficient unity and further coefficients polynomials in* $x_1, \ldots, x_n$ *with coefficients in $\mathfrak{F}$. Then if*

$$f(x) = g(x) \cdot h(x)$$

where $g(x)$ *and* $h(x)$ *are monic polynomials with coefficients rational functions of* $x_1, \ldots, x_n$, *these coefficients are all polynomials in* $x_1, \ldots, x_n$.

EXERCISES

1. State the Gauss Lemma for the case where $\mathfrak{J}$ is the domain of all ordinary integers, of all polynomials in an indeterminate t with rational coefficients, of all polynomials in t with integral coefficients.

2. State the meaning of Theorem 2.18 in terms of our definition of a unique factorization domain.

10. The results of the text. We have now obtained our fundamental coefficient domains, the integral domains and fields. It is important to remember that a field is simply an additive abelian group whose non-zero elements form a multiplicative abelian group, such that the distributive law holds. Fields are of course of two types. One has a prime subfield consisting of a set equivalent to the rational number field and is *non-modular*. The other and more unusual type has a prime subfield equivalent to the field of all residue classes modulo a prime p, and is called *modular of characteristic* p. The principal property of such fields is a consequence of the binomial theorem and states that $(a + b)^p = a^p + b^p$ since $p = 0$.

A field is a special case of an integral domain and every integral domain is contained in its quotient field. An integral domain is merely a commutative ring with a unity element and no divisors of zero, and the quotient field may be thought of as the natural extension to fractions of our integral elements. Divisibility properties and unique factorization will play no great role in the further work of our text but will be used at the beginning of Chapter X. It is important, however, to know that by polynomials $f(x)$ of $\mathfrak{F}[x]$ we mean polynomials in the indeterminate x with coefficients in the field. Also these polynomials have a unique factorization, apart from unit (constant) factors, into irreducible factors, and every non-zero pair f, g, has a unique greatest common divisor. We emphasize also that the ordinary division algorithm holds for polynomials in x with coefficients in a ring $\mathfrak{A}$ so long as $\mathfrak{A}$ has a unity element and we use a regular leading coefficient for our divisor. Then the remainder and factor theorems are true, but when $\mathfrak{A}$ is not a commutative ring care must be taken about the order of elements in our products.

11. Linear sets over $\mathfrak{F}$. We have defined fields and linear sets. The most interesting and important linear sets for our purposes are those of finite order n over a field $\mathfrak{F}$. We shall study the elementary properties of such sets. Their elements may be thought of as points in an n-dimensional space with coordinates in $\mathfrak{F}$. A particular example of this is ordinary two-space. Here the points form the elements of a linear set of order two over the real number field. We now define these linear sets.

Let $\mathfrak{L}$ be a linear set over $\mathfrak{F}$ as defined in Section 1.10. *We shall assume henceforth* that $xa = ax$ for every x in $\mathfrak{L}$ and a in $\mathfrak{F}$ so that $\mathfrak{L}$ is essentially merely a left linear set, right and left scalar multiplication being the same operation.

If $u_1, \ldots, u_n$ are in $\mathfrak{L}$ the quantity

$$x = a_1u_1 + \ldots + a_nu_n = u_1a_1 + \ldots + u_na_n \tag{20}$$

is in $\mathfrak{L}$ for every $a_1, \ldots, a_n$ in $\mathfrak{F}$. We call x a *linear combination* of $u_1, \ldots, u_n$ with coefficients $a_1, \ldots, a_n$ in $\mathfrak{F}$.

DEFINITION. *A set of quantities* $u_1, \ldots, u_n$ *of* $\mathfrak{L}$ *over* $\mathfrak{F}$ *are called linearly independent in* $\mathfrak{F}$ *if no linear combination* (20) *is zero unless the coefficients* a_i *are all zero.*

DEFINITION. *A linear set* $\mathfrak{L}$ *is said to have order* n *over* $\mathfrak{F}$ *if there exist* n *linearly independent quantities* $u_1, \ldots, u_n$ *of* $\mathfrak{L}$ *over* $\mathfrak{F}$ *such that every* x *of* $\mathfrak{L}$ *is a linear combination* (20). *We call any such set a basis of* $\mathfrak{L}$ *over* $\mathfrak{F}$ *and designate this property by the notation*

$$\mathfrak{L} = (u_1, \ldots, u_n) . \tag{21}$$

Our postulates imply that $1u = u$ for every u of the linear set $\mathfrak{L} = \mathfrak{L}_n$ of (21). Clearly two such linear sets over $\mathfrak{F}$ are equivalent if and only if they have the same order n. Moreover, $\mathfrak{L}_n$ is equivalent to the linear set of all sequences

$$(a_1, \ldots, a_n) = \sum_{i=1}^{n} a_i e_i \qquad (a_i \text{ in } \mathfrak{F})$$

where $u_i \longleftrightarrow e_i$,

$$e_1 = (1, 0, \ldots, 0), \quad e_2 = (0, 1, 0, \ldots, 0), \ldots, \quad e_n = (0, 0, \ldots, 1),$$

and addition and scalar multiplication are defined as for infinite sequences.

The set $\mathfrak{L}_n$ contains the linear subsets $\mathfrak{L}_r = (u_1, \ldots, u_r)$. Every $x \neq 0$ in $\mathfrak{L}$ is *linearly independent* in $\mathfrak{F}$. For this is the statement of the above definition of linear independence in case $n = 1$. Thus (x) is a linear set of order one and, as we have seen above, is equivalent to (u_1).

A set of quantities $x_1, \ldots, x_m$ of $\mathfrak{L}$ are called *linearly dependent* if they are not linearly independent. We may assume them distinct and non-zero without loss of generality. Then some $r \geqq 1$ of the x_i, say $x_1, \ldots, x_r$ are linearly independent and there must exist a_{ji} not all zero in $\mathfrak{F}$ such that

$$a_{j1}x_1 + a_{j2}x_2 + \ldots + a_{jr}x_r + b_jx_j = 0 \qquad (j = r + 1, \ldots, m).$$

When $b_j = 0$ the linear independence of $x_1, \ldots, x_r$ implies that $a_{j1} = a_{j2} = \ldots = a_{jr} = 0 = b_j$, a contradiction. Hence $b_j \neq 0$ has a multiplicative inverse c_j in the field $\mathfrak{F}$ and

$$x_j = c_{j1}x_1 + \ldots + c_{jr}x_r, \qquad c_{jk} = -a_{jk}c_j .$$

Theorem 21. *Let* m *distinct non-zero quantities of a linear set* $\mathfrak{L}$ *be linearly dependent. Then some* r *of them are linearly independent and the remaining* m − r *are linear combinations of these* r.

The set $a_1v_1 + \ldots + a_mv_m$ of all linear combinations with coefficients in $\mathfrak{F}$ of m quantities v_i of a linear set $\mathfrak{L}$ form a linear subset $\mathfrak{M}$ of $\mathfrak{L}$. This of course requires a proof which is a trivial consequence of our definitions. By Theorem 2.21 $\mathfrak{M}$ is a linear set of order $r \leqq m$ over $\mathfrak{F}$.

Theorem 22. *Any* n + 1 *quantities of a linear set* $\mathfrak{L}_n$ *of order* n *are linearly dependent.*

Let $x_1, \ldots, x_{n+1}$ be the quantities. If $x_1 = 0$ then $1 \cdot x_1 = 0$ with coefficient $1 \neq 0$, and if $x_1 = x_2$ then $-x_1 + x_2 = 0$ with coefficient $1 \neq 0$. Hence we may assume that the x_i are all distinct and not zero. The theorem is true when $n = 1$ since then $\mathfrak{L}_n = (u_1)$, $x_1 = a_1u_1$, $x_2 = a_2u_1$, $a_2x_1 - a_1x_2 = 0$ with $a_1 \neq 0$. We make an induction on n and have our theorem true in the subset $\mathfrak{L}_{n-1}$ of $\mathfrak{L}_n$. Thus our theorem is true when the x_i are all in $\mathfrak{L}_{n-1}$, and we may assume this not the case. If we write

$$x_i = \sum a_{ij}u_j \qquad (a_{ij} \text{ in } \mathfrak{F}) ,$$

we have assumed that for one x_i, say x_{n+1}, we have $a_{n+1,n} \neq 0$. If $b = a_{n+1,n}^{-1}$ then

$$y_i = x_i - ba_{i,n}x_{n+1} \qquad (i = 1, \ldots , n)$$

are n quantities of $\mathfrak{L}_{n-1}$ and are linearly dependent. This means that there exist $c_1, \ldots , c_n$ not all zero and in $\mathfrak{F}$ such that $c_1y_1 + \ldots + c_ny_n = 0$. But then

$$c_1x_1 + \ldots + c_nx_n - b\left(\sum_{i=1}^{n} c_ia_{i,n}\right)x_{n+1} = 0$$

and $x_1, \ldots , x_{n+1}$ are linearly dependent in $\mathfrak{F}$.

When $u_1, \ldots , u_n$ are any n linearly independent quantities of a linear set $\mathfrak{L}$ of order n over $\mathfrak{F}$ and x is in $\mathfrak{L}$ the $n + 1$ quantities $x, u_1, \ldots , u_n$ are linearly dependent. Our above considerations of dependent quantities and the assumption of the independence of $u_1, \ldots , u_n$ then imply the existence of $a_1, \ldots , a_n$ such that $x = a_1u_1 + \ldots + a_nu_n$. Thus $\mathfrak{L} = (u_1, \ldots , u_n)$ over $\mathfrak{F}$. Any subset $\mathfrak{L}_0$ of $\mathfrak{L}$ contains at most n linearly independent elements of $\mathfrak{L}$ and the maximum number r of such elements $v_1, \ldots , v_r$ is an integer less than or equal to n. But then $\mathfrak{L}_0 = (v_1, \ldots , v_r)$ is a linear set of order r over $\mathfrak{F}$. If $r = n$ we have shown that every x of $\mathfrak{L}$ is in $\mathfrak{L}_0$, that is, $\mathfrak{L} = \mathfrak{L}_0$, and if $r < n$ obviously $\mathfrak{L}_0 < \mathfrak{L}$. We have proved

Theorem 23. *Let $\mathfrak{L}$ be a linear set of order n over $\mathfrak{F}$ and $u_1, u_2, \ldots , u_n$ be any n linearly independent quantities of $\mathfrak{L}$. Then*

$$\mathfrak{L} = (u_1, \ldots , u_n) .$$

Every linear subset $\mathfrak{L}_0$ of $\mathfrak{L}$ over $\mathfrak{F}$ has order $r \leqq n$ and $\mathfrak{L}_0 = \mathfrak{L}$ if and only if $r = n$.

EXERCISE

Let $\mathfrak{L}$ have order n and $u_1, \ldots , u_r$ be linearly independent elements of $\mathfrak{L}$ for $r < n$. Prove the existence of $u_{r+1}, \ldots , u_n$ in $\mathfrak{L}$ such that $\mathfrak{L} = (u_1, \ldots , u_n)$.

12. Forms. Linear combinations of n independent indeterminates $x_1, \ldots, x_n$ over a field $\mathfrak{F}$ are the case $q = 1$ of what are called n-ary q-ic forms. These polynomials are defined as follows.

An expression

$$a_{i_1 \ldots i_n} x_1^{i_1} x_2^{i_2} \ldots x_n^{i_n} \qquad (a_{i_1 i_2 \ldots i_n} \text{ in } \mathfrak{F})$$

in n independent indeterminates $x_1, \ldots, x_n$ over $\mathfrak{F}$ is called a *term* of *degree* $i_1 + i_2 + \ldots + i_n$ in $x_1, \ldots, x_n$. Every quantity f of the polynomial domain $\mathfrak{F}[x_1, \ldots, x_n]$ is a sum of terms of the above form and we call the maximum degree of all of the terms of f the degree of f. Then f is called a *form* or *homogeneous polynomial* in $x_1, \ldots, x_n$ if every term of f has the same degree q. We say that f is an n-ary q-ic form. Here n is the number of indeterminates, q is the degree of each term of f. If $q = 1$ we call f a *linear* form, $q = 2$ a *quadratic* form, $q = 3$ a *cubic* form, and so forth. When $n = 1$ we call f a *unary* form, $n = 2$ a *binary*, $n = 3$ a *ternary*, $n = 4$ a *quaternary*, and so on. We shall later speak of quadratic forms and shall of course mean n-ary quadratic forms.

13. Algebras over $\mathfrak{F}$. We have discussed rings and have shown how to make certain sets into rings of polynomials by defining multiplication. Every linear set is an additive abelian group and may of course be made into a ring. For example, the linear set of order two over the field of real numbers consists of points (x, y) in the real plane. We make this set into the complex number field by defining $1 = (1, 0)$ as the unity element of our ring, $i = (0, 1)$ such that $i^2 = -(1, 0)$. Thus

$$(x, y)(a, b) = (xa - yb, xb + ya)$$

defines the product of two complex numbers.

The above case led Hamilton to the discovery of a certain division ring which is a ring made from a linear set of order four over the real number system. This so-called *algebra* of *real quaternions* is but a special case of *linear associative algebras* of order n over $\mathfrak{F}$ which are defined by making a linear set $\mathfrak{L}$ of order n over $\mathfrak{F}$ into a ring. This of course is done by defining multiplication of elements of $\mathfrak{L}$ so that the associative and distributive laws hold. We shall call rings of this type *algebras* over $\mathfrak{F}$.

Let $u_1, \ldots, u_n$ be a basis of $\mathfrak{L}$ over $\mathfrak{F}$. If $\mathfrak{L}$ is an algebra we must have

$$(22) \qquad u_i u_j = \sum_{k=1}^{n} \gamma_{ijk} u_k \qquad (\gamma_{ijk} \text{ in } \mathfrak{F} \text{ for } i, j, k = 1, \ldots, n),$$

and

$$(23) \qquad \left(\sum_{i=1}^{n} \xi_i u_i\right)\left(\sum_{j=1}^{n} \eta_j u_j\right) = \sum_{i,j=1}^{n} \xi_i \eta_j u_i u_j \qquad (\xi_i, \eta_j \text{ in } \mathfrak{F}).$$

Conversely it is easy to see that (22) and (23) define multiplication in $\mathfrak{L}$ such that $\mathfrak{L}$ is closed with respect to multiplication, $a(b+c) = ab + ac$, $(b+c)a = ba + ca$ for every a, b, c of $\mathfrak{L}$. But $\mathfrak{L}$ is still not an algebra unless the *multiplication constants* γ_{ijk} are so chosen that the associative law holds in $\mathfrak{L}$. A trivial computation shows that this law holds if and only if

$$u_i(u_j u_k) = (u_i u_j)u_k\,.$$

By the use of (22) we have the equivalent condition

$$\sum_s \gamma_{ijs}\gamma_{skl} = \sum_s \gamma_{jks}\gamma_{isl} \qquad (i, j, k, l = 1, \ldots, n).$$

We have defined equivalence for any rings. If however $\mathfrak{A}$ and $\mathfrak{A}'$ are algebras over $\mathfrak{F}$ we usually wish the elements of $\mathfrak{F}$ to be self-corresponding. We therefore make the

DEFINITION. *Let $\mathfrak{A}$ and $\mathfrak{A}'$ be linear algebras of order* n *over $\mathfrak{F}$ and let there be a* (1–1) *correspondence* $\mathrm{a} \longleftrightarrow \mathrm{a}'$ *between $\mathfrak{A}$ and $\mathfrak{A}'$ such that*

$$(\mathrm{a}+\mathrm{b})' = \mathrm{a}' + \mathrm{b}', \quad (\mathrm{ab})' = \mathrm{a}'\mathrm{b}', \quad (\lambda \mathrm{a})' = \lambda \mathrm{a}'$$

for every a *and* b *of $\mathfrak{A}$ and λ of $\mathfrak{F}$. Then $\mathfrak{A}$ and $\mathfrak{A}'$ are called equivalent over $\mathfrak{F}$.*

Notice that in the above definition we have made $(\lambda a)' = \lambda a'$ and not $\lambda' a'$. It is this property that implies that the quantities of $\mathfrak{F}$ are left invariant under $a \longleftrightarrow a'$ and then we do not have merely an equivalence between $\mathfrak{A}$ and $\mathfrak{A}'$ but *an equivalence over* $\mathfrak{F}$. We usually apply the definition as follows. Let $\mathfrak{A} = (u_1, \ldots, u_n)$ over $\mathfrak{F}$, $\mathfrak{A}' = (u_1', \ldots, u_n')$ over $\mathfrak{F}$ so that $\mathfrak{A} \cong \mathfrak{A}'$ would imply the existence of a particular $u_1', \ldots, u_n'$ such that $u_i \longleftrightarrow u_i'$. But then we must have $(u_i u_j)' = u_i' u_j'$ so that $\gamma_{ijk}' = \gamma_{ijk}$ in the corresponding equation (22) for $\mathfrak{A}$, $\mathfrak{A}'$. Conversely if $\mathfrak{A}$ and $\mathfrak{A}'$ have bases such that $\gamma_{ijk} = \gamma_{ijk}'$ the correspondence $\Sigma\lambda_i u_i \longleftrightarrow \Sigma\lambda_i u_i'$, λ_i in $\mathfrak{F}$, may be trivially seen to define an equivalence over $\mathfrak{F}$ of $\mathfrak{A}$ and $\mathfrak{A}'$.

Theorem 24. *Let $\mathfrak{A}$ and $\mathfrak{A}'$ be algebras of order* n *over $\mathfrak{F}$. Then $\mathfrak{A} \cong \mathfrak{A}'$ if and only if there exist bases* $(\mathrm{u}_1, \ldots, \mathrm{u}_n)$, $(\mathrm{u}_1', \ldots, \mathrm{u}_n')$ *for $\mathfrak{A}$, $\mathfrak{A}'$ respectively such that $\gamma_{ijk} = \gamma_{ijk}'$ in the corresponding multiplication tables* (22).

The notion of a subalgebra now arises. Evidently a subalgebra $\mathfrak{B}$ of an algebra $\mathfrak{A}$ over $\mathfrak{F}$ is defined as a linear subset $\mathfrak{B}$ over $\mathfrak{F}$ of $\mathfrak{A}$ which is an algebra (i.e., a ring) with respect to the operations of $\mathfrak{A}$. But then $\mathfrak{B} = (v_1, \ldots, v_m)$ over $\mathfrak{F}$ is an algebra if and only if $\mathfrak{B}$ is closed with respect to multiplication, that is $v_i v_j = \Sigma\delta_{ijk}v_k$, δ_{ijk} in $\mathfrak{F}$. We also notice that $\mathfrak{B} = \mathfrak{A}$ if and only if $m = n$. Thus we shall consider linear subsets $\mathfrak{B}$ of $\mathfrak{A}$ and shall test $\mathfrak{B}$ for closure under multiplication in order to find out whether or not $\mathfrak{B}$ is a subalgebra of $\mathfrak{A}$, and shall compare the orders of $\mathfrak{B}$ and $\mathfrak{A}$ in order to determine whether or not $\mathfrak{B}$ is a proper subalgebra of $\mathfrak{A}$.

Algebras over $\mathfrak{F}$ which are division rings are called *division algebras.* Similarly, if an algebra is a commutative ring we call it a *commutative algebra.* Finally, in a large part of the literature the distinction we have made by calling only rings which are linear sets over $\mathfrak{F}$ algebras is not made. Thus many authors have used our definition of rings as a definition of algebras and this is, in fact, the older usage.

The theory of what are called square matrices is a theory of quantities of an *algebra.* The operations on matrices are algebraic and we have now obtained what is really necessary in the study of matrices, an algebraic technique and mode of thought. We shall not study square matrices immediately but shall first consider a more general situation.

CHAPTER III

MATRICES

1. Rectangular matrices. Matrices are elements of a particular type of linear set of finite order. Instead of writing a sequence of n elements in a field $\mathfrak{F}$, we write an *array* of m such sequences. Any matrix is such an array and we shall study the consequences of certain algebraic operations on such arrays.

To be more explicit, we define an $m \times n$ (read m by n) matrix to be a rectangular array

$$A = \begin{pmatrix} a_{11} & a_{12} & \cdots & a_{1n} \\ a_{21} & a_{22} & \cdots & a_{2n} \\ \cdot & \cdot & \cdots & \cdot \\ a_{m1} & a_{m2} & \cdots & a_{mn} \end{pmatrix} \tag{1}$$

with elements a_{ij} in a field $\mathfrak{F}$. This is usually abbreviated as

$$A = (a_{ij}) \qquad (i = 1, \ldots, m; \; j = 1, \ldots, n).$$

We shall henceforth in our study of matrices speak of the matrices as *quantities* in our algebraic work and shall restrict the use of the term *element* to mean an element a_{ij} of a matrix.

The matrix A has m rows

$$a_i = (a_{i1}, a_{i2}, \ldots, a_{in}) \qquad (i = 1, \ldots, m) \tag{2}$$

and n columns

$$a^{(j)} = \begin{pmatrix} a_{1j} \\ a_{2j} \\ \cdot \\ \cdot \\ \cdot \\ a_{mj} \end{pmatrix} \qquad (j = 1, \ldots, n) \tag{3}$$

and a_{ij} is the element which is in the ith row and jth column. We may, if we wish, write

$$A = \begin{pmatrix} a_1 \\ a_2 \\ \cdot \\ \cdot \\ \cdot \\ a_m \end{pmatrix} = (a^{(1)}, a^{(2)}, \ldots, a^{(n)}),$$

where the a_i and $a^{(j)}$ are then matrices. Then any $m \times n$ matrix A is composed of m $1 \times n$ matrices or n $m \times 1$ matrices.

The *transpose* A' of A is the $n \times m$ matrix whose rows are the columns of A, that is,

$$(4) \qquad A' = (a'_{ji}) \qquad a'_{ji} = a_{ij} \qquad (j = 1, \ldots, n; \; i = 1, \ldots, m).$$

The operation of transposition will provide a duality among our theorems about matrices A by which any theorem on the rows of A has a dual on the columns of A, that is, the rows of A'.

If $A = (a_{ij})$ and $B = (b_{ij})$ are any $m \times n$ matrices, and λ and μ are in $\mathfrak{F}$, then we define

$$(5) \qquad \lambda A + \mu B = (c_{ij}), \qquad A\lambda + B\mu = (d_{ij})$$

where (actually for $\mathfrak{F}$ any ring)

$$(6) \qquad c_{ij} = \lambda a_{ij} + \mu b_{ij}, \qquad d_{ij} = a_{ij}\lambda + b_{ij}\mu \qquad (i = 1, \ldots, m; \; j = 1, \ldots, n).$$

Then the set $\mathfrak{M}_{m,n}$ of all $m \times n$ matrices with elements in $\mathfrak{F}$ is a linear set of order mn over $\mathfrak{F}$. In fact, if e_{ij} is the $m \times n$ matrix with 1 in the ith row and jth column and zeros elsewhere, we use our linear set notation to write

$$(7) \qquad \mathfrak{M}_{m,n} = (e_{11}, \ldots, e_{mn}) \text{ over } \mathfrak{F},$$

and

$$(8) \qquad A = \sum_{\substack{i = 1, \ldots, m \\ = 1, \ldots, n}} a_{ij} e_{ij}.$$

The zero quantity of $\mathfrak{M}_{m,n}$ is the $m \times n$ matrix $0_{m,n}$ whose elements are all equal to zero. Two matrices A and B are equal, $A - B = 0$, if and only if $a_{ij} = b_{ij}$ $(i = 1, \ldots, m; j = 1, \ldots, n)$.

EXERCISES

1. Prove that the 2×3 matrices

$$\begin{pmatrix} 1 & 1 & 1 \\ 2 & 0 & 1 \end{pmatrix}, \quad \begin{pmatrix} 1 & 1 & 3 \\ -1 & 0 & 3 \end{pmatrix}, \quad \begin{pmatrix} -2 & -2 & 1 \\ 0 & 0 & 1 \end{pmatrix}$$

are linearly independent in a non-modular field $\mathfrak{F}$ and generate a linear subset $\mathfrak{L}_3$ of $\mathfrak{M}_{2,3}$. Show, in fact, that $\mathfrak{L}_3 = (u_1, u_2, u_3)$ over $\mathfrak{F}$ where

$$u_1 = \begin{pmatrix} 1 & 1 & 0 \\ 0 & 0 & 0 \end{pmatrix}, \quad u_2 = \begin{pmatrix} 0 & 0 & 0 \\ 1 & 0 & 0 \end{pmatrix}, \quad u_3 = \begin{pmatrix} 0 & 0 & 1 \\ 0 & 0 & 1 \end{pmatrix}$$

are evidently linearly independent. Why is this evident?

2. Let $A_1, \ldots, A_r$ be a set of r linearly independent $m \times n$ matrices with elements in $\mathfrak{F}$ and $\mathfrak{L}_r$ be the linear set which they generate. Prove by induction that $\mathfrak{L}_r = (B_1, \ldots, B_r)$ where there exist integers i_s, j_s such that B_s has unity in the i_sth row and j_sth column while B_t has zero in the i_sth row and j_sth column for $t \neq s$.

3. What is the element in the i_sth row and j_sth column of $a_1B_1 + \ldots + a_rB_r$ where the B_i are as in Ex. 2? Use this fact to prove that if $A_1, \ldots, A_r$ are r linearly independent $m \times n$ matrices with elements in $\mathfrak{F}$ and if $\mathfrak{K}$ is a field containing $\mathfrak{F}$ then the $A_1, \ldots, A_r$ considered as matrices with elements in $\mathfrak{K}$, are linearly independent in $\mathfrak{K}$.

2. Multiplication of matrices. We have seen the properties of the linear set $\mathfrak{M}_{m,n}$ of order mn over $\mathfrak{F}$ which consists of all $m \times n$ matrices. We shall not define multiplication in $\mathfrak{M}_{m,n}$ except when $m = n$. For a more general product we define an operation on $\mathfrak{M}_{m,n}\mathfrak{M}_{n,s}$ to $\mathfrak{M}_{m,s}$ as follows. Let $A = (a_{ij})$ be any $m \times n$ matrix, $B = (b_{jk})$ be any $n \times s$ matrix, so that $i = 1, \ldots, m; j = 1, \ldots, n; k = 1, \ldots, s$. We define the product

$$C = AB = (c_{ik}), \qquad c_{ik} = \sum_{j=1}^{n} a_{ij}b_{jk} \qquad (9)$$

$$(i = 1, \ldots, m; k = 1, \ldots, s).$$

Notice that we write $(a_{ij})(b_{jk}) = (c_{ik})$ and that the inner subscript is used as a summation index. Thus the product of an $m \times n$ matrix by an $n \times s$ matrix is an $m \times s$ matrix whose element in the ith row and kth column is the sum of the products of the successive elements in the ith row of the left-hand factor by the corresponding elements in the kth column of the right-hand factor. We call this the *row by column rule* for multiplying matrices.

Notice that the number of columns in the left-hand factor A *must be the same as the number of rows in the right-hand factor* B *of* AB.

An $m \times n$ matrix A is composed of an $m \times t$ matrix A_1, whose columns are the first t columns of A, and an $m \times (n - t)$ matrix A_2 whose columns are the remaining columns of A. Thus A has the representation

$$A = (A_1\ A_2)\,, \tag{10}$$

while similarly we may write

$$B = \begin{pmatrix} B_1 \\ B_2 \end{pmatrix}, \tag{11}$$

where B_1 has t rows and s columns, B_2 has $n - t$ rows and s columns. Since

$$\sum_{j=1}^{n} a_{ij}b_{jk} = \sum_{j=1}^{t} a_{ij}b_{jk} + \sum_{j=t+1}^{n} a_{ij}b_{jk}\,,$$

we have

$$AB = (A_1B_1 + A_2B_2)\,. \tag{12}$$

We have multiplied A by B as if A_1, A_2, B_1, B_2 were elements of $\mathfrak{F}$.

We may also write

$$A = \begin{pmatrix} A_1 & A_2 \\ A_3 & A_4 \end{pmatrix}, \tag{13}$$

where A_1 and A_2 have r rows, A_3 and A_4 have $m - r$ rows, A_1 and A_3 have t columns, A_2 and A_4 have $n - t$ columns. Then (12) implies immediately

Theorem 1. *Let* A *be an* $\mathrm{m} \times \mathrm{n}$ *matrix and* B *be an* $\mathrm{n} \times \mathrm{s}$ *matrix and write*

$$\mathrm{A} = \begin{pmatrix} \mathrm{A}_1 & \mathrm{A}_2 \\ \mathrm{A}_3 & \mathrm{A}_4 \end{pmatrix}, \qquad \mathrm{B} = \begin{pmatrix} \mathrm{B}_1 & \mathrm{B}_2 \\ \mathrm{B}_3 & \mathrm{B}_4 \end{pmatrix} \tag{14}$$

where A_1 *has* r *rows and* t *columns,* B_1 *has* t *rows and* q *columns. Then*

$$\mathrm{AB} = \begin{pmatrix} \mathrm{A}_1\mathrm{B}_1 + \mathrm{A}_2\mathrm{B}_3 & \mathrm{A}_1\mathrm{B}_2 + \mathrm{A}_2\mathrm{B}_4 \\ \mathrm{A}_3\mathrm{B}_1 + \mathrm{A}_4\mathrm{B}_3 & \mathrm{A}_3\mathrm{B}_2 + \mathrm{A}_4\mathrm{B}_4 \end{pmatrix}. \tag{15}$$

This theorem states that the multiplication of matrices may be reduced to the row by column multiplication of two rowed and columned matrices *whose elements are themselves matrices.* This form (15) of AB is often much more important than the explicit form (9). Notice also that we do not assume that any of $r, t, q, m - r, n - t, s - q$ are not zero and that then (12) is a special case ($r = m, q = s$) of (15).

We may easily see that

Theorem 2. *Multiplication of matrices is associative.*

For if $A = (a_{ij})$, $B = (b_{jk})$, $C = (c_{kl})$ where $i = 1, \ldots, m; j = 1, \ldots, n; k = 1, \ldots, s; l = 1, \ldots, t$, then $A(BC) = (d_{il})$ where

$$d_{il} = \sum_{j=1}^{n} a_{ij}\left(\sum_{k=1}^{s} b_{jk}c_{kl}\right), \tag{16}$$

and

$$(AB)C = (d_{il}^{0}), \tag{17}$$

with

$$d_{il}^{0} = \sum_{k=1}^{s}\left(\sum_{j=1}^{n} a_{ij}b_{jk}\right)c_{kl}. \tag{18}$$

Multiplication in a field $\mathfrak{F}$ is associative and distributive with respect to addition so that

$$\left(\sum_{j=1}^{n} a_{ij}b_{jk}\right)c_{kl} = \sum_{j=1}^{n} (a_{ij}b_{jk})c_{kl} = \sum_{j=1}^{n} a_{ij}(b_{jk}c_{kl}). \tag{19}$$

Any finite double sum is independent of the order of summation and, by using the distributive law,

$$\begin{cases} d_{il}^{0} = \sum_{k=1}^{s}\sum_{j=1}^{n} a_{ij}(b_{jk}c_{kl}) = \sum_{j=1}^{n}\sum_{k=1}^{s} a_{ij}(b_{jk}c_{kl}) \\ \qquad = \sum_{j=1}^{n} a_{ij}\left(\sum_{k=1}^{s} b_{jk}c_{kl}\right) = d_{il}, \end{cases} \tag{20}$$

as desired. Theorem 3.2 is thus a consequence of the associative and distributive laws of the field $\mathfrak{F}$.

If A and B are $m \times n$ matrices, and C is an $n \times s$ matrix, the element in the ith row and jth column of $(A + B)C$ is

$$(21) \quad \begin{cases} \sum_{j=1}^{n} (a_{ij} + b_{ij})c_{jk} = \sum_{j=1}^{n} (a_{ij}c_{jk} + b_{ij}c_{jk}) \\ \qquad = \left(\sum_{j=1}^{n} a_{ij}c_{jk}\right) + \left(\sum_{j=1}^{n} b_{ij}c_{jk}\right), \end{cases}$$

which is the element in the ith row and kth column of $AC + BC$. A similar discussion of $C(A + B)$ gives

Theorem 3. *Multiplication of matrices is distributive with addition, that is,*

$$(22) \qquad (A + B)C = AC + BC, \qquad C(A + B) = CA + CB.$$

Let A, B, $C = AB$ be given as in (9). By (4) we have

$$(23) \qquad C' = (c'_{ki}), \qquad c'_{ki} = c_{ik} = \sum_j a_{ij}b_{jk} = \sum_j b'_{kj}a'_{ji},$$

whence $C' = B'A'$. We state this result as

Theorem 4. *The transpose of a product of two matrices is the product of their transposes in reverse order.*

A matrix is called a *square matrix* if the number of its rows is equal to the number of its columns. The set $\mathfrak{M}_n$ of all n-rowed square matrices is an algebra of order n^2 over $\mathfrak{F}$. For we have seen that $\mathfrak{M}_n$ is a linear set of order n^2 over $\mathfrak{F}$. But now we have defined multiplication in $\mathfrak{M}_n$ and $\mathfrak{M}_n$ is an algebra by Theorems 3.2 and 3.3.

EXERCISES

1. Use Theorem 3.1 to compute

$$\begin{pmatrix} 1 & 0 & 0 & 0 \\ 0 & 1 & 0 & 0 \\ -1 & 0 & 1 & 0 \\ 0 & -1 & 0 & 1 \end{pmatrix} \begin{pmatrix} 2 & 0 \\ 0 & 2 \\ 1 & 2 \\ 3 & 4 \end{pmatrix} = \begin{pmatrix} 2 & 0 \\ 0 & 2 \\ -1 & 2 \\ 3 & 2 \end{pmatrix}$$

as a product of matrices whose elements are two-rowed square matrices.

2. Show that the two-rowed matrices

$$e_{11} = \begin{pmatrix} 1 & 0 \\ 0 & 0 \end{pmatrix}, \quad e_{12} = \begin{pmatrix} 0 & 1 \\ 0 & 0 \end{pmatrix}, \quad e_{21} = \begin{pmatrix} 0 & 0 \\ 1 & 0 \end{pmatrix}, \quad e_{22} = \begin{pmatrix} 0 & 0 \\ 0 & 1 \end{pmatrix},$$

which are a basis of the linear set of all two-rowed square matrices, have the properties

$$e_{11}^2 = e_{11}, \quad e_{22}^2 = e_{22}, \quad e_{12}e_{21} = e_{11}, \quad e_{21}e_{12} = e_{22},$$

$$0 = e_{11}e_{21} = e_{12}e_{11} = e_{11}e_{22} = e_{22}e_{11} = e_{22}e_{12} = e_{21}e_{22}.$$

Show also that $IA = AI = A$ for any two-rowed matrix A and $I = e_{11}+e_{22}$.

3. Let I be the two-rowed identity matrix above, and E_{ij} be the four-rowed square matrices

$$E_{11} = \begin{pmatrix} I & 0 \\ 0 & 0 \end{pmatrix}, \quad E_{12} = \begin{pmatrix} 0 & I \\ 0 & 0 \end{pmatrix}, \quad E_{21} = \begin{pmatrix} 0 & 0 \\ I & 0 \end{pmatrix}, \quad E_{22} = \begin{pmatrix} 0 & 0 \\ 0 & I \end{pmatrix},$$

where the zeros are the two-rowed zero matrices. Show that Theorem 3.1 implies that the E_{ij} have the same multiplicative properties as the e_{ij} and that the algebra of order 4 over $\mathfrak{F}$ with basis E_{11}, E_{12}, E_{21}, E_{22} is therefore equivalent to the algebra of all two-rowed square matrices.

4. Obtain an algebra of six-rowed square matrices equivalent to the algebra of all two-rowed square matrices by a method analogous to Ex. 3.

5. Verify Theorem 3.2 by direct multiplication for the matrix product

$$\begin{pmatrix} 3 & -2 \\ -2 & 1 \end{pmatrix} \begin{pmatrix} 2 & 1 & 4 \\ 6 & 5 & -3 \end{pmatrix} \begin{pmatrix} 2 \\ -2 \\ 1 \end{pmatrix}.$$

3. Determinants. Any square matrix D with elements a_{ij} $(i, j = 1, \ldots, n)$ in a field $\mathfrak{F}$ has a determinant

$$\Delta = \Sigma(-1)^i a_{1i_1}a_{2i_2} \ldots a_{ni_n}$$

in $\mathfrak{F}$, where the sum is taken over all permutations $i_1 \ldots i_n$ of $1, 2, \ldots, n$, and i is the number of interchanges carrying the permutation into $12 \ldots n$. This definition is shown in the elementary theory of determinants to be independent of the particular interchanges used and to define a unique function Δ of the matrix D. The following elementary properties are also proved in that theory:

LEMMA 1. *The determinant of any square matrix* D *is equal to the determinant of its transpose* D′.

LEMMA 2. *If any two rows (or columns) of a square matrix* D *are interchanged the corresponding determinant is changed in sign.*

LEMMA 3. *If two rows (or columns) of a square matrix D are identical the determinant of D is zero.*

LEMMA 4. *Let k be a factor of every element of a row (or column) of a square matrix D and let D_0 be the matrix obtained from D by removing this factor k. Then the determinant of D is k times that of D_0.*

LEMMA 5. *A determinant is not changed in value if any multiple of a row (or column) is added to any other row (column).*

One of the rules for multiplying determinants is our rule for multiplication of matrices. The consequent theorem is

LEMMA 6. *The determinant of a product of two square matrices is the product of their determinants.*

All of the proofs of the above lemmas of elementary determinant theory, with the exception of Lemma 3, may be easily verified to hold for any field $\mathfrak{F}$. In this exceptional case it is usual to apply Lemma 2 to get $2\Delta = 0$ and thus $\Delta = 0$. When $\mathfrak{F}$ is a field of characteristic two we have $2\Delta = 0$ for any Δ of $\mathfrak{F}$ and the proof fails.

A simple argument completes the proof. We let $D = (d_{ij})$ where the d_{ij} in certain two columns are identical but otherwise the d_{ij} are independent indeterminates over the field $\mathfrak{R}$ of all rational numbers. Then Δ is a polynomial in the d_{ij} with integral coefficients which are all zero by the above elementary proof. The prime subfield $\mathfrak{P}_2$ of any field $\mathfrak{F}$ of characteristic two is the field of residue classes 0, 1 modulo 2. The coefficients of the d_{ij} in Δ are zero and certainly divisible by two, and hence if the d_{ij} are our restricted independent indeterminates over $\mathfrak{P}_2$ we have Δ identically zero in the d_{ij}. Replace the d_{ij} by any elements of a field $\mathfrak{F}$ of characteristic two. Evidently Δ remains zero.

The above Lemmas 2, 4, 5 are sometimes used to simplify the work of computing determinants and the process is essentially that used in solving linear equations by elimination. We shall later apply these processes to transform matrices. We shall also use the usual methods of expanding determinants and in particular the so-called *Laplace* expansion.*

We have now seen that all of the properties of determinants studied in elementary theory of equations are valid here and that these theorems may be used in our further work.

4. The rank of matrices. When $m - r$ rows and $n - s$ columns of a matrix A are omitted we obtain an $r \times s$ matrix B called a *submatrix* of A. The elements which are at the same time in the omitted rows and in the omitted columns also form a submatrix C called the *complementary* submatrix of B.

* E.g., see L. E. Dickson, *First Course in the Theory of Equations*, Chap. VIII.

An $r \times r$ submatrix B is square and has a determinant

$$\det B = |B| = |b_{ij}| \qquad (i, j = 1, \ldots, r) \tag{24}$$

which we call an r-rowed *minor* of A. We also say that B has order r.

The complementary submatrices of the elements of B are $(r-1)$-rowed square matrices whose determinants are called the minors of the corresponding elements. We designate the minor of b_{ij} by

$$(-1)^{i+j}B_{ji}, \tag{25}$$

so that B_{ji} is the *cofactor* of b_{ij}. The ordinary expansion of $|B|$ according to the elements of its ith row gives

$$|B| = \sum_{j=1}^{r} b_{ij}B_{ji} \qquad (i = 1, \ldots, r) \tag{26}$$

and the expansion of $|B|$ according to its jth column gives

$$|B| = \sum_{i=1}^{r} B_{ji}b_{ij} \qquad (j = 1, \ldots, r). \tag{27}$$

When B has identical ith and kth rows we have $|B| = 0$. The expansion of such a $|B|$ according to its ith row is then $\Sigma b_{kj}B_{ji} = 0$. But this latter sum is independent of the ith row of B so that for any B we have

$$\sum_{j=1}^{r} b_{kj}B_{ji} = 0 \qquad (k \neq i;\ i, k = 1, \ldots, r), \tag{28}$$

and, similarly,

$$\sum_{i=1}^{r} B_{ji}b_{ik} = 0 \qquad (j \neq k;\ j, k = 1, \ldots, r). \tag{29}$$

Every r-rowed minor of A is a linear combination of its $(r-1)$-rowed minors, and when these latter minors are all zero the former must also be zero. If $A = 0$ we say that A has *rank* zero. When $A \neq 0$ it has the property that some j-rowed minor of A is not zero for $j = 1, \ldots, r$ and every $(r+1), \ldots, l$-rowed minor of A is zero, where we write

$$l \leqq m, \qquad l \leqq n, \qquad l = m \text{ or } n,$$

to display the property that l is the lesser of m and n. The integer r is called the *rank* of A. We have seen that

$$r \leqq l .$$

When $r = l$ we can choose l columns of A which form a non-zero determinant or l rows according as $l = m$ or $l = n$. This is called the case where A has *maximum rank*.

The *diagonal* of a matrix is the line of elements $a_{11}, a_{22}, \ldots, a_{ll}$. The elements are called the *diagonal elements* of A and

$$A = (A_1\, A_2) \qquad \text{or} \qquad A = \begin{pmatrix} A_1 \\ A_2 \end{pmatrix}$$

where A_1 is an l-rowed *square* matrix whose diagonal elements and diagonal (line) are those of A. This line is sometimes referred to as the *main diagonal* of the square matrix A_1 and the other diagonal of the square formed by the array A_1 is called its *secondary diagonal.* We shall not use this latter concept and shall refer only to the main diagonal of A.

The elements below the diagonal in A_1 are called the elements below the diagonal in A. We shall use this terminology only where it is clear that we are referring to a specified l-rowed square submatrix of A.

5. The algebra $\mathfrak{M}_n$. The set $\mathfrak{M}_n$ of all n-rowed square matrices with elements in $\mathfrak{F}$ is an algebra of order n^2 over $\mathfrak{F}$. This is our most important example of an algebra not a field.

A square matrix A is said to be *diagonal* if $a_{ij} = 0$ for $i \neq j$. Also A is *scalar* if A is diagonal and all the a_{ii} are equal. In this case $a_{ii} = a$ in $\mathfrak{F}$ and we may write

$$A = aI, \qquad I = I_n , \tag{30}$$

where I_n is the n-rowed identity matrix, a diagonal matrix whose diagonal elements are all unity.

The set $\mathfrak{F}_n$ of all n-rowed scalar matrices is a field equivalent to $\mathfrak{F}$ under the correspondence

$$a \longleftrightarrow aI_n .$$

If A is any $m \times n$ matrix then

$$(aI_m)A = A(aI_n) = aA ,$$

where by aI_m we mean the scalar matrix whose diagonal elements are all a. Thus scalar multiplication is a special case of multiplication of matrices. Notice also that

$$I_mA = AI_n = A \tag{31}$$

for every $m \times n$ matrix A. Thus for $m = n$ $\mathfrak{F}_n[A]$ consists of all polynomials in A and I_n with coefficients in $\mathfrak{F}$ and we shall usually write instead $\mathfrak{F}[A]$.

We now consider square matrices A of order n. Define the n^2 matrices e_{ij} as in Section 3.1 so that $\mathfrak{M}_n$ is an algebra with the basis $(e_{11}, \ldots, e_{ij}, \ldots, e_{nn})$. By actual multiplication we see that

$$e_{ij}e_{jk} = e_{ik}, \qquad e_{ij}e_{tk} = 0\,, \qquad (j \neq t;\ i, j, k, t = 1, \ldots, n). \tag{32}$$

We now prove

Theorem 5. *Let* f_{ij} *be any* n^2 *quantities of* $\mathfrak{M}_n$ *not all zero and such that*

$$f_{ij}f_{jk} = f_{ik}\,, \qquad f_{ij}f_{tk} = 0 \qquad (j \neq t;\ i, j, k, t = 1, \ldots, n). \tag{33}$$

Then the f_{ij} *form what we shall call an* ***ordinary matric basis*** *of*

$$\mathfrak{M}_n = (f_{11}, \ldots, f_{ij}, \ldots, f_{nn})\,. \tag{34}$$

For if $f_{pq} \neq 0$ then all the $f_{ij} \neq 0$ since $f_{pq} = f_{pi}f_{ij}f_{jq} = 0$ when any $f_{ij} = 0$. If $\sum_{i,j=1}^{n} a_{ij}f_{ij} = 0$ then $f_{pp}\left(\sum_{i,j} a_{ij}f_{ij}\right)f_{qq} = a_{pq}f_{pq} = 0$ whence all the $a_{pq} = 0$. Thus the f_{ij} are linearly independent and form a basis of $\mathfrak{M}_n$ by Theorem 2.14.

The *centrum* of $\mathfrak{M}_n$ is the set of all matrices C of $\mathfrak{M}_n$ such that $CA = AC$ for every A of $\mathfrak{M}_n$. The matrix $e_{pq}C$ is a matrix whose pth row is the qth row $(c_{q1}, \ldots, c_{qn})$ of C and whose remaining rows are all zero. The qth column of Ce_{pq} is the pth column

$$\begin{pmatrix} c_{1p} \\ \cdot \\ \cdot \\ \cdot \\ c_{np} \end{pmatrix} \tag{35}$$

of C and its remaining columns are all zero. Then the pth row of Ce_{pq} is $(0, \ldots, c_{pp}, \ldots)$ where c_{pp} is in the qth column. Hence $c_{qj} = 0$ for $j \neq q$ and $c_{qq} = c_{pp}$, that is, C is a scalar matrix. We have proved

Theorem 6. *The centrum of* $\mathfrak{M}_n$ *is the field* $\mathfrak{F}_n$ *of all scalar matrices of* $\mathfrak{M}_n$.

The *adjoint* of any square matrix a_{ij} is the matrix

$$\text{adj}\ A = (A_{ij}) \tag{36}$$

where A_{ji} is the cofactor of a_{ij}. By (26)–(29) we have

$$A(\text{adj}\ A) = (\text{adj}\ A)A = |A|I\,. \tag{37}$$

Thus if $|A| = a \neq 0$ the matrix

$$A^{-1} = (B_{ij}), \qquad B_{ij} = a^{-1}A_{ij} \qquad (i, j = 1, \ldots, n) \tag{38}$$

has the property

$$AA^{-1} = A^{-1}A = I\,. \tag{39}$$

If also $DA = I$ then $(DA)A^{-1} = A^{-1} = D(AA^{-1}) = DI = D$. We call A^{-1} the *inverse* of A and have proved it unique.

When conversely A^{-1} exists then $AA^{-1} = I$, $|I| = 1 = |A| \cdot |A^{-1}|$ and $|A| \neq 0$. Thus a matrix has a (unique) inverse if and only if $|A| \neq 0$. We call A *non-singular* or *singular* according as $|A| \neq 0$ or $|A| = 0$.

The equation $(AB)(B^{-1}A^{-1}) = I$ and the uniqueness of the inverse gives

$$(AB)^{-1} = (B^{-1}A^{-1}) \tag{40}$$

when AB is non-singular. Also $I' = I = (AA^{-1})' = (A^{-1})'A'$ so that

$$(A^{-1})' = (A')^{-1}\,. \tag{41}$$

Equation (37) gives $|A| \cdot |\text{adj}\ A| = a^n$ so that we have proved

$$|\text{adj}\ A| = |A|^{n-1} \tag{42}$$

if $|A| \neq 0$. When $a = |A| = 0$ then (42) is equivalent to $|\text{adj}\ A| = 0$. This is evident when $A = 0$ since then adj $A = 0$. When $A \neq 0$ we use (37) to imply that $A \cdot \text{adj}\ A = 0$. If adj A were non-singular we would have $A(\text{adj}\ A)(\text{adj}\ A)^{-1} = A = 0$, a contradiction. Hence $|\text{adj}\ A| = 0$ as desired.

Before leaving our present introduction to the properties of square matrices and determinants we should notice certain elementary properties of determinants which we shall use repeatedly. The first states that *if* A *and* B *are square matrices the determinant of both*

$$\begin{pmatrix} A & C \\ 0 & B \end{pmatrix}, \quad \begin{pmatrix} A & 0 \\ D & B \end{pmatrix}$$

is the product $|A| \cdot |B|$. This is due to the Laplace expansion of a determinant. It implies also that *if the elements above or below the diagonal of a square matrix* A *are all zero then its determinant is the product of these diagonal elements.*

EXERCISES

1. Prove the last-mentioned property above by an induction on the number of rows of A.

2. Find the inverses of the rational matrices

$$\begin{pmatrix} 4 & 7 \\ 1 & 2 \end{pmatrix}, \quad \begin{pmatrix} 2 & -4 & 1 \\ 0 & 3 & 2 \\ 0 & 1 & 1 \end{pmatrix}$$

by the use of (36), (38).

3. Compute the adjoint of

$$\begin{pmatrix} 2 & 2 & 1 \\ -2 & 0 & 1 \\ 0 & 2 & 2 \end{pmatrix}$$

and show by direct multiplication that $A \cdot \text{adj } A = 0$.

4. Show that the matrix

$$A = \begin{pmatrix} 0 & 1 \\ a & 0 \end{pmatrix}$$

satisfies the equation $A^2 = aI$, where I is the two-rowed identity matrix and a is in a field $\mathfrak{F}$. Prove also that $cI + dA \neq 0$ for every non-zero d and c in $\mathfrak{F}$ so that A satisfies no equation of lower degree than two.

5. Let $a = -1$, $\mathfrak{F}$ be the field of all real numbers in Ex. 4. Show that the set $\mathfrak{F}[A]$ is equivalent to the field of all complex numbers and thus give a two-rowed square real matrix representation of any complex number $c + di$.

6. Prove that the polynomial domain $\mathfrak{F}[A]$ of Ex. 4 is a linear set of order two over $\mathfrak{F}$ and is a field over $\mathfrak{F}$ if and only if $a \neq b^2$ for any b in $\mathfrak{F}$.

7. Show that the linear combinations of the two-rowed matrices

$$I = \begin{pmatrix} 1 & 0 \\ 0 & 1 \end{pmatrix}, \quad u = \begin{pmatrix} \alpha & 0 \\ 0 & -\alpha \end{pmatrix}, \quad v = \begin{pmatrix} 0 & 1 \\ b & 0 \end{pmatrix}, \quad uv$$

form a linear set of order 4 over $\mathfrak{F}(\alpha)$ where $\alpha^2 = a \neq 0$ in $\mathfrak{F}$ of characteristic not two, $b \neq 0$ in $\mathfrak{F}$. This set is then equivalent to the set of all two-rowed square matrices with elements in $\mathfrak{F}(\alpha)$.

8. Verify that $\mathfrak{Q} = (I, u, v, uv)$ of Ex. 7 is an algebra of order four over $\mathfrak{F}$ with multiplication table given by

$$u^2 = aI, \qquad v^2 = bI, \qquad vu = -uv.$$

Such an algebra is called a *generalized quaternion algebra.* It becomes the algebra of *real* quaternions when $a = b = -1$ and $\mathfrak{F}$ is the real number field.

9. Show that the general element of $\mathfrak{Q}$ of Ex. 8 is

$$A = g_1I + g_2u + g_3v + g_4uv = \begin{pmatrix} g_1 + g_2a & g_3 + g_4a \\ (g_3 - g_4a)b & g_1 - g_2a \end{pmatrix},$$

and that

$$A^2 - 2g_1A + N(A)I = 0$$

where $N(A)$, called the *norm* of A, is its determinant,

$$N(A) = (g_1^2 - g_2^2a - g_3^2b + g_4^2ab).$$

10. Show that $\mathfrak{Q}$ is a division algebra if and only if there exist no $g_1, \ldots, g_4$ not all zero in $\mathfrak{F}$ such that $g_1^2 - g_2^2a - g_3^2b + g_4^2ab = 0$. Hence show that the algebra of real quaternions is a division algebra.

11. Use Ex. 4 to replace the matrices of Ex. 7 by an equivalent algebra of four-rowed square matrices with elements in $\mathfrak{F}$.

6. Elementary transformations over $\mathfrak{J}$. Let $\mathfrak{J}$ be any integral domain so that $\mathfrak{J}$ is either a field or a subset of a field. Thus the $m \times n$ matrices A with elements in $\mathfrak{J}$ are ordinary matrices already defined and all of the definitions and properties we have already given are still applicable.

A matrix B is said to be equivalent to A, and we write

$$B \cong A,$$

if B is obtainable from A by a finite sequence of the following three types of operations on the rows or columns of A called *elementary transformations.*

I. The interchange of any two rows (columns) of A.

II. The addition of k times the jth row (column) of A to its ith row (column) for any k of $\mathfrak{J}$ and $i \neq j$.

III. The multiplication of the ith row (column) of A by a unit a of $\mathfrak{J}$.

It is evident that each elementary transformation has an inverse transformation which is an elementary transformation of the same type. This implies that $A \cong B$ if and only if $B \cong A$. If $C \cong B$ and $B \cong A$ then C is obtainable from A by a finite sequence of elementary transformations which is the result of the two sequences by which we passed from A to B and then from B to C. The identity transformation is either a transformation of type I with the two rows (columns) the same or of type III with $a = 1$ so that any $A \cong A$. Hence the relation $A \cong B$ is an equivalence relation.

A determinant D is replaced by D, $-D$, aD by any elementary transformation on its matrix, where a is a unit of $\mathfrak{J}$. If D is an r-rowed minor of A then the only elementary transformations on A which do not induce elementary transformations on the matrix of D are those of type I which replace D by $\pm E$ where E is an r-rowed minor of A or those of type II which replace D by $D + kE$. This proves

Theorem 7. *The* r*-rowed minors of any* $B \cong A$ *are linear combinations of the* r*-rowed minors of* A *with coefficients in* $\mathfrak{J}$.

We now prove the

COROLLARY. *Equivalent matrices have the same rank.*

For if every $(r + 1)$-rowed minor of A is zero so is every $(r + 1)$-rowed minor of $B \cong A$. Thus the rank of $B \cong A$ is at most the rank of A. But $A \cong B$ and the rank of A is at most the rank of B, the ranks of A and B must be equal.

The reader is already accustomed to applying elementary transformations of type II to determinants in their computation. Similarly, transformations of types I and III were frequently used in the theory of determinants.

Elementary transformations on A *may be accomplished algebraically by the multiplication of* A *by square matrices.* We let

$$E_{ij} \tag{43}$$

be the m-rowed square matrix obtained by interchanging the ith and jth rows of I_m. Then $E_{ij}A$ is obtained from A by elementary transformation I on its ith and jth rows. The elementary transformation II is accomplished by forming the product

$$P_{ij}(k)A \tag{44}$$

where $P_{ij}(k)$ has k in its ith row and jth column, unity on the main diagonal, zeros elsewhere. Finally, we accomplish III by forming

$$R_{ii}(a)A\,, \tag{45}$$

where $R_{ii}(a)$ is a diagonal matrix with a in the ith place and unity elsewhere on the diagonal.

The corresponding column transformations are obtained by transposition. We note that

$$E'_{ij} = E_{ij}, \qquad P_{ij}(k)' = P_{ji}(k), \qquad R_{ii}(a)' = R_{ii}(a)\,. \tag{46}$$

Thus we accomplish I, II, III on the columns of A by forming

$$AE_{ij}, \qquad AP_{ji}(k), \qquad AR_{ii}(a), \tag{47}$$

where E_{ij}, $P_{ji}(k)$, $R_{ii}(a)$ are now necessarily n-rowed instead of m-rowed square matrices. Also notice that

$$E_{ij}^{-1} = E_{ij}, \qquad [P_{ij}(k)]^{-1} = P_{ij}(-k), \qquad [R_{ii}(a)]^{-1} = R_{ii}(a^{-1}), \tag{48}$$

so that, as we noticed before, the inverses of elementary transformations are elementary transformations.

A square matrix P is called an *elementary matrix* if P is a product of a finite number of the matrices E_{ij}, $P_{ij}(k)$, $R_{ii}(a)$. The product of two elementary matrices is an elementary matrix, the inverse of an elementary matrix is an elementary matrix. This definition then gives

Theorem 8. *Two* m $\times$ n *matrices* A *and* B *are equivalent in* $\mathfrak{F}$ *if and only if there exist elementary matrices* P, Q *such that*

$$B = PAQ. \tag{49}$$

In our later work we shall make several induction proofs wherein it will be convenient to use the trivial

Theorem 9. *Let*

$$A = \begin{pmatrix} A_1 & 0_{r,n-s} \\ 0_{m-r,s} & A_2 \end{pmatrix} \tag{50}$$

for zero matrices $0 = 0_{i,j}$. *Then*

$$A \cong B = \begin{pmatrix} B_1 & 0 \\ 0 & B_2 \end{pmatrix} \tag{51}$$

for any $B_1 \cong A_1$ *and* $B_2 \cong A_2$.

For the elementary transformations on the rows (columns) of A replace these rows (columns) by linear combinations of them and hence transformations on the first r rows (first s columns) do not alter the matrix $0_{r,n-s}$. But these transformations replace A_1 by any B_1 equivalent to A_1. Similarly, we replace A_2 by $B_2 \cong A_2$ when we perform elementary transformations on A involving only its last $m - r$ rows and last $n - s$ columns.

EXERCISES

1. Show that every elementary transformation on the first r rows and s columns of any matrix A replaces A by PAQ, where

$$P = \begin{pmatrix} P_1 & 0 \\ 0 & I_{m-r} \end{pmatrix}, \qquad Q = \begin{pmatrix} Q_1 & 0 \\ 0 & I_{n-s} \end{pmatrix},$$

and P_1 and Q_1 are elementary matrices of r and s rows, respectively. Thus show that if A is given by (50) we have

$$PAQ = \begin{pmatrix} B_1 & 0 \\ 0 & A_2 \end{pmatrix}, \qquad B_1 = P_1A_1Q_1 .$$

2. Let $\mathfrak{J}$ be the field $\mathfrak{R}$ of all rational numbers. Apply elementary transformations on the rows of A given by

$$A = \begin{pmatrix} 3 & 1 & 1 \\ 1 & 2 & 2 \\ 0 & -5 & -5 \end{pmatrix} \cong B = \begin{pmatrix} 1 & 0 & 0 \\ 0 & 1 & 1 \\ 0 & 0 & 0 \end{pmatrix}$$

to reduce it to B.

3. Use elementary row transformations in the integral domain $\mathfrak{J}$ of all rational integers to show that

$$\begin{pmatrix} 1 & 2 & 1 \\ 1 & 3 & -1 \\ 1 & 4 & -2 \end{pmatrix}$$

is an elementary matrix.

4. Let $\mathfrak{J} = \mathfrak{R}[x]$, $\mathfrak{R}$ the rational number field and x an indeterminate over $\mathfrak{J}$. Show that

$$\begin{pmatrix} -2x - 5 & -1 \\ 2x^2 + 5x + 6 & x \end{pmatrix}$$

is an elementary matrix.

7. Equivalence of matrices in a field $\mathfrak{F}$. We shall study the particular case where $\mathfrak{J} = \mathfrak{F}$ is a field. Then we see that every non-zero element of $\mathfrak{F}$ is a unit of $\mathfrak{F}$. We consider any $m \times n$ matrix $A = (a_{ij})$ with elements a_{ij} in $\mathfrak{F}$ and assume that $A \neq 0$.

At least one $a_{ij} \neq 0$ and a sequence of elementary transformations of type I replaces A by an equivalent matrix with $a_{11} \neq 0$. By a transformation of type III we make $a_{11} = 1$ and, by a transformation of type II with $k = -a_{i1}$ for rows, $k = -a_{1i}$ for columns, we make $a_{i1} = a_{1j} = 0$ for $i \neq 1, j \neq 1$. Thus every A is equivalent to

$$\begin{pmatrix} 1 & 0_{1,n-1} \\ 0_{m-1,1} & A_2 \end{pmatrix} \tag{52}$$

where A_2 has $m - 1$ rows and $n - 1$ columns. By Theorem 3.9 A is equivalent also to

$$\begin{pmatrix} 1 & 0 \\ 0 & B_2 \end{pmatrix}, \qquad B_2 = \begin{pmatrix} 1 & 0_{1,n-2} \\ 0_{m-2,1} & B_3 \end{pmatrix} \tag{53}$$

unless $B_2 = 0$. Hence every A is equivalent after a finite number of such steps to

$$B = \begin{pmatrix} I_r & 0 \\ 0 & 0 \end{pmatrix} = PAQ\,. \tag{54}$$

The matrix B has rank r, the rank of A, and in our notation I_r is an r-rowed identity matrix.

Theorem 10. *Every matrix of rank* r *is equivalent to* (54). *Hence two matrices* A, B *are equivalent in* $\mathfrak{F}$ *if and only if they have the same rank.*

In particular let A be an m-rowed square matrix of rank m. Then the matrix B of (54) is I_m and $PAQ = I_m$, $A = P^{-1}Q^{-1}$ is an elementary matrix. Every elementary matrix is non-singular and we have

Theorem 11. *A square matrix in* $\mathfrak{F}$ *is elementary if and only if it is non-singular. Thus* A *and* B *are equivalent if and only if there exist non-singular matrices* P *and* Q *such that*

$$\mathrm{PAQ} = \mathrm{B}\,. \tag{55}$$

This property is sometimes used as the definition of equivalence.

The rows a_i of an $m \times n$ matrix A are elements of the linear set $\mathfrak{L}_n = (u_1, \ldots, u_n)$, where u_i is a $1 \times n$ matrix with unity in the ith column and zeros elsewhere. The set

$$\mathfrak{R}_A = [a_1, \ldots, a_m]$$

of all linear combinations of the a_i with coefficients in $\mathfrak{F}$ forms a subset of $\mathfrak{L}_n$. We call $\mathfrak{R}_A$ the *row space* of A and its order is the *row rank* of A. The *column space* and *column rank* of A are defined similarly.

The rows of PA are linear combinations of the rows of A for any $p \times m$ matrix P. This gives

$$\mathfrak{R}_{PA} \leqq \mathfrak{R}_A\,. \tag{56}$$

If P is a non-singular $m \times m$ matrix the matrix $A = P^{-1}(PA)$, $\mathfrak{R}_A \leqq \mathfrak{R}_{PA}$ and thus $\mathfrak{R}_A = \mathfrak{R}_{PA}$. *Thus* A *and* PA *have the same row spaces for any non-singular* P.

The rows of AQ are the products a_iQ where

$$A = \begin{pmatrix} a_1 \\ \cdot \\ \cdot \\ \cdot \\ a_n \end{pmatrix}, \qquad AQ = \begin{pmatrix} a_1Q \\ \cdot \\ \cdot \\ \cdot \\ a_nQ \end{pmatrix}. \tag{57}$$

When Q is non-singular a linear combination

$$\sum a_i(a_iQ) = (\sum a_ia_i)Q = 0$$

if and only if $\sum a_ia_i = 0$. Thus the number of linearly independent a_i is the number of linearly independent a_iQ and AQ has the same row rank as A. This gives the

LEMMA. *Equivalent matrices have the same row rank.*

The rows of the matrix B of (54) are $u_1, \ldots, u_r, 0, \ldots, 0$. The row space of B is evidently $\mathfrak{L}_r = (u_1, \ldots, u_r)$ of order r, the rank of A. But B has the same row rank as A. A similar consideration of the columns of A gives the results:

Theorem 12. *The row rank and column rank and rank of any matrix are all equal.*

Theorem 13. *A square matrix is non-singular if and only if its rows (columns) are linearly independent.*

Theorem 3.12 may be proved without the use of (54) and the proof we are about to give is preferable for some purposes. We let s be the row rank of A and t its column rank. By elementary transformations which merely permute the rows and columns of A we replace A by

$$P_0AQ_0 = \begin{pmatrix} A_1 & A_2 \\ A_3 & A_4 \end{pmatrix}, \tag{58}$$

where the rows of (A_1A_2) are linearly independent, the columns of $\begin{pmatrix} A_1 \\ A_3 \end{pmatrix}$ are linearly independent, and A_1 is an $s \times t$ submatrix of A. The rows of $(A_1\ A_2)$ are linearly independent and Theorem 2.23 states that the rows of $(A_3\ A_4)$ are linear combinations of the rows of $(A_1\ A_2)$. We may thus write

$$(A_1\ A_2) = \begin{pmatrix} b_1 \\ \cdot \\ \cdot \\ \cdot \\ b_s \end{pmatrix} \qquad (A_3\ A_4) = \begin{pmatrix} b_{s+1} \\ \cdot \\ \cdot \\ \cdot \\ b_m \end{pmatrix}, \tag{59}$$

and have

$$b_j = -\sum_{i=1}^{s} \gamma_{ji}b_i \qquad (j = s+1, \ldots, m). \tag{60}$$

Then if

$$P_1 = \begin{pmatrix} I_s & 0 \\ \Gamma & I_{m-s} \end{pmatrix}, \qquad \Gamma = (\gamma_{ji}) \tag{61}$$

$$(i = 1, \ldots, s;\ j = s+1, \ldots, m)$$

we have

$$P_1P_0AQ_0 = \begin{pmatrix} A_1 & A_2 \\ 0 & 0 \end{pmatrix}. \tag{62}$$

The last $n - t$ columns of P_0AQ_0 are linearly dependent on the first t columns and this must be true of the partial columns which are the $n - t$ columns of A_2 and the t columns of A_1, respectively. Hence

$$\begin{cases} A_1 = (b^{(1)} \ldots b^{(t)}), \qquad A_2 = (b^{(t+1)} \ldots b^{(n)}), \\ b^{(j)} = -\sum_{i=1}^{t} b^{(i)}\delta_{ij}, \end{cases} \tag{63}$$

and if

$$Q_1 = \begin{pmatrix} I_t & \Omega \\ 0 & I_{n-t} \end{pmatrix}, \qquad \Omega = (\delta_{ij}) \qquad (i = 1, \ldots, t; j = t + 1, \ldots, n) \tag{64}$$

then

$$B = (P_1P_0)A(Q_0Q_1) = \begin{pmatrix} A_1 & 0 \\ 0 & 0 \end{pmatrix}. \tag{65}$$

The matrix B has the same row rank s and the same column rank t as A, and A_1 is an $s \times t$ matrix. It follows that the rows of A_1 are linearly independent and so are its columns. The rows of A_1 are s linearly independent $1 \times t$ matrices and are in the linear set of order t of all $1 \times t$ matrices. By Theorem 2.23 $s \leqq t$. Similarly, a discussion of columns gives $t \leqq s$ so that $s = t$. Moreover, the row space of A_1 has a basis

$$e_1 = (1, 0, \ldots, 0), \quad e_2 = (0, 1, 0, \ldots 0), \ldots, \quad e_t = (0, 0, \ldots, 1)$$

and by Theorem 2.23 we may write

$$A_1 = \begin{pmatrix} d_1 \\ \cdot \\ \cdot \\ \cdot \\ d_t \end{pmatrix}, \qquad e_i = \sum_{j=1}^{t} \pi_{ij}d_j, \qquad \pi = (\pi_{ij}), \qquad P_2 = \begin{pmatrix} \pi & 0 \\ 0 & I_{r-s} \end{pmatrix}$$

and obtain $I_t = \pi A_1$, π, A_1, and P_2 are non-singular,

$$P_2B = \begin{pmatrix} I_t & 0 \\ 0 & 0 \end{pmatrix}, \tag{66}$$

the matrix of (54). Then $t = s = r$ and we have Theorem 3.12.

We now see that (54) is an essentially unsymmetric form since our final step in (66) replaces A by PAQ with $P = P_2P_1P_0$ but only $Q = Q_0Q_1$. In a later chapter we shall wish to make the same transformations on the columns of A as on its rows, and it is clear that the matrix of (65) will be preferable to (54) for this purpose.

Only transformations of types I and II were used in obtaining (65). These are, in fact, the only essential transformations for theorems about rank. We shall use this fact again in the proof of an important result which is generally used to shorten the work of finding the rank of a matrix by actual computation of its $r + 1$ rowed minors.

Theorem 14. *Let* $\Delta \neq 0$ *be an* r*-rowed minor of* A *and let every* (r + 1)*-rowed minor of* A *with* Δ *as subdeterminant be zero. Then* A *has rank* r.

Interchanges of rows and columns of A replace its $(r + 1)$-rowed minors by other $(r + 1)$-rowed minors or their negatives, and the property of A assumed as hypothesis of our theorem is retained if we permute the rows and columns of A so as to make

$$\Delta = |D|, \qquad D = \begin{pmatrix} a_{11} & \dots & a_{1r} \\ \cdot & \cdot\ \cdot\ \cdot & \cdot \\ a_{r1} & \dots & a_{rr} \end{pmatrix},$$

where A is given by (1). By Theorem 3.13 the r rows of the matrix of D are linearly independent and form a basis of the set of all $1 \times r$ matrices. Thus a sequence of transformations of type II replaces A by a matrix

$$PA = \begin{pmatrix} D & A_2 \\ 0 & A_3 \end{pmatrix}.$$

But PA is a matrix obtained from A by subtracting linear combinations of its first r rows from its other rows, and every $(r + 1)$-rowed minor of A with D as subdeterminant is unchanged in value by the above operations. Hence every minor

$$\begin{vmatrix} a_{11} & \dots & a_{1r} & a_{1j} \\ \cdot & & \cdot & \cdot \\ \cdot & & \cdot & \cdot \\ \cdot & & \cdot & \cdot \\ a_{r1} & \dots & a_{rr} & a_{rj} \\ 0 & \dots & 0 & a_{ij} \end{vmatrix} = \Delta \cdot a_{ij} = 0,$$

and $\Delta \cdot a_{ij} = 0$, $\Delta \neq 0$, $a_{ij} = 0$, $A_3 = 0$ and A has row rank r. Then A has rank r.

EXERCISES

1. Use rational elementary transformations on the rows of the matrices

$$\begin{pmatrix} 1 & 3 & 1 & 3 \\ 1 & 7 & -3 & -5 \\ 1 & -2 & 3 & 4 \\ 1 & 3 & -1 & -3 \end{pmatrix}, \quad \begin{pmatrix} 1 & -3 & 3 & -1 \\ -3 & 5 & -7 & 1 \\ -1 & -5 & 1 & -3 \\ 0 & 2 & -1 & 1 \end{pmatrix}, \quad \begin{pmatrix} 1 & 2 & -1 & 3 \\ -2 & -4 & 2 & -6 \\ -1 & -2 & 1 & -3 \\ 3 & 6 & -3 & -9 \end{pmatrix},$$

and reduce them to rationally equivalent matrices with elements below the diagonal all zero. What is the rank of each matrix (by Theorem 3.12)?

2. In Ex. 1 find the rank of each matrix by the use of Theorem 3.14.

3. Let A be a non-singular square matrix whose elements below the diagonal are all zero and whose diagonal elements are all not zero. Show that A may be reduced by row transformations of type II in the field of its elements to a diagonal matrix.

4. Let A be a non-singular matrix with elements in a field $\mathfrak{F}$. Show that A may be reduced to the identity matrix by a sequence of elementary transformations (with coefficients in $\mathfrak{F}$) on its rows alone. Hint: The first column of A has a non-zero element which we carry into the first row and then apply elementary row transformations and Ex. 3.

5. Apply Theorem 3.12 to show that r in Theorems 2.21, 2.23 is the rank of the matrix of the coefficients in an expression of m given quantities of a linear set in terms of a basis.

6. Prove that an $m \times n$ matrix A has rank one if and only if $A = BC$ where $B \neq 0$ is an $m \times 1$ matrix, $C \neq 0$ is a $1 \times n$ matrix.

8. Linear equations. A set of m indeterminates $x_1, \ldots, x_m$ over $\mathfrak{F}$ are called *variables* or *unknowns* over $\mathfrak{F}$ if, during the course of our discussion, we allow ourselves to replace $x_1, \ldots, x_m$ by any quantities of $\mathfrak{F}$. This is of course the usual terminology.

The system of equations

$$x_i = \sum_{j=1}^{n} a_{ij} y_j \qquad (i = 1, \ldots, m) \tag{67}$$

may be thought of as a linear transformation which carries the variables $x_1, \ldots, x_m$ into new variables $y_1, \ldots, y_n$. The matrix

$$A = (a_{ij})$$

is called the *matrix of the linear transformation* and if x and y are the one-columned matrices

$$(68) \qquad x = \begin{pmatrix} x_1 \\ \cdot \\ \cdot \\ \cdot \\ x_m \end{pmatrix}, \qquad y = \begin{pmatrix} y_1 \\ \cdot \\ \cdot \\ \cdot \\ y_n \end{pmatrix},$$

then the transformation (67) may be written as either

$$a: \qquad x = Ay\,, \qquad \text{or} \qquad x' = y'A'\,.$$

If we make a second linear transformation

$$b: \qquad y = Bz$$

the product $c = ab$ of the linear transformations a and b is defined as the resultant linear transformation

$$c: \qquad x = Cz\,.$$

But $x = Ay = A(Bz) = (AB)z$, $C = AB$ by Theorem 3.2.

Theorem 15. *The matrix of a product* ab *of linear transformations* a, b *is the product* AB *of their matrices.*

This result is actually the main motive for our definition of matrix multiplication.

We call (67) a *non-singular* transformation if $m = n$ and A is non-singular. Then $y = A^{-1}x$. Thus if we think of (67) as a system of n linear equations in n unknowns y_i with $|A| \neq 0$, the system has the unique solution $y = A^{-1}x$. Here A^{-1} is of course presumably computed by the use of (36), (38). These considerations may of course also be made when A is singular or when $m \neq n$. We thus consider the system $x = Ay$ and let A have rank r. As in (58), we may permute the rows and columns of A and assume that A_1 is non-singular of order r. The corresponding transformations on x_i, y_j are merely changes of notation. We then define P_1 as in (61) and have

$$(69) \qquad \xi = P_1x = P_1Ay = \begin{pmatrix} A_1 & A_2 \\ 0 & 0 \end{pmatrix} y$$

so that necessarily

$$(70)\qquad \xi = \begin{pmatrix} x_1 \\ \cdot \\ \cdot \\ \cdot \\ x_r \\ 0 \\ \cdot \\ \cdot \\ \cdot \\ 0 \end{pmatrix}, \qquad \sum_{i=1}^{r} \gamma_{ji} x_i + x_j = 0 \quad (j = r+1, \ldots, m).$$

Write

$$X = \begin{pmatrix} x_1 \\ \cdot \\ \cdot \\ \cdot \\ x_r \end{pmatrix}, \qquad Y_1 = \begin{pmatrix} y_1 \\ \cdot \\ \cdot \\ \cdot \\ y_r \end{pmatrix}, \qquad Y_2 = \begin{pmatrix} y_{r+1} \\ \cdot \\ \cdot \\ \cdot \\ y_n \end{pmatrix}$$

and (69) reduces to

$$X - A_2 Y_2 = A_1 Y_1 .$$

But A_1 has an inverse and $x = Ay$ has the unique solution

$$(71)\qquad Y_1 = A_1^{-1}(X - A_2 Y_2) .$$

Thus after a preliminary permutation of the equations and the variables y_i we drop down to r equations in the unknowns $y_1, \ldots, y_r$, and solve for these uniquely in terms of the remaining $n - r$ unknowns y_j and the x_i. These results then satisfy the $m - r$ further equations since by (70) these $m - r$ equations are linear combinations of the first r equations.

The conditions given by (70) may be stated in terms of the rank of the $m \times (n + 1)$ matrix

$$(72)\qquad A^* = (x\ A) .$$

The rank of A^* is the same as the rank of

$$(73)\qquad P_1 A^* = (\xi\ P_1 A) = \begin{pmatrix} X & A_1 & A_2 \\ 0 & 0 & 0 \end{pmatrix}$$

when (70) is satisfied, and this matrix evidently has rank r, the rank of A. Conversely, if A^* has rank r and A has the form given by (58), then the last $m - r$ rows of A^* are linearly dependent on its first r rows and we may define a matrix P_1 for A^* such that P_1A^* is given by (73). We have proved

Theorem 16. *A system* (67) *of* m *equations in* n *unknowns has a solution if and only if the rank of the augmented matrix* (72) *is the same as the rank* r *of* A. *Then we solve the system by choosing* r *linearly independent right members of* (67) *and solving for* r *of the unknowns* y_i *uniquely in terms of the* x_i *and the remaining* $n - r$ *unknowns, the solution being given by* (71) *and an appropriate permutation of the* x_i *and* y_i.

9. Bilinear forms. Let x and y be given by (68). Then the form

$$f = x'Ay = \sum_{i=1,\ldots,m}^{j=1,\ldots,n} x_i a_{ij} y_j$$

in the variables x_i, y_i is called a *bilinear form* with matrix A. The rank of A is called the *rank* of f. A pair of non-singular linear transformations

$$x = B'\xi, \qquad y = C\eta$$

replace $f = x'Ay = \xi'BAC\eta$ by a new form with matrix BAC. We say that this new form is *equivalent to* f *in* $\mathfrak{F}$. Thus two bilinear forms with coefficients in $\mathfrak{F}$ are equivalent in $\mathfrak{F}$ if and only if their matrices are equivalent in $\mathfrak{F}$. Theorem 3.10 may now be restated as

Theorem 17. *Two bilinear forms are equivalent in* $\mathfrak{F}$ *if and only if they have the same rank.*

We of course have the immediate

COROLLARY. *Let* f *be a bilinear form of rank* r. *Then* f *is equivalent in* $\mathfrak{F}$ *to*

$$\xi_1\eta_1 + \ldots + \xi_r\eta_r .$$

The results above show that the equivalence theory of bilinear forms is identical with the theory of equivalence of matrices. The exposition of the matrix theory is simpler however and we have made the theory of forms a consequence of the matrix theory.

10. Equivalence in $\mathfrak{F}[\lambda]$. We shall consider the equivalence of matrices in the case where $\mathfrak{J} = \mathfrak{F}[\lambda]$ is the integral domain of all polynomials in λ with coefficients in a field $\mathfrak{J}$. The units of $\mathfrak{J}$ are the non-zero constant polynomials, that is, the non-zero quantities of $\mathfrak{F}$, and every quantity of $\mathfrak{J}$, is associated with a monic quantity, that is, a polynomial with leading coefficient unity. We shall prove the

LEMMA. *Let* A *be any* m × n *matrix with elements in* $\mathfrak{J}$ *and rank* r. *Then* A *is equivalent in* $\mathfrak{J}$ *to*

$$(74) \qquad \begin{pmatrix} G & 0_{r,n-r} \\ 0_{m-r,r} & 0_{m-r,n-r} \end{pmatrix}, \qquad G = \begin{pmatrix} a_1 & 0 & \dots & 0 \\ 0 & a_2 & \dots & 0 \\ . & . & \dots & . \\ 0 & 0 & \dots & a_r \end{pmatrix}$$

with G *a diagonal matrix whose* i*th diagonal element* $a_i \neq 0$ *is a monic quantity of* $\mathfrak{J}$ *such that* a_i *divides* a_{i+1} (i = 1, . . . , r − 1).

Let Σ be the set of all non-zero elements of all matrices equivalent to A. These elements have a minimum degree and there is a monic b_{11} in Σ of this degree and a corresponding matrix $B = (b_{ij})$ equivalent to A. If b_{11} does not divide b_{1j} we may write $b_{1j} = -q_j b_{11} + r_j$ where $r_j \neq 0$ has degree less than the degree of b_{11}. We add q_j times the first column of B to its jth column and obtain a matrix equivalent to A and with r_j as an element. This contradicts our hypothesis on b_{11}. Hence b_{11} divides every b_{1j} and similarly every b_{i1}. We write $b_{1j} = b_j b_{11}$, $b_{i1} = c_i b_{11}$. If b_{11} does not divide b_{ij} then $b_{ij} = -q_j b_{11} + r_j$ where $r_j \neq 0$ has degree less than the degree of b_{11}. Add $1 - b_j$ times the first column of B to its jth column and replace b_{1j} by $b_{1j}^0 = (1 - b_j)b_{11} + b_{1j} = b_{11}$, b_{ij} by $b_{ij}^0 = (1 - b_j)b_{i1} + b_{ij} = -[q_j + (b_j - 1)c_i]b_{11} + r_j = -t_{ij}b_{11} + r_j$. Then add t_{ij} times the first row of the resulting matrix to its ith row and obtain a matrix equivalent to A and with b_{ij}^0 replaced by $t_{ij}b_{1j}^0 + b_{ij}^0 = t_{ij}b_{11} - t_{ij}b_{11} + r_j = r_j$. This new matrix is equivalent to A and has r_j as an element. This is impossible and we have proved b_{11} divides every element of B.

Define b_j, c_j as above and add $-c_i$ times the first row of B to its ith row, $-b_j$ times its first column to its jth column and replace B by a matrix

$$(75) \qquad A_1 = \begin{pmatrix} b_{11} & 0 \\ 0 & B_1 \end{pmatrix}$$

where B_1 has $m - 1$ rows and $n - 1$ columns. Our theorem has then been proved if either $m = 1$ or $n = 1$. We make an induction on the number of rows and columns of A_1 and assume our theorem true for matrices with $m - 1$ rows and $n - 1$ columns. But A_1 is equivalent to

$$(76) \qquad \begin{pmatrix} b_{11} & 0 \\ 0 & B_{10} \end{pmatrix}$$

for any B_{10} equivalent to B_1 by Theorem 3.9. Moreover, the greatest common divisor of the elements of B_{10} is that of the elements of B_1, and this latter quantity of $\mathfrak{J}$ is divisible by b_{11}. Hence every element of B_{10} is divisible by b_{11}.

By the hypothesis of our induction either B_1 is zero and our theorem is true for $r = 1$, $a_1 = b_{11}$, or B_1 is equivalent to the reduced matrix

$$(77) \qquad \begin{pmatrix} G_1 & 0_{r-1,n-r} \\ 0_{m-r,r-1} & 0_{m-r,n-r} \end{pmatrix}, \qquad G_1 = \begin{pmatrix} a_2 & 0 & \dots & 0 \\ 0 & a_3 & \dots & 0 \\ . & . & \dots & . \\ 0 & 0 & \dots & a_r \end{pmatrix},$$

where the a_i are monic polynomials such that a_i divides a_{i+1}, and B_1 has rank $r - 1$. Then A is equivalent to (74) with $a_1 = b_{11}$ and A has rank r, a_1 divides a_2, a_i divides a_{i+1} as desired.

The greatest common divisor $D_s(A)$ of all s-rowed minors of A is a monic quantity which is divisible by $D_{s-1}(A)$ for $s = 2, 3, \dots, r$. If $B \cong A$ then $D_s(B) = D_s(A)$. For $D_s(A)$ divides all s-rowed minors of B by Theorem 3.7 and hence divides $D_s(B)$. Similarly, $D_s(B)$ divides $D_s(A)$ and, since both are monic quantities of $\mathfrak{J}$, $D_s(B) = D_s(A)$. We now define $D_0(A) = 1$,

$$(78) \qquad \phi_j = \frac{D_{r-j+1}(A)}{D_{r-j}(A)} \qquad (j = 1, \dots, r),$$

and call ϕ_j the jth *invariant factor* of A. It is uniquely determined by A and is the same for all matrices B equivalent to A. Moreover, as we shall see, ϕ_j is divisible by ϕ_{j+1} for $j = 1, \dots, r - 1$.

The matrix of (74) has $D_1 = a_1$ since a_1 is an element which divides each a_i. A similar computation gives $D_s = a_1 a_2 \dots a_s$, so that

$$\phi_j = a_{r-j+1} \qquad (j = 1, \dots, r).$$

Thus ϕ_j is divisible by the polynomial ϕ_{j+1}. By a permutation of the rows and columns in (74) we obtain

Theorem 18. *Let* A *be any* m $\times$ n *matrix with elements in* $\mathfrak{J}$ *and invariant factors* $\phi_1, \dots, \phi_r$. *Then* A *is equivalent to*

$$(79) \qquad \begin{pmatrix} \phi_1 & & & & \\ & \cdot & & & \\ & & \cdot & & \\ & & & \phi_r & \\ & & & & 0 \end{pmatrix}.$$

As an immediate corollary of Theorem 3.18 and the fact that we are dealing with an equivalence relation, we have

Theorem 19. *Two matrices with elements in* $\mathfrak{J} = \mathfrak{F}[\lambda]$ *are equivalent if and only if they have the same invariant factors.*

A non-singular m-rowed matrix A with determinant in $\mathfrak{F}$ has $\phi_1 = 1$ so that the matrix of (79) is I_m and $PI_mQ = A$ where P and Q are elementary matrices. Then $A = PQ$ is an elementary matrix. Conversely, every elementary matrix has determinant in $\mathfrak{F}$ and we have proved

Theorem 20. *A square matrix* A *with elements in* $\mathfrak{F}[\lambda]$ *is elementary if and only if the determinant of* A *is a non-zero quantity of* $\mathfrak{F}$.

As a special application of Theorem 3.20 we have

Theorem 21. *Let* $\mathrm{A} = \mathrm{a}\lambda + \mathrm{c}$ *and* $\mathrm{B} = \mathrm{b}\lambda + \mathrm{d}$ *where* a, b, c, d *have elements in* $\mathfrak{F}$, b *is non-singular. Then* A *and* B *are equivalent if and only if there exist non-singular matrices* p, q *with elements in* $\mathfrak{F}$ *such that*

$$\mathrm{pAq} = \mathrm{B} \,.$$

For if $pAq = B$, then p and q are elementary by Theorem 3.20 and A and B are equivalent. Conversely, let $PAQ = B$ for elementary P, Q. By Theorem 2.7 we may write

$$P = Bp_1 + p, \qquad Q = q_1B + q \tag{80}$$

where p and q have degree zero, that is, have elements in $\mathfrak{F}$. We also write

$$AQ = P_0B\,, \qquad PA = BQ_0$$

where $P_0 = P^{-1}$ and $Q_0 = Q^{-1}$ are elementary. Then

$$\begin{aligned} PAQ &= Bp_1AQ + pAQ = Bp_1P_0B + pAq_1B + pAq \\ &= B(p_1P_0)B + (P - Bp_1)Aq_1B + pAq \\ &= BRB + pAq = B\,, \end{aligned}$$

where $R = p_1P_0 + Q_0q_1 - p_1Aq_1$. Hence $p(a\lambda + c)q = B(1 - RB)$. The degree of $B(1 - RB)$ is at least two if $R \not\equiv 0$ since $B = b\lambda + d$ has non-singular leading coefficient, the degree of BRB is two plus the degree of R when $R \not\equiv 0$. But pAq has degree unity so that $R \equiv 0$, $pAq = B$. In particular $\lambda = 0$ gives $pcq = d$ so that $paq = b$. Since b is non-singular so are p and q.

The results above and those of previous sections on rectangular matrices are the principal results on such matrices. We shall leave this subject now and study square matrices.

EXERCISES

1. State the lemma and Theorem 3.18 for the case where $\mathfrak{J}$ is the domain of all rational integers. Verify that our proof is valid for this domain.

2. Let $A = \lambda I - A_0$, $B = \lambda I - B_0$ where A_0 and B_0 are square matrices with elements in $\mathfrak{F}$. Prove that if $P(\lambda)AQ(\lambda) = B$, then $q^{-1}A_0q = B_0$ where if

$$Q(\lambda) = a_0\lambda^t + a_1\lambda^{t-1} + \dots + a_t \qquad (a_i \text{ matrices in } \mathfrak{F}),$$

then

$$q = a_0B_0^t + a_1B_0^{t-1} + \dots + a_t .$$

Hint: Use (80) and Theorem 2.7, our Remainder Theorem of Algebra.

CHAPTER IV

SIMILARITY OF SQUARE MATRICES

1. Linear transformations and similar matrices. Let $x_1, \ldots, x_n, y_1, \ldots, y_n$ be symbols which are either in $\mathfrak{F}$ or are indeterminates over $\mathfrak{F}$, x and y be the $n \times 1$ matrices

$$x = \begin{pmatrix} x_1 \\ \cdot \\ \cdot \\ \cdot \\ x_n \end{pmatrix}, \quad y = \begin{pmatrix} y_1 \\ \cdot \\ \cdot \\ \cdot \\ y_n \end{pmatrix},$$

and A be any n-rowed square matrix with elements in $\mathfrak{F}$. Then the matrix equation

$$y = Ax$$

is called a linear transformation with matrix A carrying the x_j into the y_i.

A linear transformation with matrix A is called *non-singular* if A is non-singular. Then the linear transformation $x = A^{-1}y$ carries the y_i back into the x_j.

Two linear transformations are called *equivalent* if we may carry one into the other by applying the *same* non-singular linear transformation to the x_i and y_j. Thus let P be a non-singular square matrix with elements in $\mathfrak{F}$ and define ξ, η by

$$\xi = \begin{pmatrix} \xi_1 \\ \cdot \\ \cdot \\ \cdot \\ \xi_n \end{pmatrix}, \quad \eta = \begin{pmatrix} \eta_1 \\ \cdot \\ \cdot \\ \cdot \\ \eta_n \end{pmatrix}, \quad \xi = Px, \quad \eta = Py\,.$$

Then $x = P^{-1}\xi$ and the linear transformation $y = Ax$ is carried by P into the equivalent transformation

$$\eta = Py = PAx = B\xi$$

with matrix

$$B = PAP^{-1}\,.$$

A question arises which is of importance in the many branches of mathematics where linear transformations occur. *What properties of a linear transformation are invariant under equivalence?* This question is abstractly identical with the study of the invariants of a matrix A under the operations replacing A by any PAP^{-1}. We shall study this latter problem and shall make the

DEFINITION. *Two n-rowed square matrices* A *and* B *with elements in a field* $\mathfrak{F}$ *are said to be similar in* $\mathfrak{F}$ *and we write*

$$A \sim B\,, \tag{1}$$

if there exists a non-singular square matrix P *with elements in* $\mathfrak{F}$ *such that*

$$B = PAP^{-1}\,. \tag{2}$$

The similarity of matrices is a formal equivalence relation. For $A \sim A = IAI^{-1}$, $B \sim A = P^{-1}B(P^{-1})^{-1}$, and if $C = RBR^{-1}$ then $C = (RP)A(RP)^{-1} \sim A$.

EXERCISES

1. Prove that the set of all n-rowed non-singular matrices P forms a multiplicative group. Thus we are studying the invariants of matrices A under a group of transformations $A \longleftrightarrow PAP^{-1}$ called the *similarity group*.

2. Use the elementary matrix E_{ij} to prove that every matrix A is similar to a matrix obtained from A by an arbitrary permutation on the rows and the same permutation on the columns of A.

3. Let

$$A = \begin{pmatrix} P & 0 \\ 0 & Q \end{pmatrix}, \quad B = \begin{pmatrix} S & 0 \\ 0 & T \end{pmatrix}$$

where P, Q, S, T are square matrices. Show that if S is similar to P and T is similar to Q, then B is similar to A. Show also that A is similar to

$$\begin{pmatrix} Q & 0 \\ 0 & P \end{pmatrix}$$

by the use of Ex. 2. We shall use these operations later.

2. The meaning of canonical forms. The concept of the similarity of matrices is our third equivalence relation. In every such relation we define a set of matrices to be a *canonical set* if

I. *Every matrix is equivalent to a canonical matrix.*

II. *Two distinct canonical matrices are not equivalent.*

We shall obtain canonical forms of matrices under the similarity group by first proving

Theorem 1. *A matrix* $A \sim B$ *in* $\mathfrak{F}$ *if and only if their respective* λ*-matrices* $\lambda I - A$ *and* $\lambda I - B$ *have the same invariant factors.*

We are of course assuming that A and B are n-rowed square matrices with elements in $\mathfrak{F}$, I is the n-rowed identity matrix, λ is an indeterminate over $\mathfrak{F}$. Then $\lambda I - A$ and $\lambda I - B$ are non-singular matrices with elements in $\mathfrak{F}[\lambda]$ and if they have the same invariant factors we apply Theorem 3.21 to obtain

$$P(\lambda I - A)Q = \lambda I - B\,. \tag{3}$$

The hypothesis that λ is an indeterminate implies that $PQ = I$, $PAQ = B$, $Q = P^{-1}$, $B \sim A$. Conversely, $PAP^{-1} = B$ implies (3) for $Q = P^{-1}$ and, by Theorem 3.19, the matrices $\lambda I - A$ and $\lambda I - B$ have the same invariant factors.

The criterion of Theorem 4.1 will be used to obtain a set of canonical matrices. These will be matrices A with the property that to every set of invariant factors there corresponds a unique A, the invariant factors of A are uniquely determined by the form of A. We shall later obtain another canonical set depending on the factorization in $\mathfrak{F}[\lambda]$ of the invariant factors of A.

The second of the above-mentioned canonical sets depends of course on the field $\mathfrak{F}$. The first, however, does not. For the invariant factors of $\lambda I - A$ are the polynomials

$$\phi_{n-j+1}(\lambda) = \frac{D_j(\lambda)}{D_{j-1}(\lambda)} \qquad (j = 1, \ldots, n)$$

where $D_j(\lambda)$ is the greatest common divisor of all j-rowed minors of $\lambda I - A$. The polynomials $D_j(\lambda)$ are independent of the field $\mathfrak{F}$ and depend only on the elements of the matrix A. Hence the invariant factors of $\lambda I - A$ depend only on the elements of A. We may use Theorem 4.1 to prove the

Corollary. *Let* $\mathfrak{K}$ *be a field containing* $\mathfrak{F}$, A *and* B *be* n*-rowed square matrices whose elements (in* $\mathfrak{K}$*) are in* $\mathfrak{F}$*. Then* A *and* B *are similar in* $\mathfrak{K}$ *if and only if they are similar in* $\mathfrak{F}$.

The above corollary implies that we may omit the words "in $\mathfrak{F}$" from the phrase "similar in $\mathfrak{F}$" and say simply that A and B are similar. We shall do this.

EXERCISES

1. What is a set of canonical $m \times n$ matrices under ordinary equivalence in a field $\mathfrak{F}$; in the integral domain $\mathfrak{F}[\lambda]$?

2. Can we replace the words equivalent in $\mathfrak{F}$, $\mathfrak{F}[\lambda]$ by equivalent? Show that these concepts are independent of $\mathfrak{F}$ and state the corresponding results.

3. Use Theorem 4.1 to prove that every square matrix is similar to its transpose.

3. Scalar polynomials. The algebra $\mathfrak{F}_n$ of all n-rowed scalar matrices is a subalgebra of the algebra $\mathfrak{M}_n$ of all n-rowed square matrices. By Theorem 1.9 we may replace $\mathfrak{F}_n$, which is equivalent to $\mathfrak{F}$, by $\mathfrak{F}$. Thus we shall henceforth identify scalar matrices with elements of $\mathfrak{F}$.

Consider the set $\mathfrak{M}[\lambda]$ of all n-rowed square matrices with elements in $\mathfrak{F}[\lambda]$. This set is the algebra of all polynomials in λ with coefficients in $\mathfrak{M}_n$ and the scalar matrices are the polynomials.

$$f(\lambda)I$$

where $f(\lambda)$ is in $\mathfrak{F}[\lambda]$. We have already agreed to replace such a polynomial by $f(\lambda)$.

A particularly important scalar polynomial is the matrix

$$f(\lambda) = |\lambda I - A|I$$

where A is an n-rowed square matrix with elements in $\mathfrak{F}$. We call $f(\lambda)$ the *characteristic function* of A, $f(\lambda) = 0$ the *characteristic equation* of A. It is a polynomial in λ with coefficients in $\mathfrak{F}$ or, as we have seen, may be thought of as a scalar polynomial in λ. In any case $f(A)$ is defined and is a matrix of the form $A^n + a_1A^{n-1} + \ldots + a_nI$ where a_n is the determinant of $-A$. If B is any matrix we showed in Section 3.5 that $B \operatorname{adj} B = |B| \cdot I$. Take $B = \lambda I - A$ and obtain

$$f(\lambda) = (\lambda I - A) \operatorname{adj} (\lambda I - A) = (\lambda I - A)q(\lambda)$$

where $q(\lambda) = \operatorname{adj} (\lambda I - A)$ is in $\mathfrak{M}[\lambda]$. Applying our Factor Theorem of Algebra, we have $f(A) = 0$, that is:

Theorem 2. *Every square matrix satisfies its characteristic equation.*

The first invariant factor $\phi_1(\lambda)$ is the quotient

$$\frac{f(\lambda)}{D_{n-1}(\lambda)},$$

where $D_{n-1}(\lambda)$ is evidently the greatest common divisor of the elements of the matrix $\operatorname{adj} (\lambda I - A)$. We may then write

$$\operatorname{adj} (\lambda I - A) = D_{n-1}(\lambda)B \tag{4}$$

where B has elements in $\mathfrak{F}[\lambda]$ whose g.c.d. is unity. Then

$$(5) \qquad (\lambda I - A)B = \phi_1(\lambda)$$

so that, by Theorem 2.8, $\phi_1(A) = 0$.

DEFINITION. *The minimum function of a matrix* A *is the monic polynomial* $\phi(\lambda)$ *of least degree for which* $\phi(A) = 0$.

The coefficients of $\phi(\lambda)$ are of course assumed to lie in any field $\mathfrak{K}$ containing the elements of the matrix A, where $\mathfrak{K}$ is a *scalar extension* of our fundamental field $\mathfrak{F}$. This means that A *is considered as a matrix with elements in* $\mathfrak{K}$. This notion of $\mathfrak{K}$ is important as the reader might attempt to take a field $\mathfrak{K}$ of matrices in $\mathfrak{M}_n$ over $\mathfrak{F}$. We are not allowing this but are always insisting that $\mathfrak{M}_n$ over $\mathfrak{F}$ containing A is contained in $\mathfrak{M}_n$ over $\mathfrak{K}$.

The polynomial $\phi(\lambda)$ of our definition has not yet been proved unique. This follows from

Theorem 3. *The first invariant factor* $\phi_1(\lambda)$ *of the* λ*-matrix of a matrix* A *is the minimum function* $\phi(\lambda)$ *of* A. *Every scalar polynomial* $\psi(\lambda)$ *such that* $\psi(A) = 0$ *is divisible by* $\phi_1(\lambda)$.

For let $\psi(\lambda)$ be any scalar polynomial with coefficients in a field $\mathfrak{K}_n$ containing the coefficients of a minimum function $\phi(\lambda)$ of A, and such that $\psi(A) = 0$. Then

$$\psi(\lambda) = \phi(\lambda)g(\lambda) + r(\lambda), \qquad r(A) = 0$$

where $r(\lambda)$ is a scalar polynomial of degree less than that of $\phi(\lambda)$. Hence $r(\lambda) \equiv 0$, $\psi(\lambda) = \phi(\lambda)g(\lambda)$. In particular every $\mathfrak{K}_n$ contains $\mathfrak{F}_n$ and $\phi_1(A) = 0$ so that $\phi_1(\lambda) = \phi(\lambda)g(\lambda)$. But by Theorem 2.8

$$\phi(\lambda) = (\lambda I - A)C(\lambda),$$

and hence by (5)

$$(\lambda I - A)B = (\lambda I - A)C(\lambda) \cdot g(\lambda).$$

We apply the uniqueness of the quotient s_1 in Theorem 2.6 on the division of $(\lambda I - A)B$ on the left by $(\lambda I - A)$ and obtain $B = C(\lambda)g(\lambda)$. Since B has elements in $\mathfrak{K}(\lambda)$ of g.c.d. unity and $g(\lambda)$ is a scalar polynomial with coefficients in $\mathfrak{K}_n$, we must have $g = g(\lambda)$ in $\mathfrak{K}_n$. But then $\phi_1(\lambda) = g\phi(\lambda)$ where $\phi_1(\lambda)$ and $\phi(\lambda)$ are both monic polynomials. Thus $g = 1$, $\phi_1(\lambda) = \phi(\lambda)$ has coefficients in $\mathfrak{F}_n$ and in any $\mathfrak{K}_n > \mathfrak{F}_n$, $\phi_1(\lambda)$ divides every scalar $\psi(\lambda)$ such that $\psi(A) = 0$.

COROLLARY I. *The minimum function of* A *with elements in* $\mathfrak{F}$ *has coefficients in* $\mathfrak{F}$.

COROLLARY II. *The characteristic function of* A *is divisible by its minimum function.*

The equation $\phi(\lambda) = 0$ is called the *minimum equation* of A. The polynomial $\phi(\lambda)$ and the equation $\phi(\lambda) = 0$ are sometimes also called the *reduced characteristic function* and *equation*, respectively, of A.

EXERCISES

1. Show that the coefficient of λ^{n-1} in the characteristic function of an n-rowed square matrix A is $-T(A)$ where $T(A)$ is the sum of the diagonal elements a_{ii} of A. We call $T(A)$ the *trace* of A.

2. Prove that similar matrices have the same characteristic function by using the definition $|\lambda I - A|$. Why does this follow from Theorem 4.1? Are the coefficients of λ in $|\lambda I - A|$ invariants of A under similarity transformations?

3. Use the definition of the minimum function of A to prove that similar matrices have the same minimum function.

4. Use the characteristic function of A to show that the inverse of any non-singular n-rowed matrix A is a polynomial in A of degree at most $n - 1$.

5. Let the minimum function $\phi(\lambda)$ of a matrix A have degree r. Prove that the constant term of $\phi(\lambda)$ is zero if and only if A is singular. When A is non-singular show that its inverse is a polynomial in A of degree at most $r - 1$. Prove also that this expression is unique, that is, $I, A, A^2, \ldots, A^{r-1}$ are linearly independent in any scalar extension of the coefficient field $\mathfrak{F}$ of A.

4. Canonical forms. The set of canonical matrices with prescribed invariant factors will be obtained by the use of the

FUNDAMENTAL LEMMA. *Let*

$$\text{(6)} \qquad A_1 = \begin{pmatrix} 0 & 1 & 0 & \ldots & 0 & 0 \\ 0 & 0 & 1 & \ldots & 0 & 0 \\ . & . & . & \ldots & . & . \\ 0 & 0 & 0 & \ldots & 0 & 1 \\ -a_n & -a_{n-1} & -a_{n-2} & \ldots & -a_2 & -a_1 \end{pmatrix}.$$

Then the minimum function of A_1 *is its characteristic function*

$$\text{(7)} \qquad \phi(\lambda) = |\lambda I - A_1| = \lambda^n + a_1\lambda^{n-1} + \ldots + a_n\,.$$

The case $n = 1$ gives $A_1 = -a_1$, $|\lambda I - A_1| = \lambda + a_1$, and we have (7). Also $\phi(\lambda)$ is the minimum function of A in this case since the degree of the minimum function of any matrix is at least unity. We make an induction on n. Expand $|\lambda I - A_1|$ according to its first column and notice that the coefficient of λ is a determinant which is the case of order $n - 1$ of our determinant (6). Hence this term of our expansion is $\lambda^n + a_1\lambda^{n-1} +$

$\ldots + a_{n-1}\lambda$. The only other non-zero element in the first column is a_n and its cofactor is $(-1)^{n+1}|B|$ where B is an $(n-1)$-rowed matrix whose diagonal elements are -1 and whose elements above the diagonal are all zero. Thus $|B| = (-1)^{n-1}$ and the cofactor of a_n is unity. This proves that $|\lambda I - A_1| = \phi(\lambda)$ and also that $D_{n-1} = 1$ since D_{n-1} must divide all $(n-1)$-rowed minors $|B|$ of $\lambda I - A_1$. But then $\phi(\lambda)$ is the minimum function of A_1.

The lemma above is applied as follows. We write

$$A = \text{diag}\,\{A_1, \ldots, A_k\}, \tag{8}$$

where the notation means that A has been broken up into a matrix whose diagonal elements are n_i-rowed square matrices A_i, and whose non-diagonal elements are zero matrices. The importance of the form above is due to the fact that if $g(A)$ is any polynomial in I, A with coefficients in $\mathfrak{F}$ then

$$g(A) = \text{diag}\,\{g(A_1), g(A_2), \ldots, g(A_k)\}. \tag{9}$$

We shall use the notation (8) very frequently both here and in Chapter X. Assume also that A_i is an n_i-rowed square matrix of the form (6) with

$$|\lambda I_{n_i} - A_i| = \phi_i(\lambda) = \lambda^{n_i} + a_1^{(i)}\lambda^{n_i-1} + \ldots + a_{n_i}^{(i)}, \tag{10}$$

and that

$$\phi_{i+1}(\lambda) \text{ divides } \phi_i(\lambda) \qquad (i = 1, \ldots, k-1). \tag{11}$$

The invariant factors of $\lambda I - A$ *distinct from unity are called its non-trivial invariant factors.* The matrices (8) then form a canonical set. We prove this in

Theorem 4. *Every* n*-rowed square matrix* A *whose* λ*-matrix* $\lambda I - A$ *has the non-trivial invariant factors* $\phi_1, \ldots, \phi_k$ *is similar in* $\mathfrak{F}$ *to a matrix of the form* (8) *defined uniquely by the* ϕ_i *and* (10).

This theorem implies that no two distinct matrices (8) are similar. For by the theorem the λ-matrices of two distinct matrices (8) have distinct invariant factors. The theorem also implies that the non-trivial invariant factors of

$$\text{diag}\,\{\lambda I_{n_1} - A_1, \lambda I_{n_2} - A_2, \ldots, \lambda I_{n_k} - A_k\} \tag{12}$$

are $\phi_1(\lambda), \ldots, \phi_k(\lambda)$. By Theorem 4.1 it is actually sufficient to prove this last fact in order to prove our result. We apply our fundamental lemma and Theorem 3.18 to show that $\lambda I_{n_i} - A_i$ is equivalent in $\mathfrak{F}[\lambda]$ to

$$B_i = \text{diag}\,\{\phi_i(\lambda), 1, \ldots, 1\} \tag{13}$$

and, by Theorem 3.9, $\lambda I - A$ is equivalent to

$$(14) \qquad B = \text{diag}\ \{B_1, \ldots, B_k\}\ .$$

By a sequence of elementary transformation of type I we replace B by

$$(15) \qquad C = \text{diag}\ \{1, 1, \ldots, \phi_k, \phi_{k-1}, \ldots, \phi_1\}\ .$$

But this is the form in Theorem 3.18 so that C and hence the equivalent matrix $\lambda I - A$ have $\phi_1, \ldots, \phi_k$ as non-trivial invariant factors.

Our canonical form (8), (6), (10) exhibits explicitly the invariant factors of the λ-matrix of the matrix (8). It may be described as follows. Let P and Q be square matrices with elements in $\mathfrak{F}$ and

$$A = \begin{pmatrix} P & 0 \\ 0 & Q \end{pmatrix}.$$

Then A is called the *direct sum* of P and Q. Hence every matrix A whose λ-matrix has the non-trivial invariant factors $\phi_1(\lambda), \ldots, \phi_k(\lambda)$ is a direct sum of matrices A_i whose characteristic and minimum functions are $\phi_i(\lambda)$, respectively. A canonical A_i is then given in (6).

EXERCISES

1. Give canonical matrices (8) for all two-rowed and three-rowed square matrices.

2. What are the possible types of invariant factors for the λ-matrix $\lambda I - A$ of a four-rowed square matrix A?

3. Let A be an n-rowed square matrix with an irreducible minimum function $\phi(\lambda)$ of degree m. Show that $n = mq$ and that A is similar to a direct sum of q equal matrices of the form (6).

4. Let A be the nq-rowed square matrix obtained from (6) by replacing each element of (6) by the corresponding q-rowed scalar matrix. Use the method of proof of Theorem 4.4 to show that the non-trivial invariant factors of $\lambda I - A$ consist of the polynomials $\phi(\lambda)$ repeated q times. Apply this fact in Ex. 3 to obtain a new canonical form of an expanded type (6) for any matrix with an irreducible minimum function.

5. Elementary divisors. Our canonical form for a square matrix under similarity transformations is a form reducing A to a direct sum of matrices. These matrices may themselves be reducible to a direct sum of further matrices of smaller order and it is desirable to obtain this additional reduction. *We shall call a matrix* A ***indecomposable in*** $\mathfrak{F}$ *if* A *is not similar to a direct sum* (16), *and shall obtain a canonical form for any* A (in the sense of similarity) *as a direct sum of indecomposable matrices.* We shall first prove

Theorem 5. *Let* $\phi(\lambda) = \mathrm{p}(\lambda) \cdot \mathrm{q}(\lambda)$ *where* $\mathrm{p}(\lambda)$ *and* $\mathrm{q}(\lambda)$ *are relatively prime monic polynomials of* $\mathfrak{F}[\lambda]$. *Then a matrix* A *has* $\phi(\lambda)$ *as its minimum function if and only if* A *is similar in* $\mathfrak{F}$ *to*

$$\begin{pmatrix} \mathrm{P} & 0 \\ 0 & \mathrm{Q} \end{pmatrix}, \tag{16}$$

where $\mathrm{p}(\lambda)$ *is the minimum function of* P *and* $\mathrm{q}(\lambda)$ *is the minimum function of* Q.

For let P and Q have relatively prime minimum functions $p(\lambda)$ and $q(\lambda)$, respectively, and suppose that A with minimum function $\phi(\lambda)$ is similar to (16). By (9) and $\phi(A) = 0$ we have $\phi(P) = 0$, $\phi(Q) = 0$. Theorem 4.3 applied to P and Q states that $\phi(\lambda)$ is divisible by $p(\lambda)$ and $q(\lambda)$ and, since they are relatively prime, by their product $\phi_0(\lambda)$. Now

$$\phi_0(A) = \begin{pmatrix} p(P) \cdot q(P) & 0 \\ 0 & p(Q) \cdot q(Q) \end{pmatrix} = 0$$

so that $\phi(\lambda)$ divides $\phi_0(\lambda)$. Hence $\phi(\lambda) = p(\lambda) \cdot q(\lambda)$.

Conversely, let $\phi(\lambda) = p(\lambda) \cdot q(\lambda)$ be the minimum function of A where $p(\lambda)$ and $q(\lambda)$ are relatively prime monic polynomials. Let $\phi_i(\lambda)$ be the ith non-trivial invariant factor of A and $p_i(\lambda)$ be the g.c.d. of $p(\lambda)$ and $\phi_i(\lambda)$, $q_i(\lambda)$ be the g.c.d. of $\phi_i(\lambda)$ and $q(\lambda)$. Then

$$\phi_i(\lambda) = p_i(\lambda) \cdot q_i(\lambda) . \tag{17}$$

For $p_i(\lambda)$ and $q_i(\lambda)$ are relatively prime and their product $\phi_{i0}(\lambda)$ divides $\phi_i(\lambda) = \psi_i(\lambda)\phi_{i0}(\lambda)$. But every irreducible factor of $\psi_i(\lambda)$ divides $\phi(\lambda) = p(\lambda) \cdot q(\lambda)$ by our definition of $\phi(\lambda) = \phi_1(\lambda)$ and (11), and must divide either $p(\lambda)$ or $q(\lambda)$. This is contrary to our hypotheses about $p_i(\lambda)$ and $q_i(\lambda)$ unless $\psi_i(\lambda) = 1$. Then $\phi_i(\lambda) = p_i(\lambda) \cdot q_i(\lambda)$.

We take A in the form (8) and use (17). By our above proof the matrix

$$A_{i0} = \begin{pmatrix} P_i & 0 \\ 0 & Q_i \end{pmatrix}, \qquad (i = 1, \ldots, k),$$

where P_i is in the form (6) for $p_i(\lambda)$, Q_i is in the form (6) for $q_i(\lambda)$, has $\phi_i(\lambda)$ as its minimum function and is similar to A_i. Then if

$$P = \operatorname{diag}\{P_1, \ldots, P_k\}, \qquad Q = \operatorname{diag}\{Q_1, \ldots, Q_k\},$$

the matrices P and Q have the form (8) and have minimum functions $p(\lambda)$ and $q(\lambda)$, respectively. We replace the A_i of (8) for A by the similar matrices A_{i0} and permute the P_i, Q_j as in Exercise 3 of Section 4.1 and have (16).

Theorem 4.5 may be stated in the following alternative form:

Theorem 6. *Let* $f(\lambda) = g(\lambda) \cdot h(\lambda)$ *where* $g(\lambda)$ *and* $h(\lambda)$ *are relatively prime monic polynomials of* $\mathfrak{F}[\lambda]$. *Then a matrix* A *has* $f(\lambda)$ *as its characteristic function if and only if* A *is similar to*

$$\begin{pmatrix} P & 0 \\ 0 & Q \end{pmatrix}$$

where $g(\lambda)$ *is the characteristic function of* P *and* $h(\lambda)$ *is the characteristic function of* Q.

For we apply Theorem 4.5 with $p(\lambda)$ the g.c.d. of $\phi(\lambda)$ and $g(\lambda)$, $q(\lambda)$ the g.c.d. of $\phi(\lambda)$ and $h(\lambda)$. Since $\phi(\lambda)$ divides $f(\lambda)$ we evidently have $\phi(\lambda) = p(\lambda) \cdot q(\lambda)$. It is also obvious that $|\lambda I - A| = f(\lambda) = |\lambda I_1 - P| \cdot |\lambda I_2 - Q|$ and this combined with Theorem 4.5 proves our alternative form.

Theorems 4.5 and 4.6 may be used to obtain the second canonical form we have already spoken of. We write

$$|\lambda I - A| = f(\lambda) = p_1^{e_1} \cdots p_t^{e_t}, \tag{18}$$

where $p_j = p_j(\lambda)$ is a monic irreducible polynomial of $\mathfrak{F}[\lambda]$ distinct from $p_k(\lambda)$, $j \neq k$. By Theorem 4.6 any square matrix A with $f(\lambda)$ as characteristic function is similar in $\mathfrak{F}$ to

$$A = \text{diag}\,\{P_1, \ldots, P_t\}, \qquad |\lambda I_{\nu_j} - P_j| = p_j(\lambda)^{e_j}, \tag{19}$$

where the degree of $[p_j(\lambda)]^{e_j}$ is of course the order ν_j of P_j. The non-trivial invariant factors of the λ-matrix of P_j are positive powers

$$\phi_i^{(j)}(\lambda) = [p_j(\lambda)]^{f_{ji}} \qquad (i = 1, \ldots, k_j) \tag{20}$$

of the polynomial $p_j(\lambda)$ and are such that

$$f_{j1} \geqq f_{j2} \geqq \cdots \geqq f_{jk_j} > 0, \qquad \sum_{i=1}^{k_j} f_{ji} = e_j \quad (j = 1, \ldots, t). \tag{21}$$

But then by Theorem 4.4, P_j is similar in $\mathfrak{F}$ to

$$\text{diag}\,\{S_1^{(j)}, S_2^{(j)}, \ldots ; S_{k_j}^{(j)}\} \qquad (j = 1, \ldots, t), \tag{22}$$

where $S_i^{(j)}$ is a matrix of the form (6) for the polynomial $[p_j(\lambda)]^{f_{ji}}$.

DEFINITION. *The polynomials* $\phi_i^{(j)}(\lambda)$ *of* (20) *are called the* $\mathfrak{F}$*-elementary divisors of* A.

The invariant factors of a matrix are obtainable from its $\mathfrak{F}$-elementary divisors by the use of

Theorem 7. *Let* A *have the form defined by* (19), (22), (6), (18) *and arrange the factors* $p_j(\lambda)$ *so that*

$$k = k_1 \geqq k_2 \geqq \ldots \geqq k_t . \tag{23}$$

Define

$$f_{ji} = 0, \quad \phi_i^{(j)}(\lambda) = 1 \qquad (i = k_j + 1, \ldots, k).$$

Then the invariant factors of A *are*

$$\phi_i(\lambda) = \prod_{j=1}^{t}[p_j(\lambda)]^{f_{ji}} = \prod_{j=1}^{t}\phi_i^{(j)}(\lambda) \qquad (i = 1, \ldots, k). \tag{24}$$

For, a permutation of the diagonal blocks in A replaces A by a similar matrix

$$B = \text{diag}\ \{B_1, \ldots, B_k\}, \qquad B_i = \text{diag}\ \{S_i^{(1)}, S_i^{(2)}, \ldots, S_i^{(t)}\}, \tag{25}$$

where if r_j is the degree of $p_j(\lambda)$ then $S_i^{(j)}$ has $f_{ji}r_j$ rows. Hence $S_i^{(j)}$ has zero rows, that is does not appear if $i > k_j$. The minimum function of $S_i^{(j)}$ is $[p_j(\lambda)]^{f_{ji}}$ and this is also its characteristic function. By Theorems 4.5 and 4.6 the minimum and characteristic functions of B_i are the polynomial $\phi_i(\lambda)$ of (24). Since $\phi_{i+1}(\lambda)$ divides $\phi_i(\lambda)$ by (21) the matrix B of (25) similar to A will be in the canonical form (8) of A when we replace the B_i by the similar corresponding matrices (6) for $\phi_i(\lambda)$ of (24). This proves Theorem 7.

The $\mathfrak{F}$-elementary divisors of A may be easily obtained from its invariant factors. For we factor the $\phi_i(\lambda)$ as in (24) and determine the $\phi_i^{(j)}$ by (20).

We have not yet shown that (19), (22) give a reduction of A as a direct sum of indecomposable matrices. To show this it is sufficient to show that the matrices $S_i^{(j)}$ are indecomposable. Thus we wish to prove that if the minimum function of a square matrix S is its characteristic function $\phi(\lambda) = [p(\lambda)]^e$, where $p(\lambda)$ is irreducible, then S is indecomposable. Let S have this form but be decomposable,

$$S \sim \begin{pmatrix} G & 0 \\ 0 & H \end{pmatrix}.$$

Then $|\lambda I - S| = |\lambda I_1 - G| \cdot |\lambda I_2 - H| = g(\lambda) \cdot h(\lambda)$, where $g(\lambda)$ and $h(\lambda)$ are the respective characteristic functions of G and H. They must divide $p(\lambda)^e$ and we can write $g(\lambda) = p(\lambda)^\gamma$, $h(\lambda) = p(\lambda)^\delta$. Without loss of generality we assume that $\gamma \geqq \delta$ and have

$$g(S) \sim \begin{pmatrix} p(G)^\gamma & 0 \\ 0 & p(H)^\gamma \end{pmatrix} = 0$$

since $p(H)^\gamma = [p(H)]^\delta \cdot [p(H)]^{\gamma-\delta} = 0$. But $\gamma + \delta = e$ and $[p(\lambda)]^e$ divides $g(\lambda)$ by Theorem 4.3. Hence $\gamma = e$, $\delta = 0$, whereas the order of H is the degree of its characteristic function $p(\lambda)^\delta$. This contradicts our hypothesis that S is decomposable. Conversely, when S is indecomposable we apply (19), (22) and have proved

Theorem 8. *A matrix* S *is indecomposable in* $\mathfrak{F}$ *if and only if the characteristic function of* S *is its minimum function and has the form* $\mathrm{p}(\lambda)^e$ *where* $\mathrm{p}(\lambda)$ *is an irreducible monic polynomial of* $\mathfrak{F}[\lambda]$. *Every matrix* A *is similar to a direct sum of indecomposable matrices* S_k *whose characteristic functions are the* $\mathfrak{F}$*-elementary divisors of* A.

In the particular case where the field $\mathfrak{F}$ is chosen so that the p_i are linear polynomials $\lambda - \lambda_i$ we call the $(\lambda - \lambda_i)^{f_{ij}}$ *the elementary divisors of* A. The λ_i are then the roots (in $\mathfrak{F}$) of $|\lambda I - A| = 0$ and are called *the characteristic roots* of A. For a special canonical form in this case see Exercise 2 of the next section.

If the elementary divisors of A all have $f_{ij} = 1, 0$ we say that A has only *simple elementary divisors.* In this case $S_j^{(i)}$ is the 1×1 matrix

$$S_j^{(i)} = (\lambda_j)$$

and A is similar in $\mathfrak{F}$ to a diagonal matrix whose diagonal elements are the characteristic roots of A. If λ_i is a root of multiplicity r_i of $|\lambda I - A| = 0$ we may use Exercise 2 of Section 4.1 to pass to a matrix

$$\text{diag}\ \{\lambda_1 I_{r_1}, \ldots, \lambda_t I_{r_t}\} \tag{26}$$

similar to A in $\mathfrak{F}$, where I_{r_i} is of course the identity matrix of r_i rows.

We shall next consider some important special types of matrices.

EXERCISES

1. What are the possible $\mathfrak{R}$-elementary divisors for two-rowed and three-rowed square matrices with elements in the field $\mathfrak{R}$ of all rational numbers? Give the corresponding direct sums of indecomposable submatrices.

2. Find the results of Ex. 1 in the case where $\mathfrak{R}$ is replaced by the field $\mathfrak{R}'$ of all real numbers, the field $\mathfrak{R}'(i)$ of all complex numbers.

6. Nilpotent matrices. A matrix A is called *nilpotent* of *index* a if $A^{a-1} \neq 0$, $A^a = 0$ for some positive integer a. The minimum equation of A divides $\lambda^a = 0$ and has the form $\lambda^\beta = 0$, $\beta \leqq a$. But $A^\beta = 0$, $A^{a-1} \neq 0$ so that $a = \beta$, $\lambda^a = 0$ is the minimum equation of A.

The invariant factors of the matrix $\lambda I - A$ are

$$\phi_i = \lambda^{a_i}, \qquad a = a_1 \geqq a_2 \geqq \ldots \geqq 0,$$

and A is similar in $\mathfrak{F}$ to (8) where A_i is the indecomposable a_i-rowed square matrix

$$A_i = \begin{pmatrix} 0 & 1 & 0 & \ldots & 0 \\ 0 & 0 & 1 & \ldots & 0 \\ . & . & . & \ldots & 0 \\ 0 & 0 & 0 & \ldots & 1 \\ 0 & 0 & 0 & \ldots & 0 \end{pmatrix}, \tag{27}$$

whose characteristic and minimum equations are $\lambda^{a_i} = 0$. We call the integers a_i the *indices* of A and have proved

Theorem 9. *Two nilpotent matrices are similar in* $\mathfrak{F}$ *if and only if they have the same indices.*

A nilpotent matrix of index two has the property $A \neq 0$, $A^2 = 0$. Then $a_i = 1$ or 2, $A_i = 0$ or

$$A_i = \begin{pmatrix} 0 & 1 \\ 0 & 0 \end{pmatrix}.$$

A permutation of the rows and corresponding columns of A given by (8) replaces A by a similar matrix

$$\begin{pmatrix} 0_r & I_r & 0 \\ 0_r & 0_r & 0 \\ 0 & 0 & 0 \end{pmatrix} \tag{28}$$

where 0_r is an r-rowed square zero matrix. Evidently r is the rank of A and we have proved that every nilpotent matrix A of index two and rank r is similar to (28).

Theorem 10. *Two nilpotent matrices of index two are similar in* $\mathfrak{F}$ *if and only if they have the same rank.*

EXERCISES

1. Write the canonical forms (8), (27) of all nilpotent matrices of orders two, three, four.

2. Let the elementary divisors of A be $(\lambda - \lambda_k)^{g_k}$. Show that A is similar to a direct sum of matrices $S_k = \lambda_k I_{g_k} + N_k$, where N_k is the g_k-rowed nilpotent matrix (27). This is called the *classical canonical form* of A. It occurs for every A when $\mathfrak{F}$ is in particular the field of all complex numbers.

7. Idempotent matrices. A matrix E is called *idempotent* if

$$E^2 = E\,. \tag{29}$$

Since E satisfies $\lambda^2 - \lambda = 0$, its minimum function divides $\lambda^2 - \lambda$ and can only be λ, $\lambda - 1$ or $\lambda^2 - \lambda$. When E is neither the zero matrix nor the n-rowed identity matrix, it satisfies neither $\lambda = 0$ nor $\lambda - 1 = 0$ and its minimum function is $\lambda^2 - \lambda$. We assume that $E \neq 0, I$.

The elementary divisors of E are λ and $\lambda - 1$ and they are relatively prime. By Theorem 4.8 the matrix E is similar in $\mathfrak{F}$ to a diagonal matrix (19) with $t = 2$, $p_1(\lambda) = \lambda - 1$, $p_2(\lambda) = \lambda$, $e_1 = 1 = e_2$. But the form (6) for $\lambda - 1$ is (1) and for λ is (0) so that E is similar to

$$E_r = \begin{pmatrix} I_r & 0 \\ 0 & 0 \end{pmatrix}. \tag{30}$$

Thus r is the rank of E. We evidently have proved that in all cases

Theorem 11. *Two idempotent matrices are similar in $\mathfrak{F}$ if and only if they have the same rank.*

Let E have the form (30) for $0 < r < n$ and write

$$A = \begin{pmatrix} A_1 & A_2 \\ A_3 & A_4 \end{pmatrix},$$

where A_1 is an r-rowed square matrix. Then

$$EA = \begin{pmatrix} A_1 & A_2 \\ 0 & 0 \end{pmatrix}, \quad AE = \begin{pmatrix} A_1 & 0 \\ A_3 & 0 \end{pmatrix}, \quad EAE = \begin{pmatrix} A_1 & 0 \\ 0 & 0 \end{pmatrix},$$

so that

$$A = EAE + RE + EL + N\,, \tag{31}$$

where

$$R = \begin{pmatrix} 0 & 0 \\ A_3 & 0 \end{pmatrix}, \quad L = \begin{pmatrix} 0 & A_2 \\ 0 & 0 \end{pmatrix}, \quad N = \begin{pmatrix} 0 & 0 \\ 0 & A_4 \end{pmatrix}. \tag{32}$$

The matrices N, L, R have the important properties

$$RE = R, \qquad EL = L, \qquad EN = NE = ER = LE = 0\,. \tag{33}$$

We call (31) *the decomposition of* A *relative to* E. When $A_1, \ldots, A_4$ range over all possible matrices we obtain the decomposition of the algebra $\mathfrak{M}_n$ relative to its idempotent element E and may write for *any* idempotent E,

$$\mathfrak{M}_n = E\mathfrak{M}_nE + \mathfrak{R}E + E\mathfrak{L} + \mathfrak{N} .$$

Then $E\mathfrak{M}_nE$ consists of all matrices A for which $EA = AE = A$, $\mathfrak{R}$ consists of all A for which $AE = A$, $EA = 0$, $\mathfrak{L}$ consists of all A for which $EA = A$, $AE = 0$, $\mathfrak{N}$ consists of all A for which $EA = AE = 0$.

Two idempotent matrices E, E_0 *are called orthogonal* (or *supplementary*) *if* $EE_0 = E_0E = 0$. Then E_0 is in the set $\mathfrak{N}$ for E and we may take

$$E = \begin{pmatrix} I_r & 0 \\ 0 & 0 \end{pmatrix}, \qquad E_0 = \begin{pmatrix} 0 & 0 \\ 0 & E_1 \end{pmatrix}, \tag{34}$$

where E_1 is an $(n - r)$-rowed idempotent matrix whose rank is the rank r_0 of E_0. But E_0 is certainly similar in $\mathfrak{F}$ to

$$DE_0D^{-1} = \begin{pmatrix} I_r & 0 \\ 0 & D_1 \end{pmatrix} E_0 \begin{pmatrix} I_r & 0 \\ 0 & D_1^{-1} \end{pmatrix} = \begin{pmatrix} 0 & 0 \\ 0 & D_1E_1D_1^{-1} \end{pmatrix},$$

where $DED^{-1} = E$ and D_1 is at our choice. Hence we may take

$$E_0 = \begin{pmatrix} 0 & 0 & 0 \\ 0 & I_{r_0} & 0 \\ 0 & 0 & 0 \end{pmatrix}, \qquad E + E_0 = \begin{pmatrix} I_{r+r_0} & 0 \\ 0 & 0 \end{pmatrix}. \tag{35}$$

Theorem 12. *The rank of the sum of two orthogonal idempotent matrices is the sum of their ranks.*

An immediate corollary of Theorem 4.12 and its proof is given by

Theorem 13. *Let* $E_1, \ldots, E_n$ *be* n *idempotent* n*-rowed matrices which are pairwise orthogonal. Then each has rank one and there exists a matrix* Q *such that* QE_iQ^{-1} *has unity in the* ith *row and column and zeros elsewhere.*

8. The automorphisms of $\mathfrak{M}_n$. Consider a transformation

$$S: \qquad A \longleftrightarrow A^S,$$

where A and A^S range over the matrices of the algebra $\mathfrak{M}_n$ of all n-rowed square matrices with elements in a field $\mathfrak{F}$. We call S an *automorphism* of $\mathfrak{M}_n$ if

$$(A + B)^S = A^S + B^S, \qquad (AB)^S = A^SB^S, \qquad (\lambda I_n)^S = \lambda I_n, \tag{36}$$

for every A and B of $\mathfrak{M}_n$ and λ of $\mathfrak{F}$. An obvious automorphism of $\mathfrak{M}_n$ is the correspondence S defined by

$$A \longleftrightarrow A^S = PAP^{-1}, \tag{37}$$

where P is any non-singular matrix of $\mathfrak{M}_n$. We call (37) an *inner automorphism* of $\mathfrak{M}_n$ and prove

Theorem 14. *Every automorphism of $\mathfrak{M}_n$ is an inner automorphism.*

For let S be an automorphism and e_{ij} be the matrix which has unity in the ith row and jth column and zeros elsewhere. Then S is generated (i.e., completely determined) by the partial correspondence

$$e_{ij} \longleftrightarrow f_{ij} \neq 0,$$

where the f_{ij} are n^2 matrices with the properties

$$f_{ij}f_{tk} = \delta_{jt}f_{ik} \qquad (\delta_{jj} = 1,\ \delta_{jt} = 0 \text{ for } j \neq t).$$

(The function δ_{jt} is called the *Kronecker* δ).

The matrices f_{ii} are n supplementary idempotent matrices and by Theorem 4.13 there exists a non-singular matrix Q such that

$$Qf_{ii}Q^{-1} = e_{ii}, \qquad Qf_{ij}Q^{-1} = g_{ij} \qquad (i \neq j).$$

The multiplication of g_{ij} on the left by e_{ii} and on the right by e_{jj} yields a matrix with zeros everywhere except in the ith row and jth column. Since $e_{ii}g_{ij} = g_{ij}e_{jj} = g_{ij} \neq 0$ we may write

$$g_{ij} = \gamma_{ij}e_{ij} \qquad (\gamma_{ij} \neq 0 \text{ in } \mathfrak{F}).$$

Now $g_{1i}g_{i1} = e_{11} = \gamma_{1i}\gamma_{i1}e_{1i}e_{i1}$ so that $\gamma_{1i}\gamma_{i1} = 1$. Also $g_{ij} = g_{i1}g_{1j} = \gamma_{i1}\gamma_{1j}e_{ij}$ whence $\gamma_{ij} = \gamma_{i1}\gamma_{1j}$ and

$$\gamma_{ij} = \gamma_{1i}^{-1}\gamma_{1j}.$$

We now put

$$H = \operatorname{diag}\{\gamma_{11}, \gamma_{12}, \ldots, \gamma_{1n}\},$$

where H is non-singular and of course $\gamma_{11} = 1$. Then $He_{ii}H^{-1} = e_{ii}$ and

$$Hg_{ij}H^{-1} = \gamma_{1i}\gamma_{ij}\gamma_{1j}^{-1}e_{ij} = e_{ij}.$$

Thus $f_{ij} = Pe_{ij}P^{-1}$ where $P = (HQ)^{-1}$, $A^S = PAP^{-1}$.

Inner automorphisms of subalgebras of $\mathfrak{M}_n$ are defined similarly. If A is a subalgebra of $\mathfrak{M}_n$ with a unity element its regular elements act as non-singular matrices. An inner automorphism of A is then a correspondence (37) with P a regular element of A. We shall however not investigate the problem of determining the automorphisms of general subalgebras of $\mathfrak{M}_n$.

9. Another basis of $\mathfrak{M}_n$. The matrix of (27) is a special case $\gamma = 0$ of

$$y = \begin{pmatrix} 0 & 1 & 0 & \dots & 0 \\ 0 & 0 & 1 & \dots & 0 \\ . & . & . & \dots & . \\ 0 & 0 & 0 & \dots & 1 \\ \gamma & 0 & 0 & \dots & 0 \end{pmatrix} \qquad (\gamma \text{ in } \mathfrak{F}). \tag{38}$$

By our Fundamental Lemma of Section 4.4 the characteristic function of y is its minimum function

$$\phi(\lambda) = \lambda^n - \gamma\,. \tag{39}$$

If e_{ij} is the matrix with unity in the ith row and jth column the form (38) of y states that

$$y = e_{12} + e_{23} + \dots + e_{n-1,n} + \gamma e_{n1}\,. \tag{40}$$

We use the abbreviation

$$e_i = e_{ii} \qquad (i = 1, \dots, n), \tag{41}$$

and obtain by direct multiplication

$$e_i y = e_{i,i+1} = y e_{i+1} \qquad (i = 1, \dots, n-1). \tag{42}$$

Also

$$e_n y = \gamma e_{n1} = y e_1\,. \tag{43}$$

We now assume that $\gamma \neq 0$ so that y has an inverse $\gamma^{-1}y^{n-1}$ and all integral powers y^t of y are defined. Every integer t has the form

$$t = qn + r \qquad (0 < r \leqq n),$$

since we may write $t - 1 = qn + r_0$ $(0 \leqq r_0 < n)$, $r = r_0 + 1$. Then (42) and (43) imply that if we define

$$e_t = e_{tt} = e_r\,, \tag{44}$$

for all integers t, we have

(45) $$y^t e_j = e_{j-t} y^t \qquad (i, j = 1, \ldots).$$

We easily compute

(46) $$e_{1j} = e_{12}e_{23} \ldots e_{j-1,j} = e_1 y e_2 y \ldots e_{j-1} y = e_1 y^{j-1} .$$

Also $y^{i-1}e_i = e_1 y^{i-1}$ so that

(47) $$e_1 = e_{1i}e_{i1} = e_1 y^{i-1} e_{i1} = y^{i-1} e_i e_{i1} = y^{i-1} e_{i1}$$

from which

(48) $$e_{i1} = y^{1-i} e_1 .$$

But $e_1 e_1 = e_1$, $y^{1-i}e_1 = e_i y^{1-i}$ so that

(49) $$e_{ij} = e_{i1}e_{1j} = y^{1-i}e_1 e_1 y^{j-1} = e_i y^{j-i} = e_i y^k .$$

Every n-rowed square matrix $A = (a_{ij})$ is expressible uniquely as a linear combination $\Sigma a_{ij}e_{ij}$ of the e_{ij} and hence as a linear combination of the quantities

(50) $$e_i y^k \qquad (i = 1, \ldots, n;\ k = 0, \ldots, n-1)$$

since $y^t = \gamma^q y^k$ for every integer t and $k = 0, 1, \ldots, n-1$. The quantities (50) are n^2 in number and by Theorem 2.5 form a basis of the linear set $\mathfrak{M}_n$ of all n-rowed matrices over $\mathfrak{F}$. But every $A = \Sigma b_{ij} e_i y^j$ with b_{ij} in $\mathfrak{F}$, $A = \Sigma d_j y^j$ where $d_j = \Sigma b_{ij} e_i$ is a diagonal matrix uniquely determined by the matrix A. We have proved

Theorem 15. *Let* $\gamma \neq 0$ *in* y *of* (38). *Then every matrix is uniquely expressible in the form*

$$\mathrm{d}_0 + \mathrm{d}_1\mathrm{y} + \ldots + \mathrm{d}_{n-1}\mathrm{y}^{n-1} ,$$

where the d_i *are diagonal matrices.*

Any two diagonal matrices are commutative. This follows by direct multiplication. We write

(51) $$d = \mathrm{diag}\ \{\delta_1, \ldots, \delta_n\}, \qquad \delta_i \text{ in } \mathfrak{F} ,$$

and define

(52) $$d^{S^j} = \mathrm{diag}\ \{\delta_{j+1}, \delta_{j+2}, \ldots, \delta_n, \delta_1, \ldots, \delta_j\} .$$

Then evidently

$$d^{S^i} = y^i d y^{-i} \tag{53}$$

and $d = d^S$ if and only if $\delta_1 = \delta_2 = \ldots = \delta_n$, that is, d is a scalar matrix. The correspondences $d \longleftrightarrow d^{S^i}$ are evidently special cases of those inner automorphisms of $\mathfrak{M}_n$ given by

$$S^i: \qquad A \longleftrightarrow A^{S^i} = y^i A y^{-i},$$

for any matrix A of $\mathfrak{M}_n$.

An inner automorphism of $\mathfrak{M}_n$ leaves some of its quantities invariant. In fact if $A^S = yAy^{-1}$ then A is invariant under S if and only if $A = yAy^{-1}$, A is commutative with y. Thus *the problem of finding the quantities invariant under an automorphism of* $\mathfrak{M}_n$ *is equivalent to the problem of finding all matrices commutative with a given non-singular matrix.* This problem will be studied in more detail in Chapter 10 and we shall there actually omit the assumption of non-singularity. The particular case where y has the above form (38) will be treated now, and we shall prove

Theorem 16. *Let* y *be given by* (38) *with* γ *any quantity of* $\mathfrak{F}$. *Then the only matrices commutative with* y *are polynomials in* y *and the identity matrix* I *with coefficients in* $\mathfrak{F}$.

It is of course quite obvious that any polynomial in I and a matrix B is commutative with B. We are proving that these are the only matrices commutative with $B = y$ of (38).

We first let $\gamma \neq 0$. Then Theorem 4.15 states that every matrix $A = \sum_{i=0}^{n-1} d_i y^i$ where the d_i are diagonal matrices. If $Ay = yA$ we use (53) and obtain $\sum_i d_i y^{i+1} = \sum_i d_i^S y^{i+1}$, the uniqueness of the expression in Theorem 4.15 implies that every $d_i = d_i^S$, and d_i is a scalar matrix. Hence A is in $\mathfrak{F}[y]$.

Next let $\gamma = 0$ and define y_1 to be the matrix (38) with $\gamma = 1$. Every matrix of $\mathfrak{M}_n$ has the unique form $A = \sum_{i=0}^{n-1} d_i y_1^i$ and in particular,

$$y = \begin{pmatrix} 0 & 1 & 0 & \ldots & 0 \\ 0 & 0 & 1 & \ldots & 0 \\ \cdot & \cdot & \cdot & \cdots & \cdot \\ 0 & 0 & 0 & \ldots & 1 \\ 0 & 0 & 0 & \ldots & 0 \end{pmatrix} = \begin{pmatrix} 1 & 0 & \ldots & 0 & 0 \\ 0 & 1 & \ldots & 0 & 0 \\ \cdot & \cdot & \cdots & \cdot & \cdot \\ 0 & 0 & \ldots & 1 & 0 \\ 0 & 0 & \ldots & 0 & 0 \end{pmatrix} y_1 = (I - e_n) y_1$$

is the product on the left of y_1 by the diagonal matrix whose diagonal elements are all unity except for the last one which is zero. This is evident above by direct matrix multiplication. Now $y^2 = (I - e_n)y_1(I - e_n)y_1 = (I - e_n)(I - e_{n-1})y_1^2 = (I - e_n - e_{n-1})y_1^2$ and an evident induction implies that

$$y^i = [I - (e_n + e_{n-1} + \dots + e_{n-i+1})]y_1^i \qquad (i = 1, \dots, n-1).$$

Hence y^i is the product of y_1^i on the left by the diagonal matrix whose first $n - i$ diagonal elements are unity and the remaining ones zero. Write $A = \sum_{i=0}^{n-1} d_i y_1^i$ and compute

$$Ay = \left(\sum_{i=0}^{n-1} d_i y_1^i\right)(I - e_n)y_1 = \sum d_i(I - e_{n-i})y_1^{i+1}$$
$$= yA = (I - e_n)\sum d_i^S y_1^{i+1}$$

by (53). Thus $d_i(I - e_{n-i}) = (I - e_n)d_i^S$ and if

$$d_i = \operatorname{diag}\{\delta_{i1}, \delta_{i2}, \dots, \delta_{in}\}$$

then

$$\operatorname{diag}\{\delta_{i1}, \delta_{i2}, \dots, \delta_{i,n-i-1}, 0, \delta_{i,n-i+1}, \dots, \delta_{in}\} = \operatorname{diag}\{\delta_{i2}, \delta_{i3}, \dots, \delta_{in}, 0\}.$$

Comparing the right-hand matrix with the left, we have

$$0 = \delta_{in}, \qquad \delta_{in} = \delta_{in-1} = \dots = \delta_{i,n-i+1} = 0$$

and

$$\delta_{i,n-i} = \delta_{i,n-i-1} = \dots = \delta_{i2} = \delta_{i1} = \delta_i .$$

But then $d_0 = \delta_0 I$,

$$d_i = \operatorname{diag}\{\delta_i, \dots, \delta_i, 0, \dots, 0\} = \delta_i[I - (e_{n-i+1} + \dots + e_n)] \qquad (i = 1, \dots, n-1),$$

with δ_j in $\mathfrak{F}$ and

$$A = \sum_{i=0}^{n-1} d_i y_1^i = \sum_{i=1}^{n-1} \delta_i y^i + \delta_0 I$$

is in the domain $\mathfrak{F}[y]$ of all polynomials in y and I with coefficients in $\mathfrak{F}$, as desired.

We now leave the theory of the similarity of matrices and shall study a different type of equivalence relation.

EXERCISES

1. Let D be an n-rowed diagonal matrix with distinct diagonal elements δ_i in $\mathfrak{F}$. Prove by direct computation that the only matrices A commutative with D are diagonal matrices.

2. Show if D is as in Ex. 1 and $f_i(D)$ is the product for $j \neq i$ of the $D - \delta_j I$, then $f_i(D)[f_i(\delta_i)]^{-1} = e_{ii}$ is the matrix with unity in the ith diagonal place and zeros elsewhere. Then every diagonal matrix is a polynomial in D.

3. Apply Ex. 2 and Theorem 4.15 to show that every matrix is uniquely expressible in the form $\sum_{i,j=0}^{n-1} a_{ij}D^i y^j$, a_{ij} in $\mathfrak{F}$ where D is any diagonal matrix with distinct diagonal elements and y is given by (38) for $\gamma \neq 0$.

CHAPTER V

SYMMETRIC AND SKEW MATRICES

1. Involutions of $\mathfrak{M}_n$. A square matrix A is called *symmetric* if $A = A'$ is its own transpose. We also say that A is *skew* (*skew-symmetric* or *alternate*) if $A' = -A$. These concepts are defined in terms of the operation of transposition, a special case of a general transformation which we shall call an involution and shall define as follows.

Let $\mathfrak{M}_n$ be the set of all n-rowed square matrices with elements in a field $\mathfrak{K}$, and U be a transformation (i.e., a one-to-one correspondence),

$$U: \qquad A \longleftrightarrow A^U \qquad (A, A^U \text{ in } \mathfrak{M}_n)$$

of $\mathfrak{M}_n$, such that

$$(1) \qquad (A + B)^U = A^U + B^U, \quad (AB)^U = B^U A^U, \quad (A^U)^U = A\,,$$

for every A and B of $\mathfrak{M}_n$. Then we call U an *involution over* $\mathfrak{F}$ of $\mathfrak{M}_n$ when $\mathfrak{F}$ is a subfield of $\mathfrak{K}$ such that

$$(2) \qquad (\lambda I_n)^U = \lambda I_n \qquad (\lambda \text{ in } \mathfrak{K}),$$

if and only if λ is in $\mathfrak{F}$.

We shall be interested in a study of the reduction of certain types of matrices, in particular symmetric and alternate matrices, to type forms. The theory is as yet incomplete when the characteristic of $\mathfrak{F}$ is 2. Notice that for this case the concepts of symmetric and skew matrices are the same. There are also other difficulties, and in particular the important Theorem 5.1 below must be modified. *Thus we shall treat only the theory over a field* $\mathfrak{F}$ *whose characteristic is not* 2. This implies that $-1 \neq 1$ and $\frac{1}{2}$ is in $\mathfrak{F}$.

We shall apply U to the centrum of $\mathfrak{M}_n$ over $\mathfrak{K}$, that is, to the set of all scalar matrices. If a is a scalar matrix we have $aA^U = A^U a$ for every A of $\mathfrak{M}_n$. Then by (1) we have $Aa^U = a^U A$ for every A, and a^U is a scalar matrix. When $a = \lambda I_n$ with λ in $\mathfrak{K}$ we have $a^U = \lambda^U I_n$, where λ^U is a uniquely defined quantity of $\mathfrak{K}$. This result combined with (1) implies that

$$\lambda \longleftrightarrow \lambda^U$$

is a transformation of $\mathfrak{K}$ such that $(\lambda + \mu)^U = \lambda^U + \mu^U$, $(\lambda\mu)^U = \mu^U \lambda^U$, $(\lambda^U)^U = \lambda$ for every λ, μ of $\mathfrak{K}$ by (1). From (2) we have $\lambda = \lambda^U$ if and only if λ is in $\mathfrak{F}$. We may write every λ of $\mathfrak{K}$ in the form $\lambda = \lambda_1 + \lambda_2$ where $\lambda_1 = \frac{1}{2}(\lambda + \lambda^U) = \lambda_1^U$ is in $\mathfrak{F}$ and $\lambda_2 = \frac{1}{2}(\lambda - \lambda^U) = -\lambda_2^U$. If every $\lambda_2 = 0$ then $\mathfrak{K} = \mathfrak{F}$. Otherwise there exists a quantity $\theta = -\theta^U \neq 0$ in $\mathfrak{K}$ and not in $\mathfrak{F}$, $\lambda = \lambda_1 + \lambda_3\theta$, $\theta^2 = \rho \neq 0$ in $\mathfrak{F}$, $\lambda_3 = \lambda_2\,\rho^{-1}\theta = \lambda_3^U$ is in $\mathfrak{F}$. The quantity $\rho \neq \sigma^2$, σ in $\mathfrak{F}$, since otherwise $\sigma^2 = \theta^2$, $\theta = \pm\sigma$ is in $\mathfrak{F}$. We state these results in

Theorem 1. *Let* U *be an involution over* $\mathfrak{F}$ *of the algebra* $\mathfrak{M}_n$ *of all* n*-rowed square matrices with elements in a field* $\mathfrak{K}$ *containing* $\mathfrak{F}$. *Then either* $\mathfrak{K} = \mathfrak{F}$ *or* $\mathfrak{K} = \mathfrak{F}(\theta)$ *where* $\theta^2 = \rho$ *in* $\mathfrak{F}$, ρ *is not the square of any quantity of* $\mathfrak{F}$. *Every quantity of* $\mathfrak{K}$ *is uniquely expressible in the form*

$$\lambda = a + b\theta \qquad (a,\ b \text{ in } \mathfrak{F})$$

and if $\alpha = \lambda I_n$ *is any scalar matrix we have*

$$\alpha^U = \lambda^U I_n, \qquad \lambda^U = a - b\theta\,.$$

Every matrix A has the form

$$A = B + C, \qquad B = \tfrac{1}{2}(A + A^U), \qquad C = \tfrac{1}{2}(A - A^U)\,, \tag{3}$$

where $B = B^U$, $C = -C^U$. We call A *U-symmetric* if $A = A^U$ and *U-skew* if $A = -A^U$. It is easily verified that if A and B are U-symmetric and D is arbitrary the matrices

$$DAD^U, \quad A + B, \quad AB + BA \tag{4}$$

are all U-symmetric, while

$$AB - BA \tag{5}$$

is U-skew. If also $AB = BA$ then AB is U-symmetric and in fact every polynomial in A with coefficients in $\mathfrak{F}$ is U-symmetric.

Theorem 2. *The minimum equation of a* U*-symmetric matrix has coefficients in* $\mathfrak{F}$.

For let $g(\lambda) = 0$ be the minimum equation of $A = A^U$. Since $0 = g(A) = 0^U$, $0 = g(A) - [g(A)]^U = \sum_{i=1}^{t} (a_i - a_i^U)A^{t-i}$, where $g(\lambda) = \lambda^t + a_1\lambda^{t-1} + \ldots + a_t$. But $g(\lambda) = 0$ is the minimum equation of A and we must have all the $a_i = a_i^U$ in $\mathfrak{F}$.

2. Involutions determined by $\mathfrak{K}$. We shall consider the relations between various involutions over $\mathfrak{F}$ of $\mathfrak{M}_n$ over $\mathfrak{K}$. Our only interest will be in the case where the involution of $\mathfrak{K}$, that is, the correspondence $\lambda \longleftrightarrow \lambda^U$ in $\mathfrak{K}$, is preserved. Thus let T be some fixed involution over $\mathfrak{F}$ of $\mathfrak{M}_n$ and consider the set of all involutions U over $\mathfrak{F}$ with the property

$$(6) \qquad a^T = a^U$$

for any scalar matrix a. We shall determine U in terms of T.

The transformation T is an element of the transformation group of $\mathfrak{M}_n$ and

$$T^2: \qquad A \longleftrightarrow (A^T)^T$$

is evidently the identical transformation. Hence $T^{-1} = T$, $U^{-1} = U$ for any U, and we define the transformation

$$(7) \qquad S \equiv TU = T^{-1}U = TU^{-1}, \qquad U = TS.$$

But $(a^T)^{U^{-1}} = a^S = a$ for every scalar matrix a,

$$(A + B)^S = A^S + B^S, \qquad (AB)^S = (AB)^{TU} = (B^TA^T)^U = A^SB^S.$$

This proves that S is an automorphism of $\mathfrak{M}_n$ over $\mathfrak{K}$. By Theorem 4.14 every automorphism is an inner automorphism, that is, there exists a non-singular matrix P such that $A^S = PAP^{-1}$ for every A of $\mathfrak{M}_n$. But

$$(8) \qquad A^U = (A^T)^S = PA^TP^{-1},$$

and

$$A = A^{U^2} = (A^U)^U = P(A^U)^TP^{-1} = P(P^T)^{-1}AP^TP^{-1}$$
$$= (P^TP^{-1})^{-1}A(P^TP^{-1}).$$

We have used the fact that $I_n^T = I_n$, $(PP^{-1})^T = P^T(P^{-1})^T = I_n$, so that $(P^T)^{-1} = (P^{-1})^T$, and have proved that $P^TP^{-1}A = AP^TP^{-1}$,

$$\beta = P^TP^{-1}$$

is commutative with every matrix of $\mathfrak{M}_n$ and is in the centrum $\mathfrak{K}_n$ of $\mathfrak{M}_n$ over $\mathfrak{K}$. Hence $P^T = \beta P$, $P_0 = P + P^T = (1 + \beta)P$ is either zero or is non-singular. In the latter case $P_0A^TP_0^{-1} = A^U$ and in the former $P = -P^T$. Hence (8) holds with $P^T = \pm P$. Conversely it is easy to verify that any U defined by (8) with $P^T = \pm P$ generates an involution of $\mathfrak{M}_n$ over $\mathfrak{F}$ satisfying (6). We have

Theorem 3. *Let* T *be an involution over* $\mathfrak{F}$ *of* $\mathfrak{M}_n$. *Then* U *is an involution over* $\mathfrak{F}$ *of* $\mathfrak{M}_n$ *with* $\alpha^T = \alpha^U$ *for every scalar matrix* α *if and only if there exists a non-singular matrix* $P = \pm P^T$ *such that*

$$A^U = PA^TP^{-1}.$$

The above theorem completely determines *all* involutions U with a certain given involution of the scalar matrices, in terms of *any desired fixed involution* T with this property. We shall later define a particular fixed involution and shall now show that the reduction theory for a general U is completely determined by that of this more conveniently studied fixed T.

EXERCISE

Prove that if U and T are given in Theorem 3 the matrix P is uniquely determined up to a scalar factor.

3. The U-equivalence of matrices. Two matrices A and B of $\mathfrak{M}_n$ are called *U-equivalent* (with respect to an involution U over $\mathfrak{F}$ of $\mathfrak{M}_n$) if there exists a non-singular matrix D of $\mathfrak{M}_n$ such that

$$DAD^U = B. \tag{9}$$

Now

$$A = I_nAI_n^U, \qquad A = (D^{-1})B(D^{-1})^U,$$

and if $C = GBG^U$ then

$$C = (GD)A(GD)^U.$$

Thus the relation defined in (9) is an equivalence relation. We shall study only the cases $A = \pm A^U$ and have A U-*symmetric* or U-*skew* respectively. Then $B = DA^UD^U = \pm B$, that is, *when* A *is* U-*symmetric so is any* B *which is* U-*equivalent to* A; *when* A *is* U-*skew so is any* B *which is* U-*equivalent to* A.

Let U be defined by $P = \pm P^T$, $A^U = PA^TP^{-1}$. Then $D^U = PD^TP^{-1}$, $B = DAPD^TP^{-1}$, and

$$BP = D(AP)D^T.$$

Hence A *and* B *are* U-*equivalent if and only if the uniquely determined matrices* AP *and* BP *are* T-*equivalent*. Moreover, if $A = \pm A^U$ then $A = \pm PA^TP^{-1}$, $AP = \pm PA^T = \pm\epsilon(AP)^T$ where $P = \epsilon P^T$ and $\epsilon = \pm 1$. This reduces the theory of U-equivalence of U-symmetric and U-alternate matrices to the theory of the T-equivalence of T-symmetric and T-skew matrices. Since T is at our choice we shall reduce our problem to the following *classical case*. We let the involution T of $\mathfrak{K}$ be given and let T be defined in $\mathfrak{M}_n$ by

$$(10) \qquad A = (a_{ij}) , \qquad A^T = (b_{ij}) , \qquad (i, j = 1, \ldots , n),$$

where

$$(11) \qquad b_{ji} = a_{ij}^T .$$

Then T is easily verified to be an involution. We shall study this most important case.

4. Elementary T-transformations. We have now left the general environment of involutions U and are studying an involution T which may be described as follows. Call $\lambda^T = a - b\theta$ in $\mathfrak{K}$ the *conjugate* of $\lambda = a + b\theta$ when $\mathfrak{K} = \mathfrak{F}(\theta)$ in Theorem 5.1. Also call $\lambda = \lambda^T$ the conjugate of λ in $\mathfrak{F}$. Then A^T is the matrix obtained by the replacement of every element of A by its conjugate and then transposing the result. Thus A^T is what we shall call the *conjugate transpose* of A, and we shall sometimes write

$$\overline{A}'$$

instead of A^T. Notice that if $\overline{A}$ means the matrix whose elements are the conjugates of those of A then

$$(\overline{A})' = \overline{(A')} = A^T .$$

Two matrices A and B are T-equivalent* if there exists a non-singular matrix D such that $B = DAD^T$. This is, of course, the special case $T = U$ of our earlier U-equivalence. But now

$$B = DA\overline{D}' .$$

The rows of DA are linear combinations of the rows of A, and the columns of B are the conjugate linear combinations of the columns of DA. Thus B is obtained from A by simultaneous operations on both the rows and columns of A. In our definition of elementary transformations in Chapter III we were able to operate on the rows and columns separately. We shall obtain analogous transformations here operating *simultaneously* on rows and columns, and shall call them *elementary* T-*transformations*.

* It is customary to call A and B *congruent* if $\mathfrak{K} = \mathfrak{F}$, $D^T = D'$, *conjunctive* if $\mathfrak{K} = \mathfrak{F}(\theta) \neq \mathfrak{F}$, that is, $D^T = \overline{D}'$. We also call $A = A^T$ *Hermitian* if $\mathfrak{K} = \mathfrak{F}(\theta) \neq \mathfrak{F}$.

In Chapter III we defined matrices $E_{ij} = E'_{ij}$ such that $E_{ij}A$ is the same as A except that its ith and jth rows are interchanged. Then $E^T_{ij} = E_{ij}$ and

$$B = E_{ij}AE^T_{ij} \tag{12}$$

is a matrix T-equivalent to A and obtained from A by the interchange of its ith and jth rows and ith and jth columns. We shall say that B is obtained from A by an *elementary* T-*transformation of type* I. Then we have

LEMMA 1. *Any square matrix* A *is* T-*equivalent to a matrix obtained from* A *by an arbitrary permutation of the rows of* A *and the same permutation of its columns.*

A submatrix of A which is obtained by deleting $n - r$ of its rows and the same $n - r$ columns is a square matrix P_r called a *principal submatrix* of A. By a sequence of elementary T-transformations of type I we may evidently permute the rows and columns of A so as to replace it by a T-equivalent matrix of the form

$$\begin{pmatrix} P_r & A_3 \\ A_2 & A_4 \end{pmatrix}. \tag{13}$$

Notice that *all of the principal submatrices of* A *which have* P_r *as principal submatrix are either unchanged by the above transformations or have their rows and columns permuted. Their determinants are then changed at most in sign.*

In Chapter III we also defined matrices $P_{ij}(\gamma)$, where γ is in the field $\mathfrak{K}$ of the elements of our matrices, and $P_{ij}(\gamma)A$ is the matrix obtained from A by adding γ times the jth row of A to its ith row. Similarly, $A[P_{ij}(\gamma)]^T = AP_{ji}(\gamma^T)$ is obtained from A by adding γ^T times the jth column of A to its ith column. We have of course used the property that the transpose of $P_{ij}(\gamma)$ is $P_{ji}(\gamma)$. But now

$$P_{ij}(\gamma)A[P_{ij}(\gamma)]^T \tag{14}$$

is obtained from A by adding γ times the jth row of A to its ith row and then γ^T times the jth column of the result to its ith column. We call this operation an *elementary* T-*transformation of type* II.

We shall not use an arbitrary sequence of elementary T-transformations of type II, but we shall apply them only as follows. We take any principal minor $p_r = |P_r|$ and replace A by a T-equivalent matrix of the form (13). The addition of any linear combination

$$\sum_{i=1}^{r} \gamma_{ji}a_i \qquad (\gamma_{ji} \text{ in } \mathfrak{F}; j = r+1, \ldots, n), \tag{15}$$

of the first r rows a_i of A to its remaining $n - r$ rows may be accomplished by a sequence of elementary transformations of type II. This operation certainly does not change p_r or the matrix A_3 of (13). It also leaves unaltered any $(r + t)$-rowed minor of A with P_r as a submatrix of its matrix. For a determinant is unaltered in value if elementary transformations of type II are applied to its matrix and this is what we have done. We now pass to the corresponding *sequence* of elementary T-transformations. This sequence still leaves p_r unaltered and replaces the jth column $b^{(j)}$ of the result DA of our row transformations by

$$b^{(j)} + \sum_i b^{(i)}\gamma_{ji}^T, \tag{16}$$

where DA has columns $b^{(i)}$, $b^{(j)}$. We see finally that p_r is unaltered, the rows of A_2 are replaced by the sums of themselves and linear combinations of the rows of P_r, the columns of A_3 are replaced by the sums of themselves and the conjugate linear combinations of the columns of P_r, the matrix A_4 is altered in both ways. However *every* $(r + t)$*-rowed minor of* A *with* $p_r = |P_r|$ *as subdeterminant is left unaltered.* We now prove

Theorem 4. *Let* $A = \pm A^T$ *have the form* (13) *with* $p_r = |P_r| \neq 0$. *Then the* γ_{ji} *of* (15) *may be so chosen that* A *is replaced by the* T*-equivalent matrix*

$$B = \begin{pmatrix} P_r & 0 \\ 0 & B_4 \end{pmatrix}, \tag{17}$$

whose $(r + t)$*-rowed minors with* p_r *as subdeterminant have the same values as those of* A, $t = 1, \ldots, n - r$.

For Theorem 3.13 states that the rows of P_r are linearly independent. They are r linearly independent $1 \times r$ matrices and every $1 \times r$ matrix is a linear combination of them. We use (13) and see that the rows b_j of A_2 with $j = r + 1 \ldots, n$ have the form $-\sum_{i=1}^{r} \gamma_{ji}b_i$ where the b_i are the rows of P_r. These values of γ_{ji} in (15) replace A by a T-equivalent matrix B in which the b_j are replaced by $b_j + \sum_{i=1}^{r} \gamma_{ji}b_i = 0$. Then

$$B = \begin{pmatrix} P_r & B_2 \\ 0 & B_4 \end{pmatrix}.$$

Now $A = \pm(\bar{A})'$ so that $B = \pm\bar{B}'$ and $B_2 = \pm B_3^T = 0$. This gives (17).

A *principal minor* of A is the determinant of a principal submatrix. We may employ elementary T-transformations of types I and II to prove

Theorem 5. *Let an* r*-rowed principal minor* p_r *of* $A = \pm A^T$ *be not zero and let every* (r + 1)*-rowed principal minor and every* (r + 2)*-rowed principal minor of* A *with* p_r *as subdeterminant vanish. Then* A *has rank* r.

For we have seen that we change at most the signs of the $(r + 1)$-rowed and $(r + 2)$-rowed principal minors of the theorem when we apply elementary transformations of type I and carry A into the form (13) with $p_r = |P_r| \neq 0$. We do this and then apply Theorem 5.4 to carry A into the form (17) whose $(r + 1)$- and $(r + 2)$-rowed principal minors with p_r as subdeterminant are still all zero. These $(r + 1)$-rowed minors have the form

$$\begin{vmatrix} P_r & 0 \\ 0 & b_{ii} \end{vmatrix} = p_r b_{ii}$$

when $B_4 = (b_{ij})$ and $p_r b_{ii} = 0$ implies that $b_{ii} = 0$. The $(r + 2)$-rowed minors in question are then

$$\begin{vmatrix} P_r & 0 & 0 \\ 0 & 0 & b_{ij} \\ 0 & b_{ji} & 0 \end{vmatrix} = -p_r b_{ij} b_{ji} = 0$$

if and only if $b_{ij} = 0$ or $b_{ji} = 0$. When $b_{ji} = \pm\, b_{ij}^T = 0$, we have $b_{ij} = 0$ and always get $b_{ij} = 0$, $B_4 = 0$, A has rank r.

If every one-rowed and two-rowed principal minor of $A = \pm A^T$ is zero, then $a_{ii} = a_{jj} = 0$,

$$\begin{vmatrix} 0 & a_{ij} \\ a_{ji} & 0 \end{vmatrix} = \mp a_{ij} a_{ij}^T = 0$$

so that $A = 0$ has rank zero. But if some t-rowed principal minor of A is not zero and every $(t + 1)$-, $(t + 2)$-, . . . , $(r + 1)$-, $(r + 2)$-rowed principal minor of A is zero the rank of A is $t \leqq r$. This result is stated in

Theorem 6. *When every* (r + 1)*- and* (r + 2)*-rowed principal minor of* $A = \pm A^T$ *is zero the rank of* A *is at most* r.

As an immediate consequence whose proof we leave to the reader we may state the

Corollary. *Let* $A = \pm A^T$ *have rank* r. *Then there exists an* r*-rowed principal non-singular submatrix of* A.

5. The first reduction of T-symmetric matrices. We shall obtain a reduction of any T-symmetric matrix to a diagonal form. *This will not be*

a canonical form since the reduction will be seen to be possible in many ways. Let $A = A^T = (a_{ij})$ be a non-zero matrix. If every diagonal element of A is zero some $a_{ij} \neq 0$ for $i \neq j$. Add γ times the jth row to the ith row, γ^T times the resulting jth column to the ith column, and the element b_{ii} of the resulting T-equivalent matrix $B = P_{ij}(\gamma)A[P_{ij}(\gamma)]^T$ is $\gamma a_{ji} + a_{ij}\gamma^T = \gamma a_{ji} + (\gamma a_{ji})^T$ since $a_{ji}^T = a_{ij}$. If $a_{ji} + a_{ji}^T \neq 0$ we take $\gamma = 1$ and have $b_{ii} \neq 0$. If $a_{ji} = -a_{ji}^T$ then $\mathfrak{K} = \mathfrak{F}(\theta)$, $(a_{ji}\theta)^T = (-a_{ji})(-\theta) = a_{ji}\theta$ so that $\gamma = \theta$ gives $b_{ii} \neq 0$.

LEMMA 2. *Every non-zero* T-*symmetric matrix with diagonal elements all zero is* T-*equivalent to a* T-*symmetric matrix with a non-zero diagonal element.*

We may now assume that $a_{ii} \neq 0$. By Lemma 1 we may interchange the first and ith rows and corresponding columns and obtain a T-equivalent T-symmetric matrix with $a_{11} \neq 0$.

Let $g_1 = a_{11} \neq 0$ and apply Theorem 5.4. Then A is equivalent to

$$\begin{pmatrix} g_1 & 0 \\ 0 & A_2 \end{pmatrix}, \qquad g_1 = g_1^T \text{ in } \mathfrak{F}. \tag{18}$$

Evidently A_2 is T-symmetric and A is also T-equivalent to

$$\begin{pmatrix} g_1 & 0 \\ 0 & B_2 \end{pmatrix} = \begin{pmatrix} 1 & 0 \\ 0 & D_2 \end{pmatrix}\begin{pmatrix} g_1 & 0 \\ 0 & A_2 \end{pmatrix}\begin{pmatrix} 1 & 0 \\ 0 & D_2^T \end{pmatrix} \tag{19}$$

for any $B_2 = D_2A_2D_2^T$ which is T-equivalent to A_2. An evident induction yields the *reduced form*

$$\operatorname{diag}\{g_1, \ldots, g_r, 0, \ldots, 0\}, \qquad g_i = g_i^T \neq 0 \text{ in } \mathfrak{F}, \tag{20}$$

where r is the rank of A.

The above form is evidently obtainable in many ways and does not yield invariants of A. No invariantive characterization is known except when the field $\mathfrak{K}$ is specified. The invariants of A depend essentially on $\mathfrak{K}$.

We can however obtain another method of reduction which at least gives elements g_i whose values have a simple expression in terms of the principal minors of A. This is not true of (20) since when the reduction (18) is accomplished the matrix A_2 is not at all easy to recognize in A.

6. The Kronecker reduction. Let $A = A^T$ have rank r so that by the corollary to Theorem 5.6 some r-rowed principal minor of A is not zero. By Section 5.4 we may assume without loss of generality that A has the form (13). We may then apply Theorem 5.4 to pass to a T-equivalent matrix (17). But A has rank r and the $(r + 1)$-rowed minors

$$\begin{vmatrix} P_r & 0 \\ 0 & b_{ij} \end{vmatrix} = p_r b_{ij}$$

must all be zero, where $B_4 = (b_{ij})$. Since $p_r \neq 0$, $b_{ij} = 0$, $B_4 = 0$.

We now see immediately that we have A also T-equivalent to

$$\begin{pmatrix} D_1 & 0 \\ 0 & I_{n-r} \end{pmatrix} \begin{pmatrix} P_r & 0 \\ 0 & 0 \end{pmatrix} \begin{pmatrix} D_1^T & 0 \\ 0 & I_{n-r} \end{pmatrix} = \begin{pmatrix} D_1 P_r D_1^T & 0 \\ 0 & 0 \end{pmatrix}$$

and have proved

Theorem 7. *Let* A *have rank* r *and* P_r *be a non-singular principal submatrix of* A. *Then* A *is* T-*equivalent to*

$$\begin{pmatrix} Q_r & 0 \\ 0 & 0 \end{pmatrix} \tag{21}$$

where Q_r *is any matrix which is* T-*equivalent to* P_r.

The considerations above have left us the problem of the reduction of a non-singular matrix $A = A^T$ to diagonal form, since the above P_r is non-singular. We now assume without loss of generality that A is non-singular and define the principal submatrices

$$P_t = \begin{pmatrix} a_{11} & \dots & a_{1t} \\ . & \dots & . \\ a_{t1} & \dots & a_{tt} \end{pmatrix}, \qquad (t = 1, \dots, n) \tag{22}$$

of A. Their determinants are

$$p_0 = 1, \qquad p_1 = a_{11}, \dots, p_t = |P_t|, \dots, p_n = |A| \neq 0$$

and we may prove

Lemma 3. *The rows and corresponding columns of* A *may be permuted so that no two consecutive* p_i *are zero.*

The lemma is trivial if $n = 1$. Assume it true for matrices of all orders less than n and let A have order n. By Theorem 5.6 at least one principal submatrix of A of order $t = n - 1$ or $n - 2$ is non-singular. This submatrix may be carried into P_t by elementary T-transformations of type I and our induction states that we may assume that no consecutive two of $p_0 = 1$, $p_1, \dots, p_t$ are zero. Then this is evidently true of $p_0, \dots, p_n$ when $t = n - 2$ or $n - 1$, and our induction is complete.

The matrix P_{t+2} contains P_{t+1} as a principal $(t + 1)$-rowed submatrix as well as the other principal submatrix

$$(23) \qquad Q_{t+1} = \begin{pmatrix} a_{11} & \cdots & a_{1t} & a_{1\,t+2} \\ \cdot & \cdots & \cdot & \cdot \\ a_{t1} & \cdots & a_{tt} & a_{t,\,t+2} \\ a_{t+2,\,1} & \cdots & a_{t+2,\,t} & a_{t+2,\,t+2} \end{pmatrix}$$

of determinant q_{t+1}. When $p_t \neq 0$, $p_{t+1} = 0$, $p_{t+2} \neq 0$ and $q_{t+1} \neq 0$ we may interchange the $(t + 1)$-st and $(t + 2)$-d rows and columns of A and have a new sequence of p_i with $p_{t+1} \neq 0$. *We shall do this and shall call* $A = A^T$ *a* ***regular matrix*** *if no consecutive two of the* p_i *are zero and if* $p_{t+1} = 0$ *implies that also* $q_{t+1} = 0$. Our result above combined with our lemma may be stated as

Theorem 8. *Every non-singular T-symmetric matrix may be transformed into a regular matrix by permutations of its rows and corresponding columns.*

We now obtain the Kronecker reduction of any T-symmetric regular matrix. Assume first that $p_{n-1} \neq 0$. Then Theorem 5.4 states that A is T-equivalent to

$$(24) \qquad \begin{pmatrix} P_{n-1} & 0 \\ 0 & g_n \end{pmatrix}$$

with the same determinant p_n as A. Hence

$$(25) \qquad g_n = p_n p_{n-1}^{-1} .$$

Next let $p_{n-1} = 0$ so that by our definition also $q_{n-1} = 0$. By Theorem 5.4 A is T-equivalent to

$$(26) \qquad A_0 = \begin{pmatrix} P_{n-2} & 0 & 0 \\ 0 & \gamma_1 & \gamma^T \\ 0 & \gamma & \gamma_2 \end{pmatrix}$$

with the same determinants p_{n-1}, q_{n-1}, p_n as A. But

$$(27) \quad p_{n-1} = \begin{vmatrix} P_{n-2} & 0 \\ 0 & \gamma_1 \end{vmatrix} = p_{n-2}\gamma_1 = 0 ,$$

$$q_{n-1} = \begin{vmatrix} P_{n-2} & 0 \\ 0 & \gamma_2 \end{vmatrix} = p_{n-2}\gamma_2 = 0 ,$$

whence $\gamma_1 = \gamma_2 = 0$, and the determinant of A_0 is

$$(28) \qquad p_n = \begin{vmatrix} P_{n-2} & 0 & 0 \\ 0 & 0 & \gamma^T \\ 0 & \gamma & 0 \end{vmatrix} = -\gamma\gamma^T p_{n-2} = |A| .$$

It follows that if

$$(29) \qquad H = \begin{pmatrix} I_{n-2} & 0 \\ 0 & Y \end{pmatrix}, \quad Y = \begin{pmatrix} 1 & \gamma^{-1} \\ -\frac{1}{2}\gamma & \frac{1}{2} \end{pmatrix},$$

then A is T-equivalent* to

$$(30) \quad HAH^T = \begin{pmatrix} P_{n-2} & 0 \\ 0 & B_0 \end{pmatrix}, \quad B_0 = Y\begin{pmatrix} 0 & \gamma^T \\ \gamma & 0 \end{pmatrix}Y^T = \begin{pmatrix} 2 & 0 \\ 0 & -\frac{1}{2}\gamma\gamma^T \end{pmatrix}.$$

By (28) we have proved

Theorem 9. *Let* A *be a regular* T-*symmetric matrix with principal minors* $p_1, p_2, \ldots, p_n \neq 0$. *Then if* $p_{n-1} \neq 0$, A *is* T-*equivalent to*

$$(31) \qquad \begin{pmatrix} P_{n-1} & 0 \\ 0 & g_n \end{pmatrix}, \qquad g_n = p_n p_{n-1}^{-1},$$

and if $p_{n-1} = 0 = q_{n-1}$, *then* $p_{n-2} \neq 0$ *and* A *is* T-*equivalent to*

$$(32) \qquad \begin{pmatrix} P_{n-2} & 0 & 0 \\ 0 & g_{n-1} & 0 \\ 0 & 0 & g_n \end{pmatrix}, \qquad g_{n-1} = 2, \qquad g_n = \tfrac{1}{2}p_{n-2}^{-1}p_n .$$

The quantity q_{n-1} in the above theorem was of course defined as the determinant of (23) for $t = n - 2$ and is zero when $p_{n-1} = 0$ by our definition of a regular A. We apply Theorem 5.9 to obtain the diagonal matrix

$$(33) \qquad \operatorname{diag} \{g_1, \ldots, g_n\} ,$$

T-equivalent to our regular A, and have $g_i = p_i p_{i-1}^{-1}$; or $g_i = 2$ and $g_{i+1} = \frac{1}{2}p_{i-1}^{-1}p_{i+1}$; or $g_i = \frac{1}{2}p_{i-2}^{-1}p_i$ and $g_{i-1} = 2$.

EXERCISES

1. Write the Kronecker reduction of the arbitrary rational three-rowed regular symmetric matrices in terms of the possible signs (zero, positive, or negative) of the principal minors.

* This reduction evidently fails in a field K of characteristic 2.

2. Let $A_{n,n-1}$ be the cofactor of the element $a_{n,n-1}$ of A. Show that $A_{n,n-1}$ has the same value in A_0 of (26) as in A and that

$$A_{n,n-1} = -\begin{vmatrix} P_{n-2} & 0 \\ 0 & \gamma^T \end{vmatrix} = -p_{n-2}\gamma^T .$$

Prove that then $\gamma = p_n A_{n,n-1}^{-1}$. This gives an alternative form for the diagonal element $g_n = -\frac{1}{2}\gamma\gamma^T$ of (32).

3. Let

$$\begin{pmatrix} A & 0 \\ 0 & 0 \end{pmatrix}, \quad \begin{pmatrix} B & 0 \\ 0 & 0 \end{pmatrix}$$

be T-equivalent where A and B are non-singular r-rowed square matrices. Prove that A and B are T-equivalent. Show also that the analogous result holds for similar matrices. Hint: By direct multiplication show that $Q_1 A Q_1^T = B$ so that the non-singularity of B implies that Q_1 is non-singular.

7. Skew matrices. A skew matrix A is an n-rowed square matrix with elements in a field $\mathfrak{K}$ with an involution T of Theorem 5.1 such that $A^T = -A$. If $\mathfrak{K} = \mathfrak{F}(\theta)$ and $\theta^T = -\theta$ then $A_0 = \theta A$ is T-symmetric and our reduction theory is that of Sections 5.5 and 5.6. New results may only be obtained if

$$\mathfrak{K} = \mathfrak{F}, \qquad A^T = A',$$

for every A. Thus we are studying the case

$$(34) \qquad A = (a_{ij}), \qquad a_{ij} = -a_{ji} \text{ in } \mathfrak{F} \qquad (i, j = 1, \ldots, n).$$

For such matrices we obtain a far simpler result than that derived for a T-symmetric matrix. We shall, in fact, prove that every such A has rank $2r$ and is equivalent to

$$(35) \qquad \begin{pmatrix} E_r & 0 \\ 0 & 0 \end{pmatrix}, \quad E_r = \begin{pmatrix} 0 & I_r \\ -I_r & 0 \end{pmatrix}.$$

This result is also stated by

Theorem 10. *Two skew matrices are* T*-equivalent if and only if they have the same rank* 2r.

For $a_{ii} = 0$ and if $A \neq 0$ there must be some $a_{ij} \neq 0$, $i \neq j$. We interchange the ith row with the first row and the jth row with the second row and of course perform the same column interchanges. Then we obtain a new matrix with $a_{12} \neq 0$,

$$A = \begin{pmatrix} A_1 & A_2 \\ A_3 & A_4 \end{pmatrix}, \quad A_1 = \begin{pmatrix} 0 & a_{12} \\ -a_{12} & 0 \end{pmatrix}, \quad a_{12} \neq 0 .$$

The rows of A_1 are evidently linearly independent and Theorem 5.4 states that A is T-equivalent to

$$\begin{pmatrix} A_1 & 0 \\ 0 & B \end{pmatrix}, \qquad B = -B'.$$

But then A is T-equivalent to

$$\begin{pmatrix} D_1 & 0 \\ 0 & I_{n-2} \end{pmatrix}\begin{pmatrix} A_1 & 0 \\ 0 & B \end{pmatrix}\begin{pmatrix} D_1' & 0 \\ 0 & I_{n-2} \end{pmatrix} = \begin{pmatrix} E & 0 \\ 0 & B \end{pmatrix},$$

where

$$D_1A_1D_1' = E = \begin{pmatrix} a_{12}^{-1} & 0 \\ 0 & 1 \end{pmatrix}\begin{pmatrix} 0 & a_{12} \\ -a_{12} & 0 \end{pmatrix}\begin{pmatrix} a_{12}^{-1} & 0 \\ 0 & 1 \end{pmatrix} = \begin{pmatrix} 0 & 1 \\ -1 & 0 \end{pmatrix}.$$

Also A is T-equivalent to any matrix of the form

$$\begin{pmatrix} E & 0 \\ 0 & B_0 \end{pmatrix}$$

where B_0 is T-equivalent to B. Hence we ultimately obtain a matrix

$$\text{(36)} \qquad \operatorname{diag}\{E_1, \ldots, E_r, 0, \ldots, 0\}, \qquad E_i = E,$$

which is T-equivalent to A and has rank $2r$. This proves that A has rank $2r$ and also that any $A = -A'$ of rank $2r$ is equivalent to (35). Hence two skew matrices are T-equivalent if and only if they have the same rank.

The matrix (35) is a skew matrix of rank $2r$ and is equivalent to (36). Thus (35) is a canonical form and is preferable to (36) because of its simplicity.

8. Matrices over an algebraically closed field. A field $\mathfrak{F}$ is said to be *algebraically closed if every equation with coefficients in* $\mathfrak{F}$ *has a root in* $\mathfrak{F}$. We now let $\mathfrak{F} = \mathfrak{K}$ be algebraically closed, so that $A = A^T$ if and only if

$$A = (a_{ij}) = A', \qquad a_{ij} = a_{ji},$$

and call A a *symmetric* matrix. Then A is T-equivalent to

$$\text{(37)} \qquad DAD' = \operatorname{diag}\{g_1, \ldots, g_r, 0, \ldots, 0\} = D_0BD_0',$$

where

$$\text{(38)} \qquad D_0 = \operatorname{diag}\{g_1^{1/2}, \ldots, g_r^{1/2}, 1, \ldots, 1\}, \qquad B = \begin{pmatrix} I_r & 0 \\ 0 & 0 \end{pmatrix}.$$

The elements of D_0 are in $\mathfrak{F}$ and A is T-equivalent to B.

Theorem 11. *Two symmetric matrices with elements in an algebraically closed field are* T*-equivalent in that field if and only if they have the same rank.*

The result above makes evident the possible role the structure of $\mathfrak{K}$ may have on a canonical form for A.

9. Ordered fields. A field $\mathfrak{F}$ is said to be *ordered* if it is possible to classify the non-zero elements of $\mathfrak{F}$ into two distinct classes $\mathfrak{F}_P$ and $\mathfrak{F}_N$ such that

I. $\mathfrak{F}_P$ and $\mathfrak{F}_N$ have no element in common.

II. If a is in $\mathfrak{F}_N$ then $-a$ is in $\mathfrak{F}_P$.

III. If a and b are in $\mathfrak{F}_P$ then $a + b$ and ab are in $\mathfrak{F}_P$.

Postulate III implies that if a is in $\mathfrak{F}_P$ then $-a$ is in $\mathfrak{F}_N$. For otherwise $-a$ is in $\mathfrak{F}_P$ and $a + (-a) = 0$ is in $\mathfrak{F}_P$ contrary to hypothesis. We may also prove the consequent property

IV. If a and b are in $\mathfrak{F}_N$ and c is in $\mathfrak{F}_P$ the product ab is in $\mathfrak{F}_P$ and ac is in $\mathfrak{F}_N$.

For $-a$ and $-b$ are in $\mathfrak{F}_P$ and so is $(-a)(-b) = ab$. Also $-a$ and c are in $\mathfrak{F}_P$ and $(-a)c = -(ac)$ is in $\mathfrak{F}_P$, ac is in $\mathfrak{F}_N$.

The elements of $\mathfrak{F}_P$ are called the *positive* elements of $\mathfrak{F}$ and those of $\mathfrak{F}_N$ are called the *negative* elements of $\mathfrak{F}$. We symbolize these respective properties in the usual way writing $a > 0$ or $a < 0$. We also write

$$a > b, \qquad b < a \tag{39}$$

if $a - b > 0$. We now have *an order relation in* $\mathfrak{F}$, that is, one and only one of the three properties $a = b$, or $a > b$, or $a < b$ holds for every a and b of $\mathfrak{F}$. Our postulates will imply some familiar properties of the relation (39).

If $a > b$ and $b > c$, then $a > c$ since

$$a - b > 0, \qquad b - c > 0, \qquad a - c = (a - b) + (b - c) > 0$$

by III. Next $a > b$ implies that

$$a + c > b + c$$

for every c of $\mathfrak{F}$ since $a + c - (b + c) = a - b > 0$. If

$$a > b, \qquad c > 0$$

then

$$ca > cb.$$

For $c > 0$, $a - b > 0$, $c(a - b) = ca - cb > 0$ by III. Finally, let

$$a > 0, \qquad b > 0, \qquad a > b.$$

Then $ab > 0$ and $a - b = ab(b^{-1} - a^{-1}) > 0$. By IV we have

$$b^{-1} > a^{-1}.$$

The absolute value $|a|$ of a is defined to be zero if $a = 0$, a if $a > 0$, and $-a$ if $-a > 0$. Then

$$|ab| = |a| \cdot |b|, \qquad |a + b| \leqq |a| + |b|.$$

For both results are trivial if either $a = 0$ or $b = 0$. Let then $ab \neq 0$ so that $|a| = \alpha a$, $|b| = \beta b$, $\alpha = \pm 1$ and $\beta = \pm 1$, $|a| \cdot |b| = \alpha\beta ab = \pm ab > 0$, $|a| \cdot |b| = |ab|$. When $a > 0$, $b > 0$ then $|a + b| = a + b = |a| + |b|$, and when $a < 0$, $b < 0$ then $|a + b| = -(a + b) = -a - b = |a| + |b|$. Hence we may assume that one of a and b is positive and the other is negative. There is evidently no loss of generality if we choose our notation so that $a > 0$, $c = -b > 0$. Then

$$a + b = a - c < a < a + c = |a| + |b|,$$
$$-(a + b) = c - a < c < a + c = |a| + |b|,$$

so that $|a + b| < |a| + |b|$ in this case.

If $a \neq 0$ then $(-a)^2 = a^2 = |a|^2 > 0$. In particular $1^2 = 1 > 0$ and $p \cdot 1 = 1^2 + 1^2 + \ldots + 1^2 > 0$. Thus *an ordered field cannot have finite characteristic.* Every ordered field now contains a subfield equivalent to the field of all rational numbers.

DEFINITION. *Two ordered fields* $\mathfrak{F}$ *and* $\mathfrak{K}$ *are said to be ordered-equivalent if they are equivalent under a correspondence*

$$\mathrm{f} \longleftrightarrow \mathrm{k} \qquad (\mathrm{f} \text{ in } \mathfrak{F},\ \mathrm{k} \text{ in } \mathfrak{K})$$

such that $\mathrm{f} > 0$ *if and only if* $\mathrm{k} > 0$.

The field $\mathfrak{R}$ of all rational numbers is ordered by the relation generated by

$$1 > 0.$$

For then $n + 1 > n$ for every integer n and the integers 1, 2, 3, . . . are all greater than zero, the integers $-1, -2, \ldots$ are all less than zero. If α is in $\mathfrak{R}$ we may write $\alpha = a/b$ where a and $b > 0$ are integers. By properties III and IV the number α has the same sign as ab and $\alpha = 0$, $\alpha > 0$, $\alpha < 0$ according as $a = 0$, $a > 0$, $a < 0$. This proves that if $\mathfrak{R}_0$ is any ordered field equivalent to the field of all rational numbers and $1 > 0$ in $\mathfrak{R}_0$ then the ordering of $\mathfrak{R}_0$ is what we may call the *natural ordering.* But the property $1 > 0$ is always true. For properties III and IV state that a^2 is in $\mathfrak{F}_P$ for every a, and $1^2 = 1$ must be in $\mathfrak{F}_P$. We have proved

Theorem 12. *Let $\mathfrak{R}_0$ be an ordered field equivalent to the field $\mathfrak{R}$ of all rational numbers. Then the only ordering of $\mathfrak{R}_0$ is that generated by $1 > 0$, that is, the natural ordering, and $\mathfrak{R}_0$ is ordered-equivalent to $\mathfrak{R}$.*

The theorem above is applied to arbitrary fields $\mathfrak{F}$. We have already shown that when $\mathfrak{F}$ is ordered it is non-modular and its prime subfield is a field $\mathfrak{R}_0$ equivalent to $\mathfrak{R}$. The ordering of $\mathfrak{F}$ imposes an ordering of $\mathfrak{R}_0$. But this latter ordering is not arbitrary. It is the fixed natural order in $\mathfrak{R}_0$.

EXERCISE

Prove that if $\mathfrak{F}$ is an ordered field and $a > 0$ is in $\mathfrak{F}$, n is an integer, then $a^n = 1$ if and only if $a = 1$. Hint: Show that if $a > 1$ then $a^n > 1$ by induction and similarly for $a < 1$.

10. Formulation of the classical theory. The classical theory of so-called real symmetric and complex Hermitian matrices is the special case of our theory where $\mathfrak{F}$ is the field of all real numbers, $\mathfrak{K} = \mathfrak{F}$ or $\mathfrak{F}(i)$, $i^2 = -1$. We shall define complex numbers in Chapter XII, but are not yet ready to give the rather complicated concepts occurring in their definition. However we may consider our matrix problem by formulating the properties of $\mathfrak{K}$ sufficient to give the theory obtained in the above special case.

We first consider the case where $\mathfrak{F}$ is ordered and $\mathfrak{K} = \mathfrak{F}$ or $\mathfrak{F}(\theta)$, $\theta^2 = \rho < 0$ in $\mathfrak{F}$. The equation $x^2 = \rho$ is irreducible in $\mathfrak{F}$ since $x^2 > 0$ for every x of $\mathfrak{F}$. Then $\mathfrak{F}(\theta)$ is called a *quadratic field* (linear set of order 2) over $\mathfrak{F}$ and every quantity of $\mathfrak{K}$ is uniquely expressible in the form

$$\alpha = a + b\theta \qquad (a, b \text{ in } \mathfrak{F}).$$

We have called

$$\alpha^T = \bar{\alpha} = a - b\theta$$

the *conjugate* of α, and now see that

$$\alpha\alpha^T = a^2 - \rho b^2 > 0$$

unless $\alpha = 0$. When $\mathfrak{K} = \mathfrak{F}$ we have $\alpha\alpha^T = \alpha^2 > 0$ unless $\alpha = 0$. The case where $\mathfrak{F}$ is ordered is called the *ordered* case and we may prove

Theorem 13. *Let $H = (h_{ij})$ be an $n \times m$ matrix in the ordered case. Then the ith diagonal element of the n-rowed square matrix $G = H \cdot H^T$ is positive or zero according as the elements in the ith row of H are not or are all zero.*

For $G = (g_{ij})$ where $g_{ii} = \sum_{j=1}^{m} h_{ij}h_{ij}^T > 0$ unless every $h_{ij} = 0$ for a given i and $j = 1, \ldots, m$.

Theorem 5.13 has an immediate corollary.

Theorem 14. *Let* $A_1, \ldots, A_q$ *be* T-*symmetric matrices in the ordered case such that* $A_1^2 + \ldots + A_q^2 = 0$. *Then* $A_1 = A_2 = \cdots = A_q = 0$.

For the $A_j^2 = A_j A_j^T$ has diagonal elements all positive or zero and they are all zero by Theorem 5.13 only when $A_j = 0$. The sum of the diagonal elements $g_{ii}^{(j)}$ of the A_j^2 is the element g_{ii} of $0 = A_1^2 + \ldots + A_q^2$ and is positive unless the $g_{ii}^{(j)}$ which are $\geqq 0$ are all zero. Then the A_j are all zero.

We are now ready to give the properties of $\mathfrak{K}$ which lead to the so-called classical theory. We assume that $\mathfrak{F} = \mathfrak{R}'$ is an ordered field with the property that

$$\mathfrak{C} = \mathfrak{R}'(i), \qquad i^2 = -1, \tag{40}$$

is algebraically closed. We take

$$\mathfrak{K} = \mathfrak{R}' \text{ or } \mathfrak{R}'(i) \tag{41}$$

and call this the *ordered closed case*. The elements of $\mathfrak{F} = \mathfrak{R}'$ will be called *real numbers* and $\alpha^T = \alpha = \bar{\alpha}$ if α is real. The elements of $\mathfrak{C} = \mathfrak{R}'(i)$ will be called *complex numbers* and those not in $\mathfrak{R}'$ will be called *imaginary numbers*.

The field $\mathfrak{R}'$ is called an ordered closed field and is thought of as an extension of the ordered field $\mathfrak{R}$ of all rational numbers to an ordered closed field. The meaning of this definition will become clearer if we make the following discussion. Let $f(x) = 0$ have coefficients in $\mathfrak{R}'$. Then

$$f(x) = (x - \alpha_1) \ldots (x - \alpha_n) \tag{42}$$

with α_i in $\mathfrak{C}$ since $\mathfrak{C}$ is algebraically closed. Now the coefficients of $f(x)$ are in $\mathfrak{R}'$ and hence

$$f(x) = \overline{f(x)} = (x - \bar{\alpha}_1) \ldots (x - \bar{\alpha}_n) . \tag{43}$$

Thus if $\alpha = a + bi$ is a root in $\mathfrak{C}$ of $f(x) = 0$ so is $\bar{\alpha}$. When $b \neq 0$

$$f(x) = (x^2 - 2ax + a^2 + b^2) f_1(x) . \tag{44}$$

Evidently $x^2 - 2ax + a^2 + b^2 = (x - a)^2 + b^2$ is irreducible in $\mathfrak{R}'$. This result implies that $f(x)$ factors in $\mathfrak{R}'$ into linear and irreducible quadratic factors, and if α is a root of any irreducible quadratic factor then $\mathfrak{R}'(\alpha) = \mathfrak{R}'(i)$ is *not* ordered (by properties III and IV). Thus $\mathfrak{R}'$ cannot be extended by a root of an irreducible equation such that an ordered field is obtained. This is the type of closure which we have properly called *ordered closure*.

In any ordered case we call $A = A'$ *symmetric* and $A = \bar{A}'$ *Hermitian*. In the ordered closed case we shall speak of *real symmetric matrices*, that is matrices $A = A^T = A'$ with elements in $\mathfrak{R}'$, and *complex Hermitian matrices*, that is matrices $A = A^T = \bar{A}'$ with elements in $\mathfrak{R}'(i)$. We shall treat both cases simultaneously, however, by the use of the notation $A = A^T$.

11. Reduction of real symmetric and complex Hermitian matrices. We shall prove

Theorem 15. *Every* T-*symmetric matrix is* T-*equivalent in the ordered closed case to a unique matrix*

$$\begin{pmatrix} I_p & 0 & 0 \\ 0 & -I_{r-p} & 0 \\ 0 & 0 & 0 \end{pmatrix}, \tag{45}$$

where p *is called the index of* A. *Hence two* T-*symmetric matrices are* T-*equivalent in the ordered closed case if and only if they have the same rank and index.*

The integer $s = p - (r - p) = 2p - r$ is the excess of the number of positive diagonal elements in the canonical form (45) over the number of negative elements and is called the *signature* of the matrix considered. Evidently the pair s and r determine the pair p and r and we have the alternative statement given in the

COROLLARY. *Two* T-*symmetric matrices are* T-*equivalent in the ordered closed case if and only if they have the same rank and signature.*

To prove Theorem 5.15 we use the form (20) with the g_i always in the subfield $\mathfrak{F}$ of $\mathfrak{K}$ and hence real. By a sequence of interchanges of the rows and corresponding columns of (20) we carry it into a like matrix A_0 with $g_1, \ldots, g_p$ positive and $g_{p+1}, \ldots, g_r$ negative. Then if D_1 is the p-rowed diagonal matrix with $g_1^{-(1/2)}, \ldots, g_p^{-(1/2)}$ as diagonal elements, and if D_2 is the $(r - p)$-rowed diagonal matrix with diagonal elements $(-g_{p+1})^{-(1/2)}$, $(-g_{p+2})^{-(1/2)}, \ldots, (-g_r)^{-(1/2)}$ we define

$$D = \begin{pmatrix} D_1 & 0 & 0 \\ 0 & D_2 & 0 \\ 0 & 0 & I_{n-r} \end{pmatrix} = D^T = D' .$$

The matrix D has elements in $\mathfrak{F}$ and DA_0D^T is T-equivalent to A and has the form (45).

Suppose now that (45) is not unique and that A_p of (45) is equivalent to a second matrix of the same type. The integer r is the rank in each case since T-equivalent matrices have the same rank. Hence

$$A_q = \begin{pmatrix} I_q & 0 & 0 \\ 0 & -I_{r-q} & 0 \\ 0 & 0 & 0 \end{pmatrix} = HA_pH^T , \tag{46}$$

where H is a non-singular n-rowed square matrix with elements in $\mathfrak{K}$. Without loss of generality we may assume that $s = p - q > 0$.

We let $x_1, \ldots, x_p$ be independent indeterminates over $\mathfrak{K}$ and define the $1 \times n$ matrices

$$x = (x_1, x_2, \ldots, x_p, 0, \ldots, 0), \qquad y = (y_1, \ldots, y_n),$$

where the y_i are uniquely determined linear forms in $x_1, \ldots, x_p$ defined by the matrix equation

$$y = xH^{-1}, \qquad x = yH. \tag{47}$$

The system of $q < p$ linear homogeneous equations

$$y_1 = 0, y_2 = 0, \ldots, y_q = 0 \tag{48}$$

in the unknowns $x_1, \ldots, x_p$ have solutions $x_1 = \xi_1, x_2 = \xi_2, \ldots, x_p = \xi_p$ not all zero and in $\mathfrak{K}$. Each $y_i = y_i(x_1, \ldots, x_p)$ has corresponding values $\eta_i = y_i(\xi_1, \ldots, \xi_p)$ where $\eta_1 = \ldots = \eta_q = 0$. Define

$$\begin{cases} X = (\xi_1, \ldots, \xi_p), \qquad Y = (\eta_{q+1}, \ldots, \eta_r), \\ \qquad\qquad Z = (\eta_{r+1}, \ldots, \eta_n), \end{cases} \tag{49}$$

so that x and y become respectively

$$\xi = (X, 0, 0), \qquad \eta = (0, Y, Z), \tag{50}$$

when we replace the x_i by the ξ_i. The zeros in (50) are of course one-rowed zero matrices of $r - p$, $n - r$, q columns, respectively. The conjugate transpose of any rectangular matrix B may be designated by B^T and (47) gives

$$\xi = \eta H, \qquad \xi^T = H^T\eta^T. \tag{51}$$

Then

$$\eta A_q \eta^T = (0, Y, Z)\begin{pmatrix} I_q & 0 & 0 \\ 0 & -I_{r-q} & 0 \\ 0 & 0 & 0 \end{pmatrix}\eta^T = (0, -Y, 0)\begin{pmatrix} 0 \\ Y^T \\ Z^T \end{pmatrix}$$

$$= -YY^T = -\sum_{i=q+1}^{r} \eta_i\eta_i^T$$

and is a zero or negative quantity of $\mathfrak{R}$. But $\eta A_q \eta^T = \eta H A_p H^T \eta^T = \xi A_p \xi^T$ which is equal to

$$(X, 0, 0)\begin{pmatrix} I_p & 0 & 0 \\ 0 & -I_{r-p} & 0 \\ 0 & 0 & 0 \end{pmatrix}\xi^T = (X, 0, 0)\begin{pmatrix} X^T \\ 0 \\ 0 \end{pmatrix} = XX^T = \sum_{i=1}^{p} \xi_i \xi_i^T > 0$$

since not all the ξ_i are zero. This is impossible.

12. Positive definite matrices. A T-symmetric matrix A with $r = p = n$ is called *positive definite.* We call A *negative definite* if $r = r - p = n$, that is $-A$ is positive definite. When A is singular we say that A is *positive* if $r = p$, A is *negative* if $r = r - p$, that is $-A$ is positive.

When A is positive definite the matrix (45) is the n-rowed identity matrix and $DAD^T = I_n$, $G = D^{-1}$, $A = GG^T$. Thus a matrix A is positive definite *if and only if there exists a non-singular matrix* G *with elements in* $\mathfrak{R}$ *such that*

$$A = GG^T = GG' \text{ or } G\bar{G}' \tag{52}$$

according as $\mathfrak{R} = \mathfrak{N}'$ *or* $\mathfrak{N}'(i)$.

If g is the determinant of G then g^T is the determinant of G^T and $gg^T > 0$ is the determinant of A. By Theorem 5.13 every diagonal element of A is positive. These are special cases of the useful criterion

Theorem 16. *A* T-*symmetric matrix* A *is positive definite if and only if every principal minor of* A *is positive.*

For if every principal minor of A is positive the quantities in the sequence $p_0 = 1, p_1, \ldots, p_n$ of Theorem 5.9 are all positive and

$$g_i = p_i p_{i-1}^{-1} > 0 \tag{53}$$

in (33) by the Kronecker reduction. Thus A is positive definite. Conversely let P be a t-rowed principal minor of A. By Lemma 1 we may carry P into P_t, A into

$$\begin{pmatrix} P_t & A_2 \\ A_3 & A_4 \end{pmatrix}, \qquad P_t = P_t^T.$$

Then A is T-equivalent to

$$A_0 = \begin{pmatrix} D_t & 0 \\ 0 & I_{n-t} \end{pmatrix}\begin{pmatrix} P_t & A_2 \\ A_3 & A_4 \end{pmatrix}\begin{pmatrix} D_t^T & 0 \\ 0 & I_{n-t} \end{pmatrix} = \begin{pmatrix} D_t P_t D_t^T & B_2 \\ B_3 & B_4 \end{pmatrix}$$

for any non-singular D_t. But P_t is T-symmetric and by Theorem 5.15 we may take

$$D_tP_tD_t^T = \begin{pmatrix} I_a & 0 & 0 \\ 0 & -I_b & 0 \\ 0 & 0 & 0 \end{pmatrix},$$

where a is the index and $a + b$ is the rank of P_t. The matrix A_0 is positive definite and the diagonal elements of $D_tP_tD_t^T$ are diagonal elements of A_0. They must therefore all be positive and $a = t$, P_t is positive definite and has a positive determinant.

A trivial consequence of Theorem 5.16 is given by the

COROLLARY. *Every principal submatrix of a positive definite* T-*symmetric matrix is positive definite.*

EXERCISES

1. Use Theorem 5.16 to prove that the rational matrices

$$\begin{pmatrix} 5 & 3 \\ 3 & 2 \end{pmatrix}, \quad \begin{pmatrix} 6 & 0 & -1 \\ 0 & 7 & 7 \\ -1 & 7 & 10 \end{pmatrix}$$

are positive definite.

2. Let B be a real symmetric or a complex Hermitian matrix. Prove that there exists a real number γ such that $\gamma I + B$ is positive definite.

13. Orthogonal T-equivalence. Let $A = A^T$ be a matrix with elements in $\mathfrak{K} = \mathfrak{R}'$ or $\mathfrak{R}'(i)$, $\mathfrak{R}'$ an ordered closed field. Then we shall prove

Theorem 17. *The elementary divisors of* $A = AT$ *in the ordered closed case are all simple and the characteristic roots are all real. Hence* A *is similar in* $\mathfrak{K}$ *to*

$$QAQ^{-1} = \text{diag}\,\{c_1I_{m_1}, \ldots, c_tI_{m_t}\} = A_0\,, \tag{54}$$

where c_i *in* $\mathfrak{R}'$ *is a root of multiplicity* m_i *of* $|\lambda I - A| = 0$.

We recall that the elementary divisors of $\lambda I - A$ are powers $(\lambda - c_j)^{e_j}$, where c_j is in the algebraically closed field $\mathfrak{R}'(i)$. Our theorem states that the c_j are in $\mathfrak{R}'$ and $e_j = 1$. It is evidently sufficient to prove this for the factors of the minimum function $g(\lambda)$ of A since each elementary divisor divides $g(\lambda)$. By Theorem 5.2 the polynomial $g(\lambda)$ has coefficients in $\mathfrak{R}' = \mathfrak{F}$. Let $c = a + bi$ be a root of $g(\lambda) = 0$ and write

$$g(\lambda) = (\lambda - c)[g_1(\lambda) + g_2(\lambda)i]$$

where $g_1(\lambda)$ and $g_2(\lambda)$ have coefficients in $\mathfrak{F}$ and degree less than $g(\lambda)$. These coefficients and those of $g(\lambda)$ are real in our generalized sense and

$$g(\lambda) = \overline{g(\lambda)} = (\lambda - \bar{c})[g_1(\lambda) - g_2(\lambda)i] .$$

The assumption that $\bar{c} \neq c$ implies that $\lambda - c$ divides $g(\lambda)$ if and only if $\lambda - c$ divides $g_1(\lambda) - ig_2(\lambda)$. But then $h(\lambda) = [g_1(\lambda)]^2 + [g_2(\lambda)]^2$ has $g(\lambda) = (\lambda - c)[g_1(\lambda) + ig_2(\lambda)]$ as a factor and $h(A) = [g_1(A)]^2 + [g_2(A)]^2 = 0$. We apply Theorem 5.14 and have $g_1(A) = g_2(A) = 0$. This is impossible since $g(\lambda) \neq 0$ is the minimum function of A, whereas $g_1(\lambda)$ and $g_2(\lambda)$ which cannot both be zero polynomials have degrees less than $g(\lambda)$. We have proved that every characteristic root of A is real.

We now write $g(\lambda) = (\lambda - c)^r q(\lambda)$, $q(c) \neq 0$. If $r > 1$ we have $g_0(\lambda) = (\lambda - c)^{r-1}q(\lambda)$, $g_0(A) \neq 0$, $[g_0(A)]^2 = (A - cI)^{2r-2}q(A)^2 = g(A) \cdot (A - cI)^{r-2} \cdot q(A) = 0$. This is again impossible by Theorem 5.14.

We have proved the first part of our theorem. The form (54) follows from Theorem 4.5. Notice that in (54)

$$A_0 = A_0^T = A_0' \tag{55}$$

since A_0 is a real diagonal matrix.

We shall study a new equivalence relation which is a combination of similarity and T-equivalence. The result we shall obtain is a consequence of (55). For $A = Q^{-1}A_0Q = A^T = Q^TA_0(Q^T)^{-1}$ and $QQ^TA_0 = A_0QQ^T$. Write

$$G = QQ^T = (G_{ij}) \qquad (i, j = 1, \ldots, t)$$

where G_{ij} has m_i rows and m_j columns. Then

$$A_0G = (c_iG_{ij}) = GA_0 = (G_{ij}c_j) ,$$

so that $(c_i - c_j)G_{ij} = 0$. But $c_i \neq c_j$ if $i \neq j$ whence $G_{ij} = 0$ for $i \neq j$, $G_{ii} = G_i$,

$$G = \operatorname{diag} \{G_1, G_2, \ldots, G_t\} .$$

The matrix $G = QQ^T$ where Q is non-singular. By our definition expressed in (52) the matrix G is positive definite. So are its principal minors G_i by Theorem 5.16. The criterion above (52) then states that $G_i = H_iH_i^T$ and

$$G = HH^T, \qquad H = \operatorname{diag} \{H_1, \ldots, H_t\} .$$

The computation

$$A_0H = \text{diag}\ \{c_1H_1, \ldots, c_tH_t\}\ , \qquad HA_0 = \text{diag}\ \{H_1c_1, \ldots, H_tc_t\}\ ,$$

shows that $A_0H = HA_0$. We have proved the existence of an H which is commutative with A_0 and such that $QQ^T = HH^T$. Then

$$(H^{-1}Q)A(H^{-1}Q)^{-1} = H^{-1}A_0H = A_0 = CAC^{-1}$$

where

$$C = H^{-1}Q\ , \qquad CC^T = H^{-1}QQ^T(H^T)^{-1} = H^{-1}HH^T(H^T)^{-1} = I_n\ .$$

Our final result is stated as

Theorem 18. *Let* $A = A^T$ *in the ordered closed case of* $\mathfrak{K}$ *so that* A *is similar to an essentially unique canonical form*

$$A_0 = \text{diag}\ \{c_1I_{m_1}, \ldots, c_tI_{m_t}\}\ ,$$

where c_i *is a characteristic root of multiplicity* m_i *of* A. *Then there exists a non-singular matrix* Q *with elements in* $\mathfrak{K}$ *such that*

$$Q^T = Q^{-1}, \quad QAQ^T = QAQ^{-1} = A_0\ ,$$

so that A *is simultaneously similar and* T*-equivalent to* A_0.

A matrix Q is called *T-orthogonal* if $Q^{-1} = Q^T$. When $\mathfrak{K} = \mathfrak{F} = \mathfrak{R}'$ so that $Q^T = Q' = Q^{-1}$ we call Q *orthogonal*, and when $\mathfrak{K} = \mathfrak{R}'(i)$ so that $Q^{-1} = \bar{Q}'$ we call Q *unitary*. Any two matrices A and B are said to be *orthogonally T-equivalent* in $\mathfrak{K}$ if there exists a T-orthogonal matrix Q such that $B = QAQ^T$. Theorem 5.18 may now be restated as

Theorem 19. *Two* T*-symmetric matrices in the ordered closed case are orthogonally* T*-equivalent if and only if they have the same elementary divisors.*

EXERCISES

1. Find the characteristic functions of the matrices

$$A = \begin{pmatrix} 1 & 2 \\ 2 & -1 \end{pmatrix}, \quad B = \begin{pmatrix} 2 & 1 \\ 1 & -2 \end{pmatrix}.$$

Use the form of these functions to prove that A and B are similar in the field $\mathfrak{R}$ of all rational numbers. Also find a rational Q such that $QAQ^{-1} = B$. Hint: Take $Q = \begin{pmatrix} 1 & 1 \\ x & y \end{pmatrix}$ and determine x and y uniquely by direct computation.

2. Find all rational matrices G in Ex. 1 such that $GAG^{-1} = B$ by proving that $G = QP$ for Q fixed, $PA = AP$, and that P necessarily has the form $a + bA$ with rational a and b.

3. Use Ex. 2 to show that A and B are not orthogonally equivalent in the field $\mathfrak{R}$ of all rational numbers.

4. Prove that A and B of Ex. 1 are orthogonally equivalent in the field obtained by adjoining either $\sqrt{2}$ or $\sqrt{10}$ to $\mathfrak{R}$. Thus show that the criterion of Theorem 5.19 depends essentially on the assumption made about $\mathfrak{K}$. Hint: Use Ex. 3.

14. Orthogonal equivalence of real skew matrices. We have seen how the theory of the T-equivalence of skew matrices $C = -C^T$ is a consequence of the theory for $A = A^T$ when $\mathfrak{K} = \mathfrak{F}(\theta)$, $\theta^T = -\theta$, and that the only case to be considered is that where $\mathfrak{K} = \mathfrak{F}$, $a = a^T$ for every a of $\mathfrak{K}$. This is also evident in our present case of orthogonal T-equivalence. We shall therefore restrict our attention to the case where $\mathfrak{K} = \mathfrak{F} = \mathfrak{R}'$ is an ordered closed field, $C = -C'$ is skew, and shall obtain a canonical form for any such matrix under real orthogonal equivalence.

The matrix iC is an Hermitian matrix with elements in $\mathfrak{C} = \mathfrak{R}'(i)$ and is similar in $\mathfrak{R}'(i)$ to a real diagonal matrix by Theorem 5.17. Call this matrix iD_0 where D_0 has diagonal elements ai, a in $\mathfrak{R}'$. Then C is similar in $\mathfrak{R}'(i)$ to D_0 and the quantities ai are the characteristic roots of C. We have incidentally proved that *the elementary divisors of any real skew matrix are all simple.*

Let ai be a characteristic root of C, a real matrix, a real. Then $\overline{ai} = -ai$ is a characteristic root of the same multiplicity as ai. Thus *the non-zero characteristic roots of a real skew matrix* C *occur in conjugate pairs ia, $-ia$, where we may choose our notation so that $a > 0$.* It follows that C is similar in $\mathfrak{R}'(i)$ to

$$(56)\qquad \begin{pmatrix} C_0 & 0 \\ 0 & 0 \end{pmatrix}, \quad C_0 = \begin{pmatrix} ia & 0 \\ 0 & -ia \end{pmatrix},$$

where a is the real diagonal matrix

$$(57)\qquad a = \operatorname{diag}\,\{a_1 I_{r_1}, \ldots, a_t I_{r_t}\}, \qquad a_j > 0,$$

and ia_j is a characteristic root of multiplicity r_j of C.

We let I be an r-rowed identity matrix and

$$(58)\qquad V = \frac{1}{\sqrt{2}}\begin{pmatrix} I & iI \\ I & -iI \end{pmatrix}$$

so that

$$(59)\qquad V\bar{V}' = \frac{1}{2}\begin{pmatrix} I & iI \\ I & -iI \end{pmatrix}\begin{pmatrix} I & I \\ -iI & iI \end{pmatrix} = \begin{pmatrix} I & 0 \\ 0 & I \end{pmatrix},$$

and V is a unitary matrix. Put

$$D_0 = \begin{pmatrix} 0 & -a \\ a & 0 \end{pmatrix}. \tag{60}$$

By direct computation

$$VD_0V^{-1} = VD_0\bar{V}' = \begin{pmatrix} ia & 0 \\ 0 & -ia \end{pmatrix} = C_0. \tag{61}$$

The matrix (56) is similar in $\mathfrak{R}'(i)$ to C and is also similar to

$$D = \begin{pmatrix} D_0 & 0 \\ 0 & 0 \end{pmatrix} = \begin{pmatrix} V^{-1} & 0 \\ 0 & I_{n-2r} \end{pmatrix}\begin{pmatrix} C_0 & 0 \\ 0 & 0 \end{pmatrix}\begin{pmatrix} V & 0 \\ 0 & I_{n-2r} \end{pmatrix}.$$

Then D and C are matrices with elements in $\mathfrak{R}'$ and are similar in $\mathfrak{R}'(i)$. By Section 4.2 they are similar in $\mathfrak{R}'$.

Theorem 20. *Every real skew matrix* C *is similar in* $\mathfrak{R}'$ *to a matrix*

$$\mathrm{QCQ^{-1}} = \begin{pmatrix} 0 & \mathrm{a} & 0 \\ -\mathrm{a} & 0 & 0 \\ 0 & 0 & 0 \end{pmatrix},$$

where a *is given by* (57) *and* $\mathrm{a}_j\mathrm{i}$ *is a root of multiplicity* r_j *of* $|\lambda \mathrm{I} - \mathrm{C}| = 0$.

The form of a is unique apart from an arbitrary permutation of the a_j. We make it absolutely unique by assuming that

$$0 < a_1 < a_2 < \ldots < a_t. \tag{62}$$

We may then call the a_i with their multiplicities r_i the *orthogonal invariants* of C, and Theorem 5.20 implies

Theorem 21. *Two real skew matrices are similar in* $\mathfrak{R}'$ *if and only if they have the same orthogonal invariants.*

We now show that in Theorem 5.21 we may replace similarity by *orthogonal equivalence.* Let G be any real square matrix of $2r$ rows and use (58) to compute

$$\left\{\begin{aligned} VGV^{-1} &= V\begin{pmatrix} G_1 & G_2 \\ G_3 & G_4 \end{pmatrix}\bar{V}' = \frac{1}{2}\begin{pmatrix} G_1 + iG_3 & G_2 + iG_4 \\ G_1 - iG_3 & G_2 - iG_4 \end{pmatrix}\begin{pmatrix} I & I \\ -iI & iI \end{pmatrix} \\ &= \frac{1}{2}\begin{pmatrix} \Gamma(i, i) & \Gamma(i, -i) \\ \Gamma(-i, i) & \Gamma(-i, -i) \end{pmatrix}, \end{aligned}\right. \tag{63}$$

where, if x and y are indeterminates over $\mathfrak{R}'$, $\Gamma(x, y)$ is the matrix polynomial

$$\Gamma(x, y) = G_1 + xG_3 - yG_2 - xyG_4. \tag{64}$$

The assumption $GD_0 = D_0G$ is equivalent to $VGV^{-1}VD_0V^{-1} = VD_0V^{-1}VGV^{-1}$. This matrix equation is equivalent by direct computation to

$$(65) \qquad a\Gamma(i, i) = \Gamma(i, i)a, \quad a\Gamma(i, -i) = -\Gamma(i, -i)a,$$

where we use the fact that $\bar{a} = a$, $\overline{\Gamma(i,i)} = \Gamma(-i, -i)$, $\overline{\Gamma(i, -i)} = \Gamma(-i, i)$. But then

$$\begin{aligned} a(G_1 + G_4) &= (G_1 + G_4)a, & a(G_3 - G_2) &= (G_3 - G_2)a, \\ a(G_1 - G_4) &= -(G_1 - G_4)a, & a(G_3 + G_2) &= -(G_3 + G_2)a \end{aligned}$$

so that

$$(66) \quad aG_1 = G_4a, \quad aG_4 = G_1a, \quad aG_3 = -G_2a, \quad aG_2 = -G_3a,$$

and hence $a^2G_1 = G_1a^2$, $a^2G_2 = G_2a^2$.

The matrix G_1 is commutative with the diagonal matrix a^2 and since $a_j^2 \neq a_k^2$ for $j \neq k$ we see easily that $G_1 = (G_{jk}^{(1)})$, $G_{jk}^{(1)} = 0$ for $j \neq k$, $G_1a = aG_1$. Similarly, G_2, G_3, G_4 are commutative with a and by (66)

$$(67) \qquad G_1 = G_4, \quad G_3 = -G_2, \quad G = \begin{pmatrix} G_1 & -G_3 \\ G_3 & G_1 \end{pmatrix}.$$

The matrix $\Gamma(i, -i)$ of (64) has the value $G_1 + iG_3 + iG_2 - G_4 = 0$ and $\Gamma(i, i) = 2G_1 + 2iG_3$. By (63)

$$(68) \qquad VGV^{-1} = \begin{pmatrix} \Gamma & 0 \\ 0 & \bar{\Gamma} \end{pmatrix}, \quad \Gamma = G_1 + iG_3, \quad \Gamma a = a\Gamma.$$

We have proved that when $GD_0 = D_0G$ the matrix VGV^{-1} has the form (68), G has the form (67).

Conversely let G have the form given by (67) for $G_ia = aG_i$, or the equivalent form (68). Then VGV^{-1} is commutative with VD_0V^{-1} by our computation and G is commutative with D_0. This criterion will be used in the proof of the

LEMMA. *Let* G *be a real positive definite symmetric matrix commutative with* D_0. *Then*

$$G = HH',$$

for a real matrix H *commutative with* D_0.

For $V G \bar{V}'$ is T-equivalent to G in $\mathfrak{R}'(i)$ and is positive definite. Also $V G \bar{V}' = V G V^{-1}$ must have the form (68), where Γ is a principal minor of $V G \bar{V}'$ and is positive definite by Theorem 5.16. But $\Gamma a = a\Gamma$, and in the proof of Theorem 5.18 we showed that $\Gamma = \Lambda\bar{\Lambda}'$ with $\Lambda a = a\Lambda$. Hence

$$VGV^{-1} = \begin{pmatrix} \Lambda\bar{\Lambda}' & 0 \\ 0 & \bar{\Lambda}\Lambda' \end{pmatrix} = \begin{pmatrix} \Lambda & 0 \\ 0 & \bar{\Lambda} \end{pmatrix}\begin{pmatrix} \bar{\Lambda}' & 0 \\ 0 & \Lambda' \end{pmatrix}.$$

Put $\Lambda = \Lambda_1 + \Lambda_3 i$ where Λ_1 and Λ_3 are real and must be commutative with the real matrix a when $\Lambda a = a\Lambda$. Define

$$H = \begin{pmatrix} \Lambda_1 & -\Lambda_3 \\ \Lambda_3 & \Lambda_1 \end{pmatrix},$$

so that by (68)

$$VHV^{-1} = \begin{pmatrix} \Lambda & 0 \\ 0 & \bar{\Lambda} \end{pmatrix}, \qquad VGV^{-1} = VHV^{-1} \cdot V\bar{H}'V^{-1}.$$

Since $\bar{H} = H$ we have $G = HH'$, $HD_0 = D_0H$ as desired.

We now prove the final result

Theorem 22. *Two real skew matrices are orthogonally equivalent in the ordered closed field $\mathfrak{R}'$ of their elements if and only if they have the same orthogonal invariants (or the same elementary divisors).*

The relation of orthogonal equivalence is a formal equivalence relation and it is sufficient to prove that every $C = -C'$ is orthogonally equivalent to a unique D of the form

$$D = \begin{pmatrix} D_0 & 0 \\ 0 & 0 \end{pmatrix}, \qquad D_0 = \begin{pmatrix} 0 & -a \\ a & 0 \end{pmatrix}. \tag{69}$$

By Theorem 5.21 we have $D = PCP^{-1}$. The form (69) of D and a diagonal imply the skewness of D, and $-C = C' = (P^{-1}DP)' = -P'D(P')^{-1}$, so that $P^{-1}DP = P'DP'^{-1}$,

$$(PP')D = D(PP').$$

Write

$$PP' = \begin{pmatrix} G & P_2 \\ P_2' & P_3 \end{pmatrix}$$

where PP' is evidently a real positive definite matrix. Then so is G and

$$DPP' = \begin{pmatrix} D_0G & D_0P_2 \\ 0 & 0 \end{pmatrix} = PP'D = \begin{pmatrix} GD_0 & 0 \\ P_2'D_0 & 0 \end{pmatrix}.$$

Hence $D_0P_2 = 0$. Since D_0 is non-singular, $P_2 = 0$. Also P_3 is positive definite and equal to R_3R_3'. We apply our above lemma to write $G = HH'$, $HD_0 = D_0H$ and have

$$PP' = RR' , \qquad R = \begin{pmatrix} H & 0 \\ 0 & R_3 \end{pmatrix}$$

where it is trivial to show that $RD = DR$. Then since $(R')^{-1} = (R^{-1}PP')^{-1} = (P')^{-1}P^{-1}R$,

$$U = R^{-1}P$$

has the property that

$$U' = P'R'^{-1} = P^{-1}R = U^{-1} .$$

The matrix U is orthogonal and $P = RU$,

$$C = P^{-1}DP = U^{-1}(R^{-1}DR)U = U^{-1}DU .$$

Thus C is orthogonally equivalent to $D = UCU^{-1}$ as desired.

EXERCISE

Find the canonical forms (69), (62), (57) of all two-, three-, and four-rowed real skew matrices.

15. Forms. Let $x_1, \ldots, x_n$ and $y_1, \ldots, y_m$ be independent variables over a field $\mathfrak{K}$ with an involution T over $\mathfrak{F}$. Then we are assuming that $\mathfrak{K} = \mathfrak{F}$ or $\mathfrak{F}(\theta)$ and know what is meant by a^T for any a of $\mathfrak{F}$. The variables x_i, y_i are independent indeterminates over $\mathfrak{K}$, and in the course of our discussion we may replace them by quantities of $\mathfrak{K}$.

Define

$$x = (x_1, \ldots, x_n) , \qquad y = (y_1, \ldots, y_m)$$

and let $y_1^T, \ldots, y_m^T$ be new variables with the property that if y_i is replaced by a in $\mathfrak{F}$ then y_i^T is replaced by a^T. We define the $m \times 1$ matrix

$$y^T = \bar{y}' = \begin{pmatrix} y_1^T \\ y_2^T \\ \cdot \\ \cdot \\ \cdot \\ y_m^T \end{pmatrix} .$$

A transformation

$$x = \xi Q\,, \qquad y = \eta Q$$

replaces y^T by η^T where

$$y^T = (\eta Q)^T = Q^T \eta^T\,.$$

Notice that we are using the property $(AB)^T = B^T A^T$ as applied to matrices which are not square. This is a consequence of $A^T = \overline{A}'$, the case of an involution we are considering, and we have not shown it to hold otherwise.

Let $A = (a_{ij})$ be any square matrix and

$$f = xAy^T = \sum_{i,j}^{1,\ldots,n} x_i a_{ij} y_j^T$$

be the corresponding bilinear form in the $2n$ variables x_i, y_j. We call f a *symmetric bilinear form* if $A = A'$, an *alternate* (or *skew*) *bilinear form* if $A = -A'$, an *Hermitian bilinear form* if $A = \overline{A}'$, $\mathfrak{K} = \mathfrak{F}(\theta)$, a *skew-Hermitian bilinear form* if $A = -A^T = -\overline{A}'$.

A non-singular matrix Q defines a linear transformation

$$x = \xi Q\,, \qquad y = \eta Q$$

carrying f into what is called a T-*equivalent form*

$$f = \xi B \eta^T\,,$$

where evidently $f = xAy^T = \xi QAQ^T\eta^T$, so that $B = QAQ^T$ is T-equivalent to A. Thus the theory of the T-equivalence of forms is the same as that of matrices.

The results above are unchanged if we consider the quadratic forms

$$f = xAx^T\,, \qquad A = A^T\,.$$

Here $f = \sum_{i,j} x_i a_{ij} x_j$ when $\mathfrak{K} = \mathfrak{F}$ and $f = \sum_{i,j} x_i a_{ij} \bar{x}_j$ when $\mathfrak{K} = \mathfrak{F}(\theta)$, $a^T = \bar{a}$ in $\mathfrak{K}$. In this latter case f is usually called an *Hermitian quadratic form.*

There is no particular reason for our restating any of our theorems on the T-equivalence of matrices in terms of theorems on the corresponding forms. We leave this as an exercise for the reader.

CHAPTER VI

FINITE GROUPS

1. Subsets. A complete exposition of the standard theory of finite groups requires a text in itself and is certainly too long to be attempted here. Certain of the simple notions of that theory are essential for our discussion of the theory of fields, however, and will be given in the present chapter.

Let $\mathfrak{G}$ be a *finite group*, that is, a group with only a finite number of elements. As is seen below, any subset $\mathfrak{B}$ of $\mathfrak{G}$ may be extended to be closed with respect to the operation of $\mathfrak{G}$, and we designate this *closed extension* of $\mathfrak{B}$ by

$$[\mathfrak{B}] .$$

It is defined as the *intersection of all subgroups of* $\mathfrak{G}$ *which contain* $\mathfrak{B}$. Evidently $[\mathfrak{B}]$ is a subgroup of $\mathfrak{G}$ and is the group of all quantities of $\mathfrak{G}$ obtained by forming products of finite numbers of quantities of $\mathfrak{B}$. Moreover, $[\mathfrak{B}] = \mathfrak{B}$ if and only if $\mathfrak{B}$ is a subgroup of $\mathfrak{G}$.

The product

$$\mathfrak{B}\mathfrak{C} , \tag{1}$$

of any two subsets $\mathfrak{B}$ and $\mathfrak{C}$ of $\mathfrak{G}$, is the set of all quantities of $\mathfrak{G}$ of the form BC with B in $\mathfrak{B}$ and C in $\mathfrak{C}$. In particular either of these sets may consist of a single element S of $\mathfrak{G}$. Then (1) also defines the sets

$$\mathfrak{B}S , \qquad S\mathfrak{B} \tag{2}$$

for S in $\mathfrak{G}$, $\mathfrak{B} \leqq \mathfrak{G}$. The operation of *multiplication of sets* defined by (1) is a function on $\Sigma\Sigma$ to Σ where Σ is the class of all subsets of $\mathfrak{G}$, and is evidently an associative operation.

The *composite* of $\mathfrak{B}$ and $\mathfrak{C}$ is the closed extension of the logical sum of $\mathfrak{B}$ and $\mathfrak{C}$, that is, the set consisting of all of the elements of $\mathfrak{B}$ taken together with all of $\mathfrak{C}$. Thus the composite of $\mathfrak{B}$ and $\mathfrak{C}$ is the group of all quantities of $\mathfrak{G}$ obtained as products of finite numbers of quantities of $\mathfrak{B}$ and $\mathfrak{C}$.

A group $\mathfrak{G}$ is said to be *generated* by a subset $\mathfrak{B}$ if $\mathfrak{G} = [\mathfrak{B}]$. In the particular case where $\mathfrak{B}$ is a set consisting of a single element S of $\mathfrak{G}$, we have $\mathfrak{G} = [S]$, $\mathfrak{G}$ is the cyclic group composed of all powers of S. We say that

S *generates* $\mathfrak{G}$ and call all elements of $\mathfrak{G}$ which generate $\mathfrak{G}$ the *generators* of $\mathfrak{G}$.

We now let $\mathfrak{H}$ and $\mathfrak{K}$ be subgroups of a finite group $\mathfrak{G}$, and suppose that $HK = KH$ for every H of $\mathfrak{H}$ and K of $\mathfrak{K}$. The product $\mathfrak{H}\mathfrak{K}$ is now a group and is actually the composite of $\mathfrak{H}$ and $\mathfrak{K}$. When also the order of $\mathfrak{H}\mathfrak{K}$ is the product of the orders of $\mathfrak{H}$ and $\mathfrak{K}$ we call $\mathfrak{H}\mathfrak{K}$ the *direct product*

$$\mathfrak{H} \times \mathfrak{K}$$

of $\mathfrak{H}$ and $\mathfrak{K}$. We restate this definition by assuming that $H_1, \ldots, H_r$ are the distinct elements of $\mathfrak{H}$ and $K_1, \ldots, K_s$ are the distinct elements of $\mathfrak{K}$. Let $H_iK_j = K_jH_i$ for all i and j and suppose that the rs elements $H_1K_1, \ldots, H_rK_s$ are all distinct. Then the set of all these elements is the group $\mathfrak{H} \times \mathfrak{K}$.

We shall usually designate the elements of our finite group $\mathfrak{G}$ by $S_1, \ldots, S_n$ where $S_1 = I$ is the identity element of $\mathfrak{G}$, n is the order of $\mathfrak{G}$, and shall indicate that the elements of $\mathfrak{G}$ are the S_i by writing

$$\mathfrak{G} = (S_1, \ldots, S_n) .$$

2. Subgroups. We let $\mathfrak{H}$ be a subgroup of $\mathfrak{G}$ and consider the sets $\mathfrak{H}S_i$, $S_i\mathfrak{H}$. They are called the *right* and *left cosets*, respectively, of $\mathfrak{H}$. *Every* $\mathfrak{H}S$ or $S\mathfrak{H}$ *has as many elements as* $\mathfrak{H}$. For if $H_1S = H_2S$ or $SH_1 = SH_2$ then always $H_1 = H_2$. We also see that if $\mathfrak{H}$ is a group then $\mathfrak{H}\mathfrak{H} = \mathfrak{H}$. But also

$$H\mathfrak{H} = \mathfrak{H}H = \mathfrak{H} , \tag{3}$$

for every H of $\mathfrak{H}$. This follows from the fact that $H\mathfrak{H} \leqq \mathfrak{H}$, $\mathfrak{H}H \leqq \mathfrak{H}$, and if H_0 is any element of $\mathfrak{H}$ then $H_0 = H(H^{-1}H_0) = (H_0H^{-1})H$ is in $H\mathfrak{H}$ and $\mathfrak{H}H$.

Two cosets $\mathfrak{H}S$ *and* $\mathfrak{H}T$ *are equal if and only if* TS^{-1} *is in* $\mathfrak{H}$. For if $T = HS$ with H in $\mathfrak{H}$ then $\mathfrak{H}T = (\mathfrak{H}H)S = \mathfrak{H}S$. Conversely if $\mathfrak{H}S = \mathfrak{H}T$ then $IT = T$ is in $\mathfrak{H}S$, $T = HS$ with H in $\mathfrak{H}$. Also if $\mathfrak{H}S \neq \mathfrak{H}T$ then they have no elements in common. For otherwise $H_1S = H_2T$, $H = H_2^{-1}H_1 = TS^{-1}$ is in $\mathfrak{H}$.

Our group $\mathfrak{G}$ has only a finite number of subsets altogether and thus only a finite number q of distinct cosets of a subgroup $\mathfrak{H}$. This integer q is called the *index* of $\mathfrak{H}$ under $\mathfrak{G}$. Since every S of $\mathfrak{G}$ is in some coset $\mathfrak{H}S$ of $\mathfrak{H}$, the q cosets each having the same number m of elements exhaust the group $\mathfrak{G}$. No two distinct cosets have an element in common and this proves

Theorem 1. *The order* m *of a subgroup* $\mathfrak{H}$ *of a finite group* $\mathfrak{G}$ *divides the order* n *of* $\mathfrak{G}$ *and* n = mq, *where* q *is the index of* $\mathfrak{H}$ *under* $\mathfrak{G}$. *Thus* $\mathfrak{G}$ *has the* ***decomposition*** *into* q *distinct cosets given by the logical sum*

$$\mathfrak{G} = \mathfrak{H} + \mathfrak{H}S_2 + \mathfrak{H}S_3 + \ldots + \mathfrak{H}S_q . \tag{4}$$

The result above may of course also be stated for left cosets and gives

$$\mathfrak{G} = \mathfrak{H} + T_2\mathfrak{H} + \ldots + T_q\mathfrak{H} . \tag{5}$$

It is of course not true in general that $S_i\mathfrak{H} = \mathfrak{H}S_i$ or even that the T_i can be taken to be the S_i. However, it is obvious that we may take $T_i = S_i^{-1}$ if we so choose. The proof is left to the reader.

A subgroup $\mathfrak{H}$ of $\mathfrak{G}$ is called a *proper* subgroup of $\mathfrak{G}$ if $\mathfrak{G} > \mathfrak{H}$. We call $\mathfrak{G}$ an improper subgroup of itself. Theorem 6.1 implies

COROLLARY. *Let* $\mathfrak{H}$ *be a proper subgroup of prime index under* $\mathfrak{G}$. *Then* $\mathfrak{H}$ *is not contained properly in any proper subgroup* $\mathfrak{K}$ *of* $\mathfrak{G}$.

We leave the proof to the reader.

3. Cyclic groups. Every element S of a finite group $\mathfrak{G}$ generates a cyclic subgroup

$$[S] = (I, S, \ldots, S^{r-1}) ,$$

where $S^r = I$ is the identity element of $\mathfrak{G}$. This group was called the *period* of S by **Gauss.** It is called the *identity group* when $S = I$. Its order r is called the *order* of S. By Theorem 6.1 we see that r divides n and thus that

$$S^n = I$$

for every S of a group $\mathfrak{G}$ of order n. The integer r is the least t for which S^t is the identity element I of $\mathfrak{G}$. The *division algorithm for integers* then implies the

LEMMA. *Let* S *have order* r. *Then* $S^t = I$ *if and only if* r *divides* t.

We now have

Theorem 2. *A group* $\mathfrak{G}$ *of prime order* p *is cyclic and is the period* [S] *of any* $S \neq I$ *of* $\mathfrak{G}$.

For if $S \neq I$ the group $[S]$ has order $r > 1$ and a divisor of p. It follows that $r = p$, $\mathfrak{G} = [S]$.

An arbitrary group $\mathfrak{G}$ may possibly have many subgroups of a prescribed order. This is not true of cyclic groups and we may in fact prove

Theorem 3. *Let* $\mathfrak{G} = [S]$ *be cyclic of order* n = mq *for integers* m *and* q. *Then the cyclic group*

$$\mathfrak{H} = [S^q] \tag{6}$$

has order m *and is the only subgroup of order* m *of* $\mathfrak{G}$.

If the order of $\mathfrak{H} = [S^q]$ is t we have $(S^q)^t = I$. By the lemma qt is divisible by $n = qm$ and t is divisible by m, $t \geqq m$. Also $S^n = (S^q)^m = I$ so that t divides m, $t \leqq m$, $t = m$. Conversely let $\mathfrak{H}_0$ be a subgroup of $\mathfrak{G} = [S]$ of order m. Every element of $\mathfrak{G}$ is a power of S with positive integral exponent and the elements of $\mathfrak{H}_0$ are in $\mathfrak{G}$. We let g be the least exponent which occurs among the elements S^h of $\mathfrak{H}_0$. The greatest common divisor d of g and n has the form $ag + bn$ and $S^n = I$, $S^d = (S^g)^a$ is in $\mathfrak{H}_0$. Hence the divisor d of g and n is greater than or equal to g and evidently $d = g$, g divides n. Let S^h be in $\mathfrak{H}_0$ and write $h = gg_0 + k$ where $0 \leqq k < g$. Then $(S^h)(S^g)^{-g_0} = S^k$ is in $\mathfrak{H}_0$ and $k < g$ implies that $k = 0$, every element of $\mathfrak{H}_0$ is a power of S^g, $\mathfrak{H}_0 = [S^g]$. By our first proof if $n = gm_0$ the order of $\mathfrak{H}_0$ is m_0. Since it is m we have $m_0 = m$, $g = q$, $\mathfrak{H}_0 = [S^q] = \mathfrak{H}$ as desired.

We defined the generators of a cyclic group in Section 6.1. They are determined in

Theorem 4. *An element* S^t *generates* $[S]$ *if and only if* t *is prime to the order* s *of* S.

For if $t = at_0$ and $s = as_0$ then $[S^t] \leqq [S^a]$ which has order $s_0 < s$ when $a > 1$ by Theorem 6.3. Thus $[S^t] = [S]$ implies that t is prime to s. Conversely t prime to s implies that $1 = tt_0 + ss_0$, $S = S^{tt_0+ss_0} = (S^t)^{t_0}$ is in $[S^t]$, $[S] \leqq [S^t]$. Since $[S^t]$ is always a subgroup of $[S]$ we have $[S] = [S^t]$.

Theorem 5. *Let* $\mathfrak{G} = [S]$ *be cyclic of order* $n = ab$, *where* a *and* b *are relatively prime. Then* $\mathfrak{G}$ *is uniquely expressible as the direct product*

$$\mathfrak{G} = \mathfrak{H} \times \mathfrak{K} \tag{7}$$

of two cyclic groups of respective orders a *and* b.

For we take $\mathfrak{H} = [S^b]$, $\mathfrak{K} = [S^a]$ and Theorem 6.3 states that $\mathfrak{H}$ has order a, $\mathfrak{K}$ has order b. Since a and b are relatively prime we have $aa_0 + bb_0 = 1$ for integers a_0 and b_0, and $S^t = H^{b_0t} \cdot K^{a_0t}$ where $H = S^b$, $K = S^a$. The elements S^t are $n = ab$ distinct elements in $\mathfrak{H}\mathfrak{K}$, they are all commutative, $\mathfrak{H}\mathfrak{K}$ has order at most ab. Hence $\mathfrak{H}\mathfrak{K}$ has order ab and is the direct product $\mathfrak{H} \times \mathfrak{K}$. If also $\mathfrak{G} = \mathfrak{H}_1 \times \mathfrak{K}_1$ the group $\mathfrak{H}_1$ has order a and is equal to $\mathfrak{H}$ by Theorem 6.3. Similarly $\mathfrak{K}_1 = \mathfrak{K}$ and our expression (7) is unique.

We may write

$$n = p_1^{e_1} \dots p_t^{e_t}, \tag{8}$$

where the p_i are distinct primes. Then Theorem 6.5 states that *every cyclic group* $\mathfrak{G}$ *of order* n *is uniquely expressible in the form*

$$\mathfrak{G} = \mathfrak{G}_1 \times \dots \times \mathfrak{G}_t, \tag{9}$$

where $\mathfrak{G}_i$ *is cyclic of order* $p_i^{e_i}$ *and in fact* $n = p_i^{e_i}q_i$, $\mathfrak{G}_i = [S^{q_i}]$.

The converse of Theorem 6.5 states that *if $\mathfrak{G}$ is the direct product of two cyclic groups $\mathfrak{H}$ and $\mathfrak{K}$ of respective relatively prime orders* a *and* b *then $\mathfrak{G}$ is cyclic of order* n = ab. In fact if $\mathfrak{H} = [T]$, $\mathfrak{K} = [U]$ then $\mathfrak{G} = [S]$ where $S = TU$. We shall prove a more comprehensive result which implies this converse in

Theorem 6. *Let* S *be an element of order* n *in any finite group $\mathfrak{G}$. Express* S *as a product* TU *of commutative elements* T *and* U *of $\mathfrak{G}$ of respective orders* a *and* b *which are relatively prime. Then* n = ab, T *and* U *are unique and are in* [S].

For $S = TU$, $S^{ab} = (T^a)^b(U^b)^a = I$ so that n divides ab. Also b is prime to a and by Theorem 6.4 the element $S^b = (TU)^b = T^b$ generates $[T]$, T is in $[S]$. Similarly U is in $[S]$. Theorem 6.3 states that the groups $[U]$ and $[T]$ are subgroups of $[S]$ uniquely determined by their order and that a and b divide n. The relatively prime integers a and b divide n and so does their product ab. Since already $n \leqq ab$ we have $n = ab$. Suppose now that $S = T_1U_1 = U_1T_1$ so that by our proof $[T_1] = [T]$, $[U_1] = [U]$. Then $T_1 = T^\beta$, $U_1 = U^\alpha$, and $T^\beta U^\alpha = TU$, $T^{\beta-1} = U^{1-\alpha}$. The order of $[T^{\beta-1}]$ divides a, the order of $[U^{1-\alpha}]$ divides b, where a and b are relatively prime integers. Hence $T^{\beta-1}$ has order one, $T^\beta = T = T_1$, $U^\alpha = U = U_1$.

DEFINITION. *Let $\mathfrak{G}$ be an abelian (commutative) group. Then the maximum of the orders of the elements of $\mathfrak{G}$ is called the exponent of $\mathfrak{G}$.*

Theorem 7. *The exponent* m *of an abelian group $\mathfrak{G}$ is divisible by the order of every element of $\mathfrak{G}$, that is*

$$S^m = I \tag{10}$$

for every S *of $\mathfrak{G}$.*

For let T in $\mathfrak{G}$ have order m, S have order s not a divisor of m. Then there exists a prime p such that $m = p^a m_0$, $s = p^b s_0$ with $b > a$ and $s_0 m_0$ prime to p. The order of $T^{p^a} = T_0$ is m_0 and of $S^{s_0} = S_0$ is p^b by Theorem 6.3 and by Theorem 6.6 the order of S_0T_0 is $p^b m_0 > m$. This contradicts our definition of m.

As an immediate corollary of the above we have

Theorem 8. *Let $\mathfrak{G}$ be an abelian group whose order* n *is the least integer* e *such that*

$$S^e = I \tag{11}$$

for every S *of $\mathfrak{G}$. Then $\mathfrak{G}$ is a cyclic group.*

For n is the exponent of $\mathfrak{G}$ and is the order of some S of $\mathfrak{G}$, $\mathfrak{G} = [S]$.

EXERCISES

1. Determine all groups of orders $n \leq 6$.

2. Prove that Theorem 6.5 holds for abelian groups $\mathfrak{G}$ where now $\mathfrak{H}$ and $\mathfrak{K}$ are abelian. Hint: Show that if p is a prime divisor of the order of an abelian group, then the group contains an element of order p. Use this with $\mathfrak{H}$, the subgroup of $\mathfrak{G}$ of all elements whose orders divide a, and similarly for $\mathfrak{K}$, and apply Theorems 6.1, 6.3, 6.6.

3. Show that the non-zero residue classes modulo a prime p form a group of order $p - 1$ and hence that $a^{p-1} \equiv 1 \pmod{p}$ for every a prime to p.

4. Conjugate groups. Two elements A and B of a group $\mathfrak{G}$ are called *conjugate* if there exists an S in $\mathfrak{G}$ such that

$$B = S^{-1}AS.$$

Now $A = I^{-1}AI$, $A = (S^{-1})^{-1}BS^{-1}$, and if $C = T^{-1}BT$ then $C = (ST)^{-1}AST$. This proves that the relation of conjugacy is an equivalence relation. We also have $S^{-1}ABS = S^{-1}ASS^{-1}BS$. Thus if $\mathfrak{H}$ is a group so is the set $S^{-1}\mathfrak{H}S$. We call this group and $\mathfrak{H}$ a pair of *conjugate subgroups of* $\mathfrak{G}$.

Conjugate groups are evidently equivalent. They are all equal to $\mathfrak{H}$ if and only if

$$\mathfrak{H}S = S\mathfrak{H}$$

for every S of $\mathfrak{G}$. When this is true we call $\mathfrak{H}$ a *normal divisor* (or invariant subgroup, or self-conjugate subgroup) of $\mathfrak{G}$.

Theorem 9. *The intersection of two normal divisors of* $\mathfrak{G}$ *is a normal divisor of* $\mathfrak{G}$.

For the intersection of two subgroups $\mathfrak{H}$ and $\mathfrak{K}$ is a subgroup $\mathfrak{D}$ of both. If S is in $\mathfrak{G}$ then $S^{-1}\mathfrak{H}S = \mathfrak{H}$ so that $S^{-1}\mathfrak{D}S$ is in $\mathfrak{H}$. But also $S^{-1}\mathfrak{D}S$ is in $\mathfrak{K}$ since $\mathfrak{D} \leq \mathfrak{K}$. Hence $S^{-1}\mathfrak{D}S$ is in the intersection $\mathfrak{D}$ of $\mathfrak{H}$ and $\mathfrak{K}$, $S^{-1}\mathfrak{D}S = \mathfrak{D}$ is a normal divisor.

The *centrum* of a group $\mathfrak{G}$ is the set $\mathfrak{C}$ of all elements C of $\mathfrak{G}$ such that $CS = SC$ for every S of $\mathfrak{G}$. Evidently $\mathfrak{C}$ is an abelian normal divisor of $\mathfrak{G}$. Every group $\mathfrak{G}$ has the *improper normal divisor* $\mathfrak{G}$. All other normal divisors are of course called *proper normal divisors* of $\mathfrak{G}$. Thus we make the

DEFINITION. *A group* $\mathfrak{G}$ *is called a simple group if it has no proper normal divisor except the identity group* [I].

A final elementary result on normal divisors is given by

Theorem 10. *Let* $\mathfrak{H}$ *and* $\mathfrak{K}$ *be two normal divisors of* $\mathfrak{G}$ *whose intersection is the identity element of* $\mathfrak{G}$. *Then* $\mathfrak{H}\mathfrak{K} = \mathfrak{H} \times \mathfrak{K}$.

For if A is in $\mathfrak{H}$ and B is in $\mathfrak{K}$ then $B^{-1}\mathfrak{H}B = \mathfrak{H}$ implies that $B^{-1}A^{-1}B$ is in $\mathfrak{H}$ and so is $S = B^{-1}A^{-1}BA = (AB)^{-1}BA$. But $S = B^{-1}(A^{-1}BA)$ is in $\mathfrak{K}$ and $S = I$, $BA = AB$ for every A of $\mathfrak{H}$ and B of $\mathfrak{K}$. We now let $AB = A_1B_1$ where A and A_1 are in $\mathfrak{H}$, B and B_1 are in $\mathfrak{K}$. Then $A_1^{-1}A = B_1B^{-1}$ is in both $\mathfrak{H}$ and $\mathfrak{K}$, $A_1^{-1}A = I = B_1B^{-1}$, $A_1 = A$, $B_1 = B$. We have proved that the products $A_iB_j = B_jA_i$ are all distinct and that $\mathfrak{H}\mathfrak{K} = \mathfrak{H} \times \mathfrak{K}$.

EXERCISES

1. Let $\mathfrak{H}$ be a subgroup of index two of $\mathfrak{G}$. Prove that $\mathfrak{H}$ is a normal divisor of $\mathfrak{G}$.
2. Find all normal divisors of the groups $\mathfrak{G}$ of order four.
3. Find the normal divisors of a cyclic group of order twelve.
4. Extend our definition of normal divisors to infinite groups.
5. What are the normal divisors of an abelian group?

5. Quotient groups. Let $\mathfrak{H}$ be a normal divisor of a group $\mathfrak{G}$ so that $\mathfrak{H}S = S\mathfrak{H}$ for every S of $\mathfrak{G}$. Then

$$(\mathfrak{H}S)(\mathfrak{H}T) = \mathfrak{H}(ST) , \tag{12}$$

and the product of two cosets of $\mathfrak{H}$ is a coset of $\mathfrak{H}$. We use this operation of multiplication of cosets and prove

Theorem 11. *The cosets of a normal divisor $\mathfrak{H}$ of $\mathfrak{G}$ form a finite group with respect to multiplication of subsets of $\mathfrak{G}$. This group is called the* ***quotient group***

$$\mathfrak{G}/\mathfrak{H} . \tag{13}$$

For the elements of $\mathfrak{G}/\mathfrak{H}$ are a finite number of cosets closed with respect to multiplication. The element $\mathfrak{H}$ is the identity element by (12) and the inverse of $\mathfrak{H}S$ is $\mathfrak{H}S^{-1}$. But multiplication of sets has already been seen to be associative and the group postulates are satisfied, $\mathfrak{G}/\mathfrak{H}$ is a group.

We notice that $\mathfrak{G}/\mathfrak{H}$ consists of sets of elements whose aggregate makes up $\mathfrak{G}$. Any subset $\mathfrak{K}_0$ of $\mathfrak{G}_0 = \mathfrak{G}/\mathfrak{H}$ consists also of some of these cosets and we may write

$$\mathfrak{K}_0 = (\mathfrak{H}S_1, \ldots , \mathfrak{H}S_u) . \tag{14}$$

The elements of the cosets in $\mathfrak{K}_0$ form a subset

$$\mathfrak{K} = \mathfrak{H}S_1 + \ldots + \mathfrak{H}S_u \tag{15}$$

of $\mathfrak{G}$. The connection between these sets is given by

Theorem 12. *The set* $\mathfrak{K}_0$ *is a subgroup of index* t *of* $\mathfrak{G}_0 = \mathfrak{G}/\mathfrak{H}$ *if and only if* $\mathfrak{K}$ *is a subgroup of index* t *of* $\mathfrak{G}$ *with* $\mathfrak{H}$ *as a normal divisor and*

$$\mathfrak{K}_0 = \mathfrak{K}/\mathfrak{H}\,. \tag{16}$$

For we let $\mathfrak{G}$ have order $n = mq$ where m is the order of $\mathfrak{H}$. If $\mathfrak{K}_0$ is a subgroup of $\mathfrak{G}_0$ it contains the identity $\mathfrak{H}$ of $\mathfrak{G}_0$ and $\mathfrak{H}S_i\mathfrak{H}S_j = \mathfrak{H}S_k$ with $i, j, k = 1, \ldots, u$. Since $\mathfrak{G}_0$ has order q we have $q = ut$ and $\mathfrak{K}$ has um elements, $n = (um)t$. But $H_1S_iH_2S_j = H_3S_k$ is in $\mathfrak{K}$ for every H_1S_i and H_2S_j of $\mathfrak{K}$. Thus $\mathfrak{K}$ is a subgroup of $\mathfrak{G}$ and t is its index. Also $\mathfrak{H} \leqq \mathfrak{K}$ is a normal divisor of $\mathfrak{G}$ and consequently of $\mathfrak{K}$. The form of $\mathfrak{K}$ in (15) implies (16). Conversely when $\mathfrak{K}$ is a subgroup of $\mathfrak{G}$ which contains $\mathfrak{H}$ we may write (15) and see that $\mathfrak{K}_0 = \mathfrak{K}/\mathfrak{H}$ is a group. But evidently (14) holds and $\mathfrak{K}_0$ is a subgroup of $\mathfrak{G}_0$ with $\mathfrak{K}$ the set defined for this $\mathfrak{K}_0$ by (15). Then the properties of our theorem hold.

Theorem 6.12 gives a direct connection between the subgroups of $\mathfrak{G}/\mathfrak{H}$ and those of $\mathfrak{G}$. We may go considerably deeper into this relation. We write

$$\mathfrak{G} = \mathfrak{K}T_1 + \mathfrak{K}T_2 + \ldots + \mathfrak{K}T_t\,, \tag{17}$$

and see that the only possible cosets of $\mathfrak{H}$ under $\mathfrak{G}$ are the $q = ut$ cosets $\mathfrak{H}S_iT_j$. Since $\mathfrak{H}$ has index q the $\mathfrak{H}S_iT_j$ are distinct for $i = 1, \ldots, u$ and $j = 1, \ldots, t$. By (15) and $\mathfrak{H}\mathfrak{H} = \mathfrak{H}$, we have $(\mathfrak{H}S_i)(\mathfrak{H}T_j) = \mathfrak{H}S_iT_j$. Then

$$\mathfrak{G}_0 = \mathfrak{K}_0(\mathfrak{H}T_1) + \mathfrak{K}_0(\mathfrak{H}T_2) + \ldots + \mathfrak{K}_0(\mathfrak{H}T_t)\,, \tag{18}$$

and the cosets of the subgroup $\mathfrak{K}_0 = \mathfrak{K}/\mathfrak{H}$ under $\mathfrak{G}_0 = \mathfrak{G}/\mathfrak{H}$ are $\mathfrak{K}_0(\mathfrak{H}T_j)$. Clearly the elements of each such coset $\mathfrak{K}_0(\mathfrak{H}T_j)$ are the cosets $\mathfrak{H}S_iT_j$ $(i = 1, \ldots, u)$.

If $\mathfrak{K}$ is a normal divisor of $\mathfrak{G}$ then $\mathfrak{K}T_j = T_j\mathfrak{K}$, $T_j\mathfrak{H}S_i \leqq \mathfrak{K}T_j$. But $\mathfrak{H}$ is a normal divisor of $\mathfrak{G}$, $T_j\mathfrak{H}S_i = \mathfrak{H}T_jS_i$ is a coset of $\mathfrak{H}$ in $\mathfrak{K}T_j$ and must be one of the $\mathfrak{H}S_kT_j$. Thus $T_j\mathfrak{H}S_i = \mathfrak{H}S_kT_j$, $(\mathfrak{H}T_j)(\mathfrak{H}S_i) = (\mathfrak{H}S_k)(\mathfrak{H}T_j)$, $(\mathfrak{H}T_j)\mathfrak{K}_0 = \mathfrak{K}_0(\mathfrak{H}T_j)$, and $\mathfrak{K}_0$ is a normal divisor of $\mathfrak{G}_0$. The converse is obtained by reversing the steps of our proof.

Suppose that $\mathfrak{K}T_i\mathfrak{K}T_j = \mathfrak{K}T_k$, which is true when $\mathfrak{K}$ is a normal divisor of $\mathfrak{G}$. Then $\mathfrak{K}_0(\mathfrak{H}T_i) \cdot \mathfrak{K}_0(\mathfrak{H}T_j) = \mathfrak{K}_0(\mathfrak{H}T_i)(\mathfrak{H}T_j) = \mathfrak{K}_0\mathfrak{H}T_iT_j = \mathfrak{K}_0\mathfrak{H}T_k$ for the same k as above. For $T_iT_j = HS_iT_k$ with H in $\mathfrak{H}$, $\mathfrak{H}T_iT_j = (\mathfrak{H}S_i)(\mathfrak{H}T_k)$, $\mathfrak{K}_0\mathfrak{H}T_iT_j = \mathfrak{K}_0(\mathfrak{H}S_i)\mathfrak{H}T_k = \mathfrak{K}_0(\mathfrak{H}T_k)$ since $\mathfrak{H}S_i$ is in $\mathfrak{K}_0$. The correspondence

$$\mathfrak{K}T_j \longleftrightarrow \mathfrak{K}_0(\mathfrak{H}T_j)$$

is a one-to-one correspondence between the elements of $\mathfrak{G}/\mathfrak{K}$ and $\mathfrak{G}_0/\mathfrak{K}_0$ which is preserved under the respective group operations. We have proved

Theorem 13. *A subgroup $\mathfrak{K}$ of $\mathfrak{G}$ is a normal divisor of $\mathfrak{G}$ if and only if $\mathfrak{K}_0 = \mathfrak{K}/\mathfrak{H}$ is a normal divisor of $\mathfrak{G}_0 = \mathfrak{G}/\mathfrak{H}$. Moreover, the quotient groups*

$$\mathfrak{G}_0/\mathfrak{K}_0\,, \qquad \mathfrak{G}/\mathfrak{K}$$

are equivalent.

A *maximal* normal divisor of $\mathfrak{G}$ is a proper normal divisor which is not a proper subgroup of any proper normal divisor of $\mathfrak{G}$. For such normal divisors we have the

Theorem 14. *Let $\mathfrak{H}$ be a normal divisor of $\mathfrak{G}$. Then $\mathfrak{G}/\mathfrak{H}$ is simple if and only if $\mathfrak{H}$ is maximal.*

For let $\mathfrak{K}$ and $\mathfrak{K}_0$ be groups of Theorem 6.12 such that $\mathfrak{G}/\mathfrak{H} > \mathfrak{K}_0$, $\mathfrak{G} > \mathfrak{K} \geqq \mathfrak{H}$. Then $\mathfrak{K}_0$ is a normal divisor of $\mathfrak{G}/\mathfrak{H}$ if and only if $\mathfrak{K}$ is a normal divisor of $\mathfrak{G}$ and by the equality of indices in Theorem 6.12 we have $\mathfrak{K} > \mathfrak{H}$ if and only if $\mathfrak{K}_0$ is not the identity element of $\mathfrak{G}/\mathfrak{H}$. When $\mathfrak{H}$ is maximal no such $\mathfrak{K}_0$ exists and $\mathfrak{G}/\mathfrak{H}$ is simple. Conversely, if $\mathfrak{G}/\mathfrak{H}$ is simple no $\mathfrak{K}$ can exist and $\mathfrak{H}$ is maximal.

COROLLARY. *Let $\mathfrak{G} > \mathfrak{K} > \mathfrak{H}$ where $\mathfrak{K}$ and $\mathfrak{H}$ are normal divisors of $\mathfrak{G}$. Then $\mathfrak{K}_0 = \mathfrak{K}/\mathfrak{H}$ is a maximal normal divisor of $\mathfrak{G}_0 = \mathfrak{G}/\mathfrak{H}$ if and only if $\mathfrak{K}$ is a maximal normal divisor of $\mathfrak{G}$.*

For we have seen that $\mathfrak{G}_0/\mathfrak{K}_0$ is equivalent to $\mathfrak{G}/\mathfrak{K}$. When either $\mathfrak{K}_0$ or $\mathfrak{K}$ is maximal both quotient groups are simple and the other is maximal.

EXERCISES

1. Find the quotient groups in Exercises 2, 3 of Section 6.4.

2. Let $\mathfrak{G}$ be the infinite additive group of all integers and $\mathfrak{H}$ be the subgroup of all even integers. Find $\mathfrak{G}/\mathfrak{H}$. Such a group is usually called the *difference group* $\mathfrak{G}-\mathfrak{H}$.

3. Let $\mathfrak{G}_1$ consist of all classes $\{a\}$ of integers a where we put a and b into the same class if $a - b$ is divisible by twelve. Take $\mathfrak{G}$ to be the multiplicative group of all classes $\{a\}$ with a prime to twelve and $\mathfrak{H}$ to consist of $\{1\}$, $\{-1\}$. Find $\mathfrak{G}/\mathfrak{H}$.

4. A group $\mathfrak{G}$ is said to be *homomorphic* (multiply isomorphic) to a group $\mathfrak{G}_0$ if there is a correspondence $g \to g_0$ on $\mathfrak{G}$ to $\mathfrak{G}_0$ such that every element of $\mathfrak{G}_0$ is a correspondent g_0 and if also $h \to h_0$ then $gh \to g_0h_0$. Show that the set of elements of $\mathfrak{G}$ which correspond to the identity element of $\mathfrak{G}_0$ forms an invariant subgroup $\mathfrak{H}$ of $\mathfrak{G}$ and that $\mathfrak{G}_0$ and $\mathfrak{G}/\mathfrak{H}$ are equivalent.

6. Composition series. We shall first prove the preliminary

Theorem 15. *Let $\mathfrak{H}$ and $\mathfrak{K}$ be subgroups of $\mathfrak{G}$, $\mathfrak{D}$ be their intersection, and*

$$K\mathfrak{H} = \mathfrak{H}K \tag{19}$$

for every K *of* $\mathfrak{K}$. *Then*

$$\mathfrak{L} = \mathfrak{H}\mathfrak{K} \tag{20}$$

is a group, $\mathfrak{H}$ *is a normal divisor of* $\mathfrak{L}$, $\mathfrak{D}$ *is a normal divisor of* $\mathfrak{K}$, *and*

$$\mathfrak{L}/\mathfrak{H} \cong \mathfrak{K}/\mathfrak{D} \, . \tag{21}$$

For (19) implies that $\mathfrak{H}\mathfrak{K} = \mathfrak{K}\mathfrak{H}$ and therefore that

$$\mathfrak{L}\mathfrak{L} = \mathfrak{H}\mathfrak{K}\mathfrak{H}\mathfrak{K} = \mathfrak{H}\mathfrak{H}\mathfrak{K}\mathfrak{K} = \mathfrak{H}\mathfrak{K} = \mathfrak{L}$$

since $\mathfrak{H}$ and $\mathfrak{K}$ are groups. Then $\mathfrak{L}$ is a group. The cosets of $\mathfrak{H}$ in $\mathfrak{L}$ have the form $\mathfrak{H}K$ with K in $\mathfrak{K}$ and (19) states that $\mathfrak{H}$ is a normal divisor of $\mathfrak{L}$. Also $K^{-1}\mathfrak{D}K$ has all of its elements in $\mathfrak{H}$ since $\mathfrak{H}$ is a normal divisor of $\mathfrak{L}$, $K^{-1}\mathfrak{D}K \leqq \mathfrak{D}$ since $\mathfrak{D} \leqq \mathfrak{K}$, $K \leqq \mathfrak{K}$. Thus $K^{-1}\mathfrak{D}K = \mathfrak{D}$ and $\mathfrak{D}$ is a normal divisor of $\mathfrak{K}$.

Two cosets $\mathfrak{H}K_i$ and $\mathfrak{H}K_j$ of $\mathfrak{H}$ in $\mathfrak{L}$ are equal if and only if the element $K_iK_j^{-1}$ of $\mathfrak{K}$ is in $\mathfrak{H}$ and hence in $\mathfrak{D}$. But then $\mathfrak{D}K_i = \mathfrak{D}K_j$ if and only if $\mathfrak{H}K_i = \mathfrak{H}K_j$, and the correspondence

$$\mathfrak{H}K_i \longleftrightarrow \mathfrak{D}K_i$$

is a (1–1) correspondence between $\mathfrak{L}/\mathfrak{H}$ and $\mathfrak{K}/\mathfrak{D}$. It is preserved under multiplication since if $K_iK_j = K_t$ then $(\mathfrak{H}K_i)(\mathfrak{H}K_j) = \mathfrak{H}K_t \longleftrightarrow \mathfrak{D}K_t = (\mathfrak{D}K_i)(\mathfrak{D}K_j)$.

An immediate corollary of the above is given by

Theorem 16. *Let* $\mathfrak{H}$ *and* $\mathfrak{K}$ *be distinct maximal normal divisors of* $\mathfrak{G}$ *and* $\mathfrak{D}$ *be their intersection. Then*

$$\mathfrak{G}/\mathfrak{H} \cong \mathfrak{K}/\mathfrak{D} \, , \qquad \mathfrak{G}/\mathfrak{K} \cong \mathfrak{H}/\mathfrak{D} \, . \tag{22}$$

For we have (19) as well as $H\mathfrak{K} = \mathfrak{K}H$ for every H of $\mathfrak{H}$, and (22) follows from (20), (21) when we prove that $\mathfrak{L} = \mathfrak{H}\mathfrak{K} = \mathfrak{G}$. Now $\mathfrak{H} \neq \mathfrak{K}$, $\mathfrak{L} > \mathfrak{H}$, $\mathfrak{G} \geqq \mathfrak{L}$, $S^{-1}\mathfrak{L}S = S^{-1}\mathfrak{H}SS^{-1}\mathfrak{K}S = \mathfrak{H}\mathfrak{K} = \mathfrak{L}$. Thus $\mathfrak{L}$ is a normal divisor of $\mathfrak{G}$ and $\mathfrak{L} = \mathfrak{G}$ since $\mathfrak{H}$ is maximal.

A sequence of subgroups

$$\mathfrak{G} = \mathfrak{H}_0 > \mathfrak{H}_1 > \ldots > \mathfrak{H}_r = I \tag{23}$$

is called a ***composition series*** *of* $\mathfrak{G}$ *if* $\mathfrak{H}_i$ *is a maximal normal divisor of* $\mathfrak{H}_{i-1}$ *for* $i = 1, \ldots, r$. *Every normal divisor* $\mathfrak{H}$ of $\mathfrak{G}$ is a member of *some composition series* of $\mathfrak{G}$. For this is evident when $\mathfrak{H}$ is maximal since we may take $\mathfrak{H} = \mathfrak{H}_1$, $\mathfrak{H}_2$ any maximal normal divisor of $\mathfrak{H}_1$ and so forth. When $\mathfrak{H}$ is not maximal it is contained in a maximal $\mathfrak{H}_1$ and is a normal divisor of $\mathfrak{H}_1$.

If $\mathfrak{H}$ is not maximal in $\mathfrak{H}_1$ it is contained as a normal divisor of a maximal $\mathfrak{H}_2$ of $\mathfrak{H}_1$. We reduce the orders of our groups $\mathfrak{G}$, $\mathfrak{H}_1$, $\mathfrak{H}_2$ and so forth and ultimately obtain an $\mathfrak{H}_t$ such that $\mathfrak{H} = \mathfrak{H}_{t+1}$ is a maximal normal divisor of $\mathfrak{H}_t$ and is a member of the corresponding composition series of $\mathfrak{G}$.

The quotient groups

(24) $$\mathfrak{G}/\mathfrak{H}_1\,, \qquad \mathfrak{H}_1/\mathfrak{H}_2, \ldots, \qquad \mathfrak{H}_{r-1}/\mathfrak{H}_r \cong \mathfrak{H}_{r-1}$$

are all simple groups by Theorem 6.14. They are called the *prime factor groups* of $\mathfrak{G}$. If

(25) $$\mathfrak{G} > \mathfrak{K}_1 > \mathfrak{K}_2 > \ldots > \mathfrak{K}_s = I$$

is any other composition series of $\mathfrak{G}$ and

(26) $$\mathfrak{G}/\mathfrak{K}_1\,, \ldots, \mathfrak{K}_{s-1}/\mathfrak{K}_s$$

are the corresponding prime factor groups it is of course evident that these are not the same groups as (24). It is also not necessarily true that $\mathfrak{K}_{i-1}/\mathfrak{K}_i$ is equivalent to $\mathfrak{H}_{i-1}/\mathfrak{H}_i$. But it is true that $r = s$ and that the groups (26) may be rearranged so that corresponding groups in the new arrangement (24) are equivalent. We state this result in the **Jordan-Hölder**

Theorem 17. *The prime factor groups* of $\mathfrak{G}$ *are unique in the sense of equivalence.*

The above theorem is true for groups of order one and we assume it true for all groups whose order is less than that of $\mathfrak{G}$. We now let (23) and (25) be two composition series of $\mathfrak{G}$. If $\mathfrak{K}_1 = \mathfrak{H}_1$ we delete $\mathfrak{G}$ in (23) and (25) and obtain two composition series of $\mathfrak{H}_1$. By our hypothesis the corresponding quotient groups are equivalent and we obtain (24), (26) by adjoining $\mathfrak{G}/\mathfrak{H}_1 = \mathfrak{G}/\mathfrak{K}_1$ to each set and have shown that (24), (26) are equivalent. Hence let $\mathfrak{H}_1 \neq \mathfrak{K}_1$ and $\mathfrak{D}$ be their intersection. Since $\mathfrak{G}/\mathfrak{H}_1$ is simple and (22) holds, the group $\mathfrak{K}_1/\mathfrak{D}$ is simple, $\mathfrak{D}$ is a maximal normal divisor of $\mathfrak{K}_1$. Similarly $\mathfrak{D}$ is a maximal normal divisor of $\mathfrak{H}_1$ and $\mathfrak{G}$ has the two composition series

(27) $$\mathfrak{G} > \mathfrak{H}_1 > \mathfrak{D} > \mathfrak{D}_3 > \ldots > \mathfrak{D}_t = I\,,$$

(28) $$\mathfrak{G} > \mathfrak{K}_1 > \mathfrak{D} > \mathfrak{D}_3 > \ldots > \mathfrak{D}_t = I\,.$$

Also the corresponding prime factor groups are

(29) $$\mathfrak{G}/\mathfrak{H}_1 \cong \mathfrak{K}_1/\mathfrak{D}\,, \qquad \mathfrak{H}_1/\mathfrak{D} \cong \mathfrak{G}/\mathfrak{K}_1\,, \qquad \mathfrak{D}/\mathfrak{D}_3\,, \ldots, \mathfrak{D}_{t-1}/\mathfrak{D}_t\,.$$

But we have already assumed in our induction that the prime factor groups of $\mathfrak{G}$ defined by (23) and (27) are equivalent, those defined by (25) and (28) are equivalent. Then (29) states that those defined by (27) and (28) are equivalent and thus that the prime factor groups defined by any two composition series (23) and (25) are equivalent in some order. Our induction is complete.

DEFINITION. *A group $\mathfrak{G}$ is called solvable if its prime factor groups all have prime order.*

The prime factor groups of a solvable group are evidently cyclic groups C_i of prime order p_i. These groups are of interest in the theory of the *solvability of equations by radicals* and we shall prove two results for later application. The first is

Theorem 18. *Every subgroup $\mathfrak{H}$ of a solvable group $\mathfrak{G}$ is solvable and its prime factor groups are equivalent to a part of those of $\mathfrak{G}$.*

For let $\mathfrak{G} > \mathfrak{H}_1 > \ldots > \mathfrak{H}_r$ be a composition series of $\mathfrak{G}$. Since $\mathfrak{G}$ contains $\mathfrak{H}$ there is some $\mathfrak{H}_{i-1}$ which contains $\mathfrak{H}$ and is such that $\mathfrak{H}_i$ does not properly contain $\mathfrak{H}$. But $\mathfrak{H}_i$ is a normal divisor of $\mathfrak{H}_{i-1}$ and is therefore commutative with every H of $\mathfrak{H} \leqq \mathfrak{H}_{i-1}$. We apply Theorem 6.15. The product $\mathfrak{K} = \mathfrak{H}\mathfrak{H}_i$ is a subgroup of $\mathfrak{H}_{i-1}$ and our result is obvious if $\mathfrak{H}_i = \mathfrak{H}$. In the contrary case $\mathfrak{H}_{i-1} \geqq \mathfrak{K} > \mathfrak{H}_i$ and we must have $\mathfrak{H}_{i-1} = \mathfrak{K}$ since the index of $\mathfrak{H}_i$ under $\mathfrak{H}_{i-1}$ is a prime. Theorem 6.15 now states that if $\mathfrak{D}$ is the intersection of $\mathfrak{H}$ and $\mathfrak{H}_i$ we have

$$\mathfrak{H}/\mathfrak{D} \cong \mathfrak{H}_{i-1}/\mathfrak{H}_i .$$

Then $\mathfrak{D}$ is a maximal normal divisor of $\mathfrak{H}$ contained in $\mathfrak{H}_i$ since $\mathfrak{H}_{i-1}/\mathfrak{H}_i$ is simple. We apply the process on $\mathfrak{H}$ and $\mathfrak{G}$ to our new groups $\mathfrak{D}$ and $\mathfrak{H}_i$ and ultimately obtain a composition series

$$\mathfrak{D}_0 = \mathfrak{H} > \mathfrak{D} = \mathfrak{D}_1 > \mathfrak{D}_2 > \ldots > \mathfrak{D}_t = I ,$$

such that

$$\mathfrak{D}_{j-1}/\mathfrak{D}_j \cong \mathfrak{H}_{k_j-1}/\mathfrak{H}_{k_j} .$$

In each case the quotient groups have prime order and $\mathfrak{H}$ is solvable, its prime factor groups are equivalent to a part of those of $\mathfrak{G}$.

We next assume that $\mathfrak{G}$ is any group and that $\mathfrak{H}$ is a normal divisor of $\mathfrak{G}$. Since $\mathfrak{H}$ is contained in some composition series of $\mathfrak{G}$ its prime factor groups are a part of those of $\mathfrak{G}$. But we may prove

Theorem 19. *The prime factor groups of $\mathfrak{G}/\mathfrak{H}$ and those of $\mathfrak{H}$ form a set equivalent to the prime factor groups of $\mathfrak{G}$. Thus $\mathfrak{G}$ is solvable if and only if both $\mathfrak{H}$ and $\mathfrak{G}/\mathfrak{H}$ are solvable.*

For let $\mathfrak{K}_0 = \mathfrak{G} > \mathfrak{K}_1 > \mathfrak{K}_2 > \ldots > \mathfrak{K}_s = \mathfrak{H} > \mathfrak{H}_1 > \ldots > \mathfrak{H}_t = I$ be a composition series of $\mathfrak{G}$. In our corollary to Theorem 6.14 we saw that $\mathfrak{K}_{0i} = \mathfrak{K}_i/\mathfrak{H}$ is a maximal normal divisor of $\mathfrak{K}_{0\,i-1}$ if and only if $\mathfrak{K}_i$ is a maximal normal divisor of $\mathfrak{K}_{i-1}$, and in Theorem 6.13 that

$$\mathfrak{K}_{0\,i-1}/\mathfrak{K}_{0i} \cong \mathfrak{K}_{i-1}/\mathfrak{K}_i\,.$$

Thus $\mathfrak{G}/\mathfrak{H} = \mathfrak{K}_{00} > \mathfrak{K}_{01} > \ldots > I_0 = \mathfrak{H}$ is a composition series of $\mathfrak{G}/\mathfrak{H}$ with prime factor groups equivalent to the corresponding prime factor groups $\mathfrak{K}_{i-1}/\mathfrak{K}_i$ of $\mathfrak{G}$. This proves our theorem.

7. The Sylow theorems. The theory of finite groups is a large subject and we have studied only some of its theorems here. These theorems were introduced because they are essential in an understanding of our later discussion of the Galois theory of Chapter VIII. Hence we do not intend the present chapter to be at all a complete discourse on finite groups. There are certain theorems whose proofs are too long to be included here but which are too important to be omitted. The results stated in these theorems are simple and easily understandable. We shall state these *Sylow* theorems without proof.*

Sylow Theorem 1. *Let* $\mathfrak{G}$ *be a group of order* $p^r q$ *where* p *is a prime which does not divide* q. *Then there exist subgroups of* $\mathfrak{G}$ *of order* p^r *called the* ***Sylow groups*** *of* $\mathfrak{G}$ *for the prime* p. *Any two such subgroups are conjugate groups.*

Sylow Theorem 2. *A Sylow group is solvable.*

The Sylow Theorem 2 states that every group of prime power order is solvable. The prime factor groups are then cyclic groups of the same order p. The theorem also states that *if* $r \geqq s$ *there is a subgroup of order* p^s *in* $\mathfrak{G}$. For we take this subgroup from the composition series defined for the group of order p^r. The converse property is then given by the

Sylow Theorem 3. *Every subgroup of order* p^s *of* $\mathfrak{G}$ *is contained in a composition series of a Sylow group of* $\mathfrak{G}$.

8. Permutation groups. In Chapter I we discussed briefly the group $\mathfrak{G}$ of all transformations on a set $\mathfrak{M}$. We shall now study in detail the case where $\mathfrak{M}$ is a finite set.

Let $\mathfrak{M}$ have n elements designated by $x_1, \ldots, x_n$ and S be a transformation

$$S: \qquad x_j \longleftrightarrow x_{i_j} \qquad (j = 1, \ldots, n)$$

* For proofs see, e.g., A. Speiser, *Theorie der Gruppen von endlicher Ordnung* (Berlin, 1927), Chaps. V and VI. They will also of course be found in the many other texts on finite groups.

on $\mathfrak{M}$. Then the integers $i_1, \ldots, i_n$ are merely a permutation of $1, 2, \ldots, n$ and we represent S by what is called its *two-rowed representation*

$$(30) \qquad S = \begin{pmatrix} 1 & 2 & \ldots & n \\ i_1 & i_2 & \ldots & i_n \end{pmatrix}.$$

It is evident that we may permute the columns of S without altering the transformation S on $\mathfrak{M}$. If T is given by

$$(31) \qquad \begin{pmatrix} 1 & 2 & \ldots & n \\ j_1 & j_2 & \ldots & j_n \end{pmatrix}$$

then we may also write T as

$$(32) \qquad \begin{pmatrix} i_1 & i_2 & \ldots & i_n \\ k_1 & k_2 & \ldots & k_n \end{pmatrix},$$

and we saw in Chapter I that ST is the resulting transformation

$$(33) \qquad \begin{pmatrix} 1 & 2 & \ldots & n \\ k_1 & k_2 & \ldots & k_n \end{pmatrix}$$

on $\mathfrak{M}$.

The set $\mathfrak{S}_n$ of *all permutations* (30) is a group called the *symmetric group* on n letters. Every permutation group $\mathfrak{G}$ on n letters is a subgroup of $\mathfrak{S}_n$ and we call n the *degree* of $\mathfrak{G}$. In courses in elementary algebra it is shown that there are

$$(34) \qquad n! = n(n-1) \ldots 3 \cdot 2$$

distinct permutations on n letters. This shows that the order of $\mathfrak{S}_n$ is $n!$ and that the order of any $\mathfrak{G}$ of degree n is a *divisor* of $n!$.

The identity transformation of $\mathfrak{G}_n$ is the permutation

$$(35) \qquad I = \begin{pmatrix} 1 & 2 & \ldots & n \\ 1 & 2 & \ldots & n \end{pmatrix},$$

and it follows that if S is given by (30) then

$$(36) \qquad S^{-1} = \begin{pmatrix} i_1 & \ldots & i_n \\ 1 & \ldots & n \end{pmatrix}.$$

A *cycle* $(12 \ldots r)$ is a permutation

$$(37) \qquad \begin{pmatrix} 1 & 2 & \ldots & r-1 & r \\ 2 & 3 & \ldots & r & 1 \end{pmatrix}.$$

Every permutation may be expressed as a product of cycles each involving letters different from any other cycle in the product.

For this is true when $n = 1$ and we make an induction on n. If $S \neq I$ it carries x_1 to some other letter which we may call x_2. Similarly, S carries x_2 to x_3, x_3 to x_4, etc., and ultimately x_r to a letter which is one of $x_1, \ldots, x_r$ which have already been used and must be x_1, so that S has the form

$$(38) \qquad \begin{pmatrix} 1 & 2 & \ldots & r-1 & r & r+1 & \ldots & n \\ 2 & 3 & \ldots & r & 1 & i_{r+1} & \ldots & i_n \end{pmatrix} = (12 \ldots r)T$$

where T is a permutation on $x_{r+1}, \ldots, x_n$. By our assumption T is a product of cycles involving distinct letters in the set $x_{r+1}, \ldots, x_n$ and hence so is S.

The permutation $S = (123)(45)$ carries x_1 to x_2, x_2 to x_3, x_3 to x_1 and interchanges x_4 and x_5. The corresponding cyclic group $\mathfrak{G} = [S]$ thus permutes the letters x_1, x_2, x_3 among themselves and x_4 with x_5. This is an example of the

DEFINITION. *A set of elements* $x_1, \ldots, x_r$ *of* $\mathfrak{M}$ *is called a transitive system of a permutation group* $\mathfrak{G}$ *if there exists an* S_i *in* $\mathfrak{G}$ *which carries* x_1 *to* x_i *for* $i = 1 \ldots, r$ *and if no* S *of* $\mathfrak{G}$ *carries* x_1 *to any* x *not one of* $x_1, \ldots, x_r$.

If $x_1, \ldots, x_r$ is a transitive system and S_i, S_j are as above then S_i^{-1} carries x_i to x_1 so that $S_i^{-1}S_j$ carries x_i to x_j. If S carries x_i to x then S_iS in $\mathfrak{G}$ carries x_1 to x and $x = x_j$ is one of $x_1, \ldots, x_r$. Thus in a transitive system the letters comprising it are merely permuted among themselves by $\mathfrak{G}$ and any one can be carried to any other by a properly chosen S of $\mathfrak{G}$.

DEFINITION. *Let* $\mathfrak{G}$ *be a permutation group on* $x_1, \ldots, x_n$. *Then* $\mathfrak{G}$ *is called a transitive group if* $x_1, \ldots, x_n$ *are a transitive system of* $\mathfrak{G}$.

In the following we shall often replace the letters $x_1, \ldots, x_n$ of the set $\mathfrak{M}$ by their subscripts $1, 2, \ldots, n$ and speak of $1, 2, \ldots, n$ as the letters *permuted* by $\mathfrak{G}$.

9. The order of a permutation. If $C = (12 \ldots r)$ is a cycle, then $C^{-1} = (r\ r-1 \ldots 1)$. Also C carries i to $i + 1$ and C^i carries i to $i + j$ where by $i + j$ we mean its positive remainder on division by r if it exceeds r. The least j for which C^j carries 1 to itself is evidently $j = r$, and evidently i is carried to $i + r = i$ in the above sense. Thus $C^r = I$, $C^{r-1} = C^{-1}$. We now write $S = C_1 \ldots C_t$ where C_i is a cycle on r_i letters all distinct from those in C_j for $j \neq i$. Since C_i transforms a subset $\mathfrak{M}_i$ of $\mathfrak{M}$ the result C_iC_j is the same as C_jC_i and $C_iC_j = C_jC_i$, $S^r = C_1^r \ldots C_t^r$ for every r.

Then $S^r = I$ if and only if every $\mathfrak{M}_i$ is unaltered by S^r, that is $C_i^r = I$. But C_i has order r_i and this proves that the order r of S is divisible by each r_i and hence by their least common multiple r_0. However $S^{r_0} = C_1^{r_0} \dots C_t^{r_0} = I$ so that r_0 is divisible by r, $r_0 = r$. *The order of* S *is the least common multiple of the numbers of letters in its cycles* C_i.

A *transposition* is a cycle on two letters. Every cycle on r letters has the form

$$(39) \qquad (12 \dots r) = (12)(13) \dots (1r)$$

as may be easily shown and is a product of transpositions. This is therefore true of every permutation. But the expression of a permutation as a product of transpositions is by no means unique since, for example,

$$(40) \qquad (123) = (12)(13) = (13)(23) = (13)(23)(34)(34) .$$

Theorem 20. *Let* $S = T_1T_2 \dots T_r = U_1U_2 \dots U_s$, *where the* T_i *and* U_j *are transpositions. Then* r *and* s *are both even and we call* S *an* ***even*** *permutation, or both are odd and we call* S *an* ***odd*** *permutation.**

For let $\mathfrak{R}$ be the field of all rational numbers and $x_1, \dots, x_n$ be indeterminates over $\mathfrak{R}$. Then the polynomial

$$(41) \quad P = (x_1 - x_2)(x_1 - x_3) \dots (x_1 - x_n)(x_2 - x_3) \dots (x_2 - x_n) \dots (x_{n-1} - x_n)$$

is called the *alternating function*. If $1 \leqq i < j \leqq n$ we may write

$$(42) \qquad P = \pm Q \cdot (x_i - x_j) \prod_{k=1,\dots,n}^{k \neq i,\ k \neq j} (x_i - x_k)(x_j - x_k) ,$$

where Q does not involve x_i or x_j. The transposition (ij) then replaces P by $-P$ since $x_i - x_j$ goes into $x_j - x_i$, $x_i - x_k$ and $x_j - x_k$ are merely interchanged, Q is unaltered. The application of S to P replaces P by

$$(43) \qquad P_S = (x_{i_1} - x_{i_2}) \dots (x_{i_{n-1}} - x_{i_n}) .$$

But S is a product of transpositions and $P_S = (-1)^r P$ since we apply $S = T_1 \dots T_r$ to P by applying $T_1, T_2, \dots, T_r$ in succession and obtain $(-1)^r P$. But also $P_S = (-1)^s P$ since we may apply $U_1, \dots, U_s$. Then $(-1)^s P = (-1)^r P$, $(-1)^s = (-1)^r$ as desired.

If $S = (a_1b_1)(a_2b_2) \dots (a_kb_k)$ is an expression of S as a product of k transpositions then $S(a_kb_k)(a_{k-1}b_{k-1}) \dots (a_2b_2)(a_1b_1) = I$. Thus S^{-1} is also

* This is the fact mentioned in the definition of a determinant in Chap. III.

a product of k transpositions and is even or odd according as S is even or odd. This result also follows from the fact that $I = (ab)(ab)$ is an even permutation and $SS^{-1} = I$ so that S and S^{-1} are both even or both odd. We obtain a more general theorem in

Theorem 21. *The set $\mathfrak{H}$ of all even permutations of a permutation group $\mathfrak{G}$ is either the whole group $\mathfrak{G}$ or is a normal divisor of index two of $\mathfrak{G}$.*

For if S_1 and S_2 are in $\mathfrak{H}$ so is S_1S_2 and since $\mathfrak{G}$ is finite this is sufficient to imply that $\mathfrak{H}$ is a group. When $\mathfrak{H}$ is a proper subgroup of $\mathfrak{G}$ there is an odd permutation T in $\mathfrak{G}$. Then T^{-1} is odd and every S of $\mathfrak{G}$ is either in $\mathfrak{H}$ or is odd and equal to $(ST^{-1})T$ where ST^{-1} is in $\mathfrak{H}$. This shows that $\mathfrak{G} = \mathfrak{H} + \mathfrak{H}T$ where $\mathfrak{H}T$ consists of all odd permutations of $\mathfrak{G}$. But the permutations of $T\mathfrak{H}$ are all odd and are the same in number as those of $\mathfrak{H}T$. Hence $\mathfrak{H}T = T\mathfrak{H}$ and $\mathfrak{H}$ is a normal divisor of $\mathfrak{G}$.

The symmetric group $\mathfrak{S}_n$ on n letters contains (12) and Theorem 6.21 states that $\mathfrak{S}_n$ contains a normal divisor $\mathfrak{A}_n$ of *all even permutations* on n letters. This group has order $\frac{1}{2}n!$ and is called *the alternating group.* Its elements are even permutations but we shall require the form given by the

LEMMA. *Every even permutation* S *can be expressed as a product of cycles on three letters.*

For $(ab)(cd) = (abd)(acd)$ and $(ab)(ac) = (abc)$. Thus if two adjacent transpositions of S have no element in common we write their product as a product of two cycles each on three letters and otherwise as a single cycle. This replaces each pair of transpositions by one or two cycles on three letters and since S is even, the lemma is proved.

10. A composition series of $\mathfrak{S}_n$. We have seen that every permutation $S = C_1 \ldots C_t$ where the C_i are cycles no two of which have a letter in common. Then

$$T^{-1}ST = (T^{-1}C_1T) \ldots (T^{-1}C_tT) \tag{44}$$

for every T of $\mathfrak{G}$. Write $C = (a_1 \ldots a_r)$ and

$$T = \begin{pmatrix} a_1 & \ldots & a_n \\ b_1 & \ldots & b_n \end{pmatrix}, \qquad T^{-1} = \begin{pmatrix} b_1 & \ldots & b_n \\ a_1 & \ldots & a_n \end{pmatrix}. \tag{45}$$

Then

$$T^{-1}C = \begin{pmatrix} b_1 & b_2 & \ldots & b_r & b_{r+1} & \ldots & b_n \\ a_2 & a_3 & \ldots & a_1 & a_{r+1} & \ldots & a_n \end{pmatrix} \tag{46}$$

and

$$T^{-1}CT = \begin{pmatrix} b_1 & b_2 & \ldots & b_r & b_{r+1} & \ldots & b_n \\ b_2 & b_3 & \ldots & b_1 & b_{r+1} & \ldots & b_n \end{pmatrix} = (b_1 \ldots b_r). \tag{47}$$

This shows that *we obtain* $T^{-1}ST$ *by applying the permutation* T *to the letters of the cycles* C_i *of* S. For example

$$(123)^{-1}(12)(34)(123) = (23)(14)\,.$$

Another illustrative example is the general one where we write

$$S = C_1 \dots C_t\,, \quad C_1 = (a_1 \dots a_r) \tag{48}$$

with $r \geqq 3$. Then if

$$T = (a_{r-2}a_{r-1}a_r) \tag{49}$$

we have

$$T^{-1}ST = U = D_1C_2 \dots C_t \tag{50}$$

where

$$D_1 = (a_1 \dots a_{r-3}a_{r-1}a_ra_{r-2}) \tag{51}$$

is obtained from C_1 by permuting its last three letters cyclically. When $r > 3$ we have

$$S(T^{-1}ST)^{-1} = SU^{-1} = C_1D_1^{-1} = (a_{r-3}a_ra_{r-2})\,. \tag{52}$$

This result will be used in the proof of

Theorem 22. *The alternating group* $\mathfrak{A}_n$ *on* $n > 4$ *letters is simple.*

For let $\mathfrak{H} < \mathfrak{A}_n$ be a normal divisor of $\mathfrak{A}_n$. If a, b, c, d, e are any five letters permuted by $\mathfrak{A}_n$ and $\mathfrak{H}$ contains $S = (\alpha\beta\gamma)$, then one of

$$T = T_1 = \begin{pmatrix} \alpha & \beta & \gamma & \delta & \epsilon & \dots \\ a & b & c & d & e & \dots \end{pmatrix},$$
$$T = (\delta\epsilon)T_1 = \begin{pmatrix} \alpha & \beta & \gamma & \epsilon & \delta & \dots \\ a & b & c & d & e & \dots \end{pmatrix}$$

is in $\mathfrak{A}_n$ and $T^{-1}ST = (abc)$ is in the normal divisor $\mathfrak{H}$ of $\mathfrak{A}_n$. But then $\mathfrak{H}$ contains every (abc) and, by our lemma above, $\mathfrak{H}$ contains $\mathfrak{A}_n$. This contradicts $\mathfrak{H} < \mathfrak{A}_n$.

Every S of $\mathfrak{H}$ has the form (48) and T of (49) is in $\mathfrak{A}_n$, $T^{-1}ST = U$ is in $\mathfrak{H}$, SU^{-1} is in $\mathfrak{H}$. Hence in (48) r cannot be > 3 for C_1, since we have just shown that $\mathfrak{H}$ cannot contain SU^{-1} of (52). The cycles C_i of (48) are thus cycles on three or two letters. If any one of them has two letters the order of $S \neq I$ is then two or six and in the latter case S^2 has order three.

In the former case S has order two and is a product with each C_i a transposition. If $t = 2$ in equation (48) we have $S = (ab)(cd)$ and let e be a fifth letter, $T = (abe)$, $T^{-1}ST = (be)(cd)$ is in $\mathfrak{H}$, $ST^{-1}ST = (bae)$ is in $\mathfrak{H}$ contrary to proof. Hence t is at least four and

$$S = (a_1b_1)(a_2b_2)(a_3b_3)(a_4b_4)C_5 \dots C_t ,$$

where the eight letters $a_1, \dots, b_4$ are all distinct. We take $T = (b_2a_3)(b_3a_4)$ in $\mathfrak{A}_n$ and have

$$T^{-1}ST = (a_1b_1)(a_2a_3)(b_2a_4)(b_3b_4)C_5 \dots C_t$$

in $\mathfrak{H}$ and $ST^{-1}ST = (a_2a_4b_3)(a_3b_4b_2)$ of order three in $\mathfrak{H}$. Thus in every case either $\mathfrak{H}$ is the identity group or contains an S of order three. This S has the form (48) with $t > 1$ and each C_i a cycle on three letters. Write $S = (a_1a_2a_3)(b_1b_2b_3)C_3 \dots C_t$ and take $T = (a_3b_1b_2)$ in $\mathfrak{A}_n$,

$$T^{-1}ST = (a_1a_2b_1)(b_2a_3b_3)C_3 \dots C_t$$

in $\mathfrak{H}$, so that

$$ST^{-1}ST = (a_1b_1a_3a_2b_3)C_3^2 \dots C_t^2$$

is in $\mathfrak{H}$. But we have shown this impossible and $\mathfrak{H} = (I)$, $\mathfrak{A}_n$ is simple.

A composition series of $\mathfrak{S}_n$ is thus given by

$$\mathfrak{S}_n > \mathfrak{A}_n > (I) ,$$

with indices

$$2, \tfrac{1}{2}n! = n(n-1) \dots 5 \cdot 4 \cdot 3 \qquad (n > 4).$$

This last integer is divisible by 60 and is not a prime. This gives

Theorem 23. *The symmetric group on* n > 4 *letters is not a solvable group.*

11. Finite permutation groups. The permutation groups are not merely interesting but have an added importance due to

Theorem 24. *Every finite group is equivalent to a permutation group.*

For let $\mathfrak{G} = (S_1, \dots, S_n)$ be a group of order n. If S is in $\mathfrak{G}$ the n elements $S_1S, \dots, S_nS$ are all distinct since, as we saw in (3), $\mathfrak{G}S = \mathfrak{G}$. They are also a permutation of $S_1, \dots, S_n$ and we represent this permutation by

$$P_S = \begin{pmatrix} S_1 & \dots & S_n \\ S_1S & \dots & S_nS \end{pmatrix} .$$

The correspondence

$$S \longleftrightarrow P_S$$

is evidently a one-to-one correspondence between the elements of $\mathfrak{G}$ and the set $\mathfrak{G}_0$ of all permutations P_S. If T is in $\mathfrak{G}$ then

$$P_T = \begin{pmatrix} S_1 & \dots & S_n \\ S_1T & \dots & S_nT \end{pmatrix} = \begin{pmatrix} S_1S & \dots & S_nS \\ S_1ST & \dots & S_nST \end{pmatrix},$$

and

$$P_{ST} = \begin{pmatrix} S_1 & \dots & S_n \\ S_1ST & \dots & S_nST \end{pmatrix} = P_SP_T .$$

This proves that $\mathfrak{G}_0$ is a group equivalent to $\mathfrak{G}$.

There are generally many other permutation groups equivalent to $\mathfrak{G}$ but we shall not pursue this study further.

EXERCISES

1. Find all permutation groups on two and three letters. Find their composition series.

2. Let $a = (12)$, $b = (13)(24)$ be permutations on four letters. Find the group generated by a and b. We call this group $\mathfrak{G}_8$.

3. Find the two groups $\mathfrak{H}_8$ and $\mathfrak{K}_8$ obtained by the respective interchanges of 2 and 3, 2 and 4, in $\mathfrak{G}_8$.

4. Show that the intersection of $\mathfrak{G}_8$, $\mathfrak{H}_8$, $\mathfrak{K}_8$ is a non-cyclic abelian group $\mathfrak{G}_4$ of order four.

5. Find the composition series of $\mathfrak{G}_8$ and of the alternating group on four letters.

6. What are all groups of order four on four letters?

7. Which of the groups of Ex. 6 are transitive?

CHAPTER VII

FIELDS OVER $\mathfrak{F}$

1. Symmetric functions. The most interesting and probably the most important branch of the theory of fields is the theory of algebraic fields $\mathfrak{K}$ over $\mathfrak{F}$. Here $\mathfrak{F}$ is an arbitrary field, $\mathfrak{K}$ is obtained from $\mathfrak{F}$ by adjoining roots ξ_i of equations with coefficients in $\mathfrak{F}$. Some of the properties of $\mathfrak{K}$ over $\mathfrak{F}$ are identically true, that is, true when the ξ_i are replaced by independent indeterminates x_i over $\mathfrak{F}$, and they are best proved by considering this latter case. We shall first study these properties and shall formulate our results as follows:

Let $\mathfrak{J}$ be an integral domain and $x_1, \ldots, x_n$ be independent indeterminates over $\mathfrak{J}$. The set

$$\mathfrak{J}_n = \mathfrak{J}[x_1, \ldots, x_n] \tag{1}$$

is the integral domain of all polynomials

$$\phi = \phi(x_1, \ldots, x_n) = \sum_{j_k=1}^{t_k} a_{j_1 j_2 \ldots j_n} x_1^{j_1} x_2^{j_2} \ldots x_n^{j_n}, \tag{2}$$

with $a_{j_1 j_2 \ldots j_n}$ in $\mathfrak{J}$. It has the automorphisms

$$S: \qquad \phi \longleftrightarrow \phi^S = \phi(x_{i_1}, \ldots, x_{i_n}) = \sum_{j_k=1}^{t_k} a_{j_1 \ldots j_n} x_{i_1}^{j_1} \ldots x_{i_n}^{j_n},$$

where S is determined by a permutation $i_1, \ldots, i_n$ of $1, \ldots, n$. We may designate this permutation by S itself.

A quantity ϕ of $\mathfrak{J}_n$ is called *symmetric* if $\phi = \phi^S$ for every permutation S of $1, 2, \ldots, n$. Then $\mathfrak{J}_n$ contains the integral domain* $\mathfrak{S}_n$ of all symmetric polynomials ϕ of $\mathfrak{J}_n$. In particular $\mathfrak{S}_n$ contains

$$\begin{cases} c_1 = x_1 + x_2 + \ldots + x_n, \\ c_2 = x_1x_2 + x_1x_3 + \ldots + x_{n-1}x_n, \\ \quad \cdots\cdots\cdots\cdots \\ c_n = x_1x_2 \ldots x_n. \end{cases} \tag{3}$$

* We are, of course, not to confuse this with our notation of Chap. VI where $\mathfrak{S}_n$ was the symmetric group. The connotations are desirable in both cases. Notice that in Sec. 8.8 we use $\mathfrak{G}^{(n)}$ for the symmetric group.

We call the c_i the *elementary symmetric functions* of $x_1, \ldots, x_n$ and have

$$(4) \qquad \mathfrak{J}_n \geqq \mathfrak{S}_n \geqq \sum\nolimits_n = \mathfrak{J}[c_1, \ldots, c_n].$$

The total degree of an expression $x_1^{j_1}x_2^{j_2} \ldots x_n^{j_n}$ was defined in Chapter II to be the sum $j_1 + j_2 + \ldots + j_n$. The total degree of ϕ given by (2) is the maximum of the total degrees of its terms $x_1^{j_1} \ldots x_n^{j_n}$ with $a_{j_1 \ldots j_n} \neq 0$. This definition implies that the total degree of a product $\phi_1\phi_2 \neq 0$ is the sum of the total degrees of ϕ_1 and ϕ_2 and that the total degree of $\phi_1 + \phi_2$ is at most the maximum of the total degree of ϕ_1 and the total degree of ϕ_2. Thus c_i has the total degree i.

Every ϕ of $\mathfrak{J}_n$ has the form

$$(5) \qquad \phi = \phi_0 + \phi_1 x_n + \ldots + \phi_r x_n^r \qquad (\phi_i \text{ in } \mathfrak{J}_{n-1}).$$

In particular ϕ_0 is obtained from ϕ by replacing x_n by zero, and we may write

$$(6) \qquad \phi_0 = \phi(x_1, \ldots, x_{n-1}, 0), \qquad c_n = x_n d_n, \qquad c_i = c_{i0} + x_n d_i \quad (i = 1, \ldots, n-1).$$

The polynomials d_i are in $\mathfrak{J}_{n-1}$ and the c_{i0} are the elementary symmetric functions of $x_1, \ldots, x_{n-1}$. Since c_i has the same total degree as c_{i0} we may replace c_{i0} in any $\psi_0 = \psi(c_{10}, \ldots, c_{n-1\,0})$ of Σ_{n-1} by c_i and obtain a polynomial $\psi = \psi(c_1, \ldots, c_{n-1})$ in Σ_n with the same total degree as ψ_0. Moreover,

$$(7) \qquad \psi = \psi_0 + x_n\psi_1 \qquad (\psi_1 \text{ in } \mathfrak{J}_n).$$

If $n = 1$ then $x_1 = c_1$ and $\mathfrak{J}_1 = \mathfrak{S}_1 = \Sigma_1$. We are trying to prove that $\mathfrak{S}_n = \Sigma_n$. Make an induction on n and assume that $\mathfrak{S}_{n-1} = \Sigma_{n-1}$. If ϕ has total degree unity and is in $\mathfrak{S}_n$ it must have the form $a_0(x_1 + \ldots + x_n) + a_1 = a_0c_1 + a_1$ with a_0, a_1 in $\mathfrak{J}$. Then ϕ is in Σ_n. We may now make an induction on the total degree k of ϕ and assume that every polynomial in $\mathfrak{S}_n$ of degree at most $k - 1$ is in Σ_n. Suppose then that ϕ is given by (5), has degree k, and is in $\mathfrak{S}_n$.

The polynomial $\phi = \phi^S$ for every permutation S of $x_1, \ldots, x_n$ and in particular for every permutation of $x_1, \ldots, x_{n-1}$. These permutations replace (5) by $\phi_0^S + \phi_1^S x_n + \ldots + \phi_r^S x_n^r = \phi$ so that every ϕ_i must be in $\mathfrak{S}_{n-1}$. By our hypothesis $\phi_0 = \psi_0(c_{10}, \ldots, c_{n-1\,0}) = \psi(c_1, \ldots, c_{n-1}) - x_n\psi_1$ where ψ has the same total degree as ϕ_0. This total degree is at most k and the total degree of

$$(8) \qquad \phi - \psi = \phi_0 + x_n\pi - \phi_0 - x_n\psi_1 = x_n\tau$$

is at most k. Since ϕ and ψ are in $\mathfrak{S}_n$ so is x_{nT} and x_{nT} must be divisible by $x_1, x_2, \ldots, x_{n-1}$ and therefore by c_n. We may now write

$$\phi - \psi = \omega c_n \tag{9}$$

with ω in $\mathfrak{S}_n$. The total degree of ω is at most $k - n$ and ω is in Σ_n, $\phi = \psi + c_n\omega$ is in Σ_n. Our induction on k is complete and every ϕ of $\mathfrak{S}_n$ is in Σ_n. This completes our induction on n and we have proved

Theorem 1. *Every symmetric polynomial in the indeterminates* $x_1, \ldots, x_n$ *with coefficients in* $\mathfrak{J}$ *is a polynomial in their elementary symmetric functions* $c_1, \ldots, c_n$ *with coefficients in* $\mathfrak{J}$, *that is,*

$$\mathfrak{S}_n = \Sigma_n \,. \tag{10}$$

The expression of any polynomial of $\mathfrak{S}_n$ as a quantity of Σ_n is unique. This is a consequence of

Theorem 2. *The quantities* $c_1, \ldots, c_n$ *are independent indeterminates over* $\mathfrak{J}$.

The result is true when $n = 1$, $c_1 = x_1$. We again make an induction on n and assume that $c_{10}, \ldots, c_{n-1\,0}$ are independent indeterminates. Any polynomial ψ in $c_1, \ldots, c_{n-1}$ with coefficients not all zero has the same coefficients as $\psi_0 = \psi(c_{10}, \ldots, c_{n-1\,0})$ and is not zero since the $c_{10}, \ldots, c_{n-1\,0}$ are independent indeterminates. Then $c_1, \ldots, c_{n-1}$ are independent indeterminates over $\mathfrak{J}$. We now write

$$\phi = \phi_0 + \phi_1 c_n + \ldots + \phi_t c_n^t \tag{11}$$

where the ϕ_i are in $\mathfrak{J}[c_1, \ldots, c_{n-1}]$. Let $\phi = 0$. Replace x_n by zero and ϕ becomes $\phi_0(c_{10}, \ldots, c_{n-1\,0})$ so that $\phi_0 = 0$ by the above argument. Hence $\phi = c_n(\phi_1 + \ldots + \phi_t c_n^{t-1}) = 0$ only if $\phi_1 + \ldots + \phi_t c_n^{t-1} = 0$. The same argument shows that $\phi_1 = 0$ and ultimately that $\phi_t = 0$, ϕ has coefficients all zero. Then $c_1, \ldots, c_n$ are independent indeterminates over $\mathfrak{J}$. This completes our induction.

The following lemma is a consequence of Theorem 7.1.

LEMMA 1. *Let* $x_1, \ldots, x_n, x$ *be independent indeterminates over* $\mathfrak{J}$, a *be in* $\mathfrak{J}[x_1, \ldots, x_n]$. *Then the coefficients of the polynomial in* x *defined by*

$$g(x) = \prod_{S}(x - a^S) \,,$$

where S *ranges over all permutations of* $x_1, \ldots, x_n$, *are in* $\mathfrak{J}[c_1, \ldots, c_n]$.

For the coefficients b_i of $g(x)$ are in $\mathfrak{J}[x_1, \ldots, x_n]$. If we apply any permutation T of $x_1, \ldots, x_n$ to the coefficients of $g(x)$ we replace $x - a^S$ by

$x - \alpha^{ST}$. But ST ranges over all permutations of $x_1, \ldots, x_n$ when S does. Thus the factors of $g(x)$ are merely permuted by T and the coefficients b_i have the property $b_i = b_i^T$, b_i is in Σ_n by Theorem 7.1.

Let $\xi_1, \ldots, \xi_n$ be in a field $\mathfrak{K}$ containing $\mathfrak{J}$ and x be an indeterminate over $\mathfrak{K}$. A polynomial $\alpha(\xi_1, \ldots, \xi_n)$ of $\mathfrak{J}[\xi_1, \ldots, \xi_n]$ is said to be *formally unaltered* by a permutation S of $\xi_1, \ldots, \xi_n$ if the corresponding polynomial

$$\alpha = \alpha(x_1, \ldots, x_n)$$

in independent indeterminates $x_1, \ldots, x_n$ over $\mathfrak{J}$ is equal to α^S. We also say that $\alpha(\xi_1, \ldots, \xi_n)$ is a *symmetric polynomial* in $\xi_1, \ldots, \xi_n$ if $\alpha = \alpha^S$ for every S. With these definitions in mind we apply Theorem 7.1 and obtain immediately

LEMMA 2. *Let $\mathfrak{J}$ be an integral domain contained in a field $\mathfrak{K}$ and*

$$f(x) = (x - \xi_1) \ldots (x - \xi_n)\,, \tag{12}$$

with ξ_i in $\mathfrak{K}$. Then every symmetric polynomial in the ξ_i with coefficients in $\mathfrak{J}$ is a polynomial in the coefficients of $f(x)$ *with coefficients in $\mathfrak{J}$.*

This result implies the useful

Theorem 3. *Let $\xi_1, \ldots, \xi_n$ be in a field $\mathfrak{K}$ containing an integral domain $\mathfrak{J}$ and $c_1, \ldots, c_n$ be the coefficients of* $f(x)$ *of* (12). *Then if*

$$\alpha(x) = a_0 + a_1 x + \ldots + a_{n-1}x^{n-1} \qquad (a_i \text{ in } \mathfrak{J}),$$

the coefficients of

$$h(x) = [x - \alpha(\xi_1)][x - \alpha(\xi_2)] \ldots [x - \alpha(\xi_n)]$$

are in $\mathfrak{J}[c_1, \ldots, c_n]$.

For the permutations of $\xi_1, \ldots, \xi_n$ merely permute the factors of $h(x)$ and do not alter its coefficients. Our result then follows from Lemma 2.

We next prove

Theorem 4. *Let $\mathfrak{J}$ be an integral domain, $f_1(x), \ldots, f_t(x)$ be monic polynomials of $\mathfrak{J}[x]$, $\mathfrak{K}$ be a field in which*

$$f(x) = f_1(x)f_2(x) \ldots f_t(x)$$

factors into linear factors. Assume also that η_i in $\mathfrak{K}$ is a root of $f_i(x) = 0$ and that

$$\theta = \theta(\eta_1, \ldots, \eta_t)$$

is any polynomial in the η_i *with coefficients in* $\mathfrak{J}$. *Then* θ *is a root of an equation*

$$g(x) = 0$$

where g(x) *is a monic polynomial of* $\mathfrak{J}[x]$.

For $f(x) = (x - \xi_1) \dots (x - \xi_n) = x^n + c_1x^{n-1} + \dots + c_n$ has coefficients c_i in $\mathfrak{J}$ and $\theta = \theta(\eta_1, \dots, \eta_t) = \Theta(\xi_1, \dots, \xi_n)$ is a root of

$$g(x) = \prod_S (x - \theta^S) = 0$$

where S ranges over all permutations of $\xi_1, \dots, \xi_n$. But $g(x)$ has coefficients in $\mathfrak{J}[c_1, \dots, c_n]$ by Lemma 2 and hence in $\mathfrak{J}$.

We shall later show that the field $\mathfrak{K}$ of Theorem 4 exists and is in fact uniquely defined by $f(x)$ in the sense of equivalence. Thus the restrictive hypothesis of the existence of $\mathfrak{K}$ in Theorem 4 is not really necessary but must be made at the present stage of our theory.

A result of more general algebraic application may also be obtained. We consider an equation $f(x) = 0$ with coefficients in a field $\mathfrak{F}$ and let $\mathfrak{K}$ be a field containing $\mathfrak{F}$ and such that $f(x)$ has the factorization (12) with ξ_i in $\mathfrak{K}$. We apply Lemma 1 and obtain without further argument

Theorem 5. *Every polynomial with coefficients in* $\mathfrak{F}$ *in the roots* $\xi_1, \dots, \xi_n$ *in* $\mathfrak{K}$ *of* $f(x) = 0$, *which has coefficients in* $\mathfrak{F} \leqq \mathfrak{K}$, *is a root of an equation*

$$g(y) = y^m + d_1y^{m-1} + \dots + d_m = 0 \qquad (d_i \textit{ in } \mathfrak{F}),$$

whose other roots are all polynomials in $\xi_1, \dots, \xi_n$ *with coefficients in* $\mathfrak{F}$.

EXERCISES

1. The discriminant Δ of $f(x)$ of (12) is defined to be the product $(\xi_1 - \xi_2)^2 \cdot (\xi_1 - \xi_3)^2 \dots (\xi_1 - \xi_n)^2(\xi_2 - \xi_3)^2 \dots (\xi_2 - \xi_n)^2 \dots (\xi_{n-1} - \xi_n)^2$. Prove that Δ is a polynomial in the coefficients of (12) with rational integral coefficients. Hint: First take the ξ_i to be independent indeterminates over the integral domain of all rational integers.

2. Show that the discriminant of $f(x) = x^2 + bx + c$ is $b^2 - 4c$. Then if b and c are in a field $\mathfrak{F}$ and $\Delta = b^2 - 4c = d^2$ for d in $\mathfrak{F}$ the quadratic is reducible. But if the characteristic of $\mathfrak{F}$ is two, $\Delta = b^2$ and the criterion fails.

3. Let x_1, x_2, x_3 be independent indeterminates over the field $\mathfrak{R}$ of all rational numbers, $f(x) = (x - x_1)(x - x_2)(x - x_3) = x^3 + bx^2 + cx + d$ and verify (as in L. E. Dickson, *First Course in the Theory of Equations*, pp. 45–47) that $\Delta = 18bcd - 4b^3d + b^2c^2 - 4c^3 - 27d^2$. Prove that then Δ has the same formula for the cubic $f(x)$ of (12) with coefficients b, c, d in any field $\mathfrak{F}$. Hint: Use the argument of Section 3.3.

4. Let $\mathfrak{F}$ be a field of characteristic two. Prove that the discriminant Δ of an equation (12) with coefficients in $\mathfrak{F}$ is the square of a quantity of $\mathfrak{F}$. Thus all algebraic criteria which may be stated in terms of Δ a square or a non-square of $\mathfrak{F}$ fail in this case.

5. Let $f(x) = x^n + c_1x^{n-1} + \ldots + c_n$ have the factorization (12) and $s_k = \xi_1^k + \ldots + \xi_n^k$. Then in the elementary theory of symmetric functions it is shown that

$$s_k + c_1s_{k-1} + \ldots + c_{k-1}s_1 + kc_k = 0 \quad (k = 1, 2, \ldots, n),$$

and

$$s_k + c_1s_{k-1} + \ldots + c_ns_{k-n} = 0 \qquad (k > n).$$

Verify that these *Newton identities* hold when $\xi_1, \ldots, \xi_n$ are independent indeterminates over the integral domain $\mathfrak{J}$ of all rational integers. Show therefore that they hold for any ξ_i in a field containing an integral domain $\mathfrak{J}$.

6. Let $f(x) = x^3 + bx^2 + cx + d$. Use the identities above to compute $s_3 = -b^3 + 3bc - 3d$. Pass to the reciprocal equation and show that

$$\xi_1^{-3} + \xi_2^{-3} + \xi_3^{-3} = -(cd^{-1})^3 + 3bcd^{-2} - 3d^{-1}.$$

7. The *cyclic function* $\mu = \xi_1^2\xi_2 + \xi_2^2\xi_3 + \xi_3^2\xi_1$ satisfies the quadratic equation $u^2 - (\mu + \bar{\mu})u + \mu\bar{\mu} = 0$ where $\bar{\mu} = \xi_1^2\xi_3 + \xi_2^2\xi_1 + \xi_3^2\xi_2$. Show that the coefficients of the quadratic are

$$\mu + \bar{\mu} = -bs_2 - s_3 = cs_1 + 3d = 3d - bc,$$

$$\begin{aligned}\mu\bar{\mu} &= -ds_3 - 3d^2 + (\xi_1^3\xi_2^3 + \xi_2^3\xi_3^3 + \xi_3^3\xi_1^3) = -ds_3 - 3d^2 \\ &\quad - d^3(\xi_1^{-3} + \xi_2^{-3} + \xi_3^{-3}) = c^3 + b^3d + 3d^2 - 6bcd.\end{aligned}$$

2. Fields over $\mathfrak{F}$. Our present chapter is being devoted to the formulation of an adequate foundation for what is called the *Galois Theory.* This theory is an algebraic discussion of fields over a fundamental coefficient field $\mathfrak{F}$. The field $\mathfrak{F}$ will occur as a subfield of all the fields we shall discuss. When $\mathfrak{K}$ is a field which contains $\mathfrak{F}$ we shall call $\mathfrak{K}$ *a field over* $\mathfrak{F}$. Since $\mathfrak{F}$ will always be a field we shall use this notation without further comment.

An important general notion which will occur frequently is that given by the

DEFINITION. *The composite in a field $\mathfrak{K}$ of two of its subfields $\mathfrak{L}$, $\mathfrak{F}$ is the unique intersection*

$$(\mathfrak{L}, \mathfrak{F} : \mathfrak{K})$$

of all subfields of $\mathfrak{K}$ which contain both $\mathfrak{L}$ and $\mathfrak{F}$.

It is evident that $(\mathfrak{L}, \mathfrak{F} : \mathfrak{K})$ is the field of all the quantities of $\mathfrak{K}$ which are obtained by a finite number of the rational operations of $\mathfrak{K}$ on the quantities of $\mathfrak{L}$ and $\mathfrak{F}$. It is a field over $\mathfrak{F}$ as well as a field over $\mathfrak{L}$.

3. Simple extensions of $\mathfrak{F}$. Consider a field $\mathfrak{K}$ over $\mathfrak{F}$. Every quantity ξ of $\mathfrak{K}$ is either an indeterminate over $\mathfrak{F}$ or algebraic over $\mathfrak{F}$. In either case $\mathfrak{K}$ contains the integral domain

$$\mathfrak{F}[\xi]$$

of all polynomials in ξ with coefficients in $\mathfrak{F}$, and its quotient field $\mathfrak{F}(\xi)$. The field $\mathfrak{F}(\xi)$ is called a *simple transcendental extension* of $\mathfrak{F}$ when ξ is an indeterminate over $\mathfrak{F}$. This case has been amply treated in Sections 2.6 and 7.1. We now study the case where ξ is algebraic over $\mathfrak{F}$ and call $\mathfrak{F}(\xi)$ *a simple algebraic extension of* $\mathfrak{F}$.

If x is an indeterminate over $\mathfrak{F}$, and $\mathfrak{L}$ is the set of all non-zero polynomials $f_0(x)$ of $\mathfrak{F}[x]$ such that $f_0(\xi) = 0$, then $\mathfrak{L}$ is a non-empty set. For ξ is algebraic over $\mathfrak{F}$. The quantities of $\mathfrak{L}$ have a minimum degree n and $\mathfrak{L}$ contains

$$f(x) = x^n + a_1x^{n-1} + \ldots + a_n \qquad (a_i \text{ in } \mathfrak{F}). \tag{13}$$

We shall prove $f(x)$ unique. We call it the *minimum function* of ξ over $\mathfrak{F}$ and call $f(x) = 0$ the *minimum equation* of ξ over $\mathfrak{F}$.

Suppose that $f(x)$ were reducible so that $f = gh$ where g and h are in $\mathfrak{F}[x]$ and each has lower degree than f. Then $g(\xi) \cdot h(\xi) = 0$ in a field $\mathfrak{F}(\xi)$ and $g(\xi) = 0$ or $h(\xi) = 0$, $g(x)$ or $h(x)$ is in $\mathfrak{L}$. This contradicts our hypothesis about the degree of $f(x)$. Thus $f(x)$ is irreducible in $\mathfrak{F}$. Theorem 2.10 now states that $f(x)$ is unique and divides every $f_0(x)$ of $\mathfrak{L}$. For $f(x)$ divides every $f_0(x)$ and when f and f_0 are both irreducible monic polynomials they must be identical.

A polynomial $g(x)$ which is not divisible by $f(x)$ is prime to $f(x)$. Let $g(\xi) \neq 0$ so that $g(x)$ is prime to $f(x)$, $g(x)\,h(x) + f(x)\,k(x) = 1$, $g(\xi)\,h(\xi) = 1$ where $h(x)$ is in $\mathfrak{F}[x]$. This proves that every non-zero quantity of $\mathfrak{F}[\xi]$ has an inverse in $\mathfrak{F}[\xi]$ and that

$$\mathfrak{F}[\xi] = \mathfrak{F}(\xi)$$

is its own quotient field. We now divide $g(x)$ by $f(x)$ and obtain $g = fq + r$, $g(\xi) = r(\xi)$ where

$$r(x) = r_0 + r_1x + \ldots + r_{n-1}x^{n-1} \qquad (r_i \text{ in } \mathfrak{F}). \tag{14}$$

No $r(\xi) = 0$ unless the r_i are all zero since $f(x)$ is the minimum function of ξ. Thus $1, \xi, \ldots, \xi^{n-1}$ are linearly independent in $\mathfrak{F}$ and we have

Theorem 6. *Let $\mathfrak{F}(\xi)$ be a simple algebraic extension of $\mathfrak{F}$. Then there*

exists a unique monic irreducible polynomial f(x) *of degree* n *and coefficients in* $\mathfrak{F}$ *called the* ***minimum function over*** $\mathfrak{F}$ *of* ξ *such that*

$$(15) \qquad f(\xi) = 0, \quad \mathfrak{F}(\xi) = \mathfrak{F}[\xi] = (1, \xi, \ldots, \xi^{n-1})$$

is a linear set of order n *over* $\mathfrak{F}$. *We call* n *the* **degree** *of* $\mathfrak{F}(\xi)$ *over* $\mathfrak{F}$, *as well as the degree of* ξ *over* $\mathfrak{F}$.

The form (15) of $\mathfrak{F}(\xi)$ may be generalized and the result is stated as

Theorem 7. *Let* $\mathfrak{K} = \mathfrak{F}(\xi_1, \ldots, \xi_s)$ *be a field and* ξ_i *be algebraic of degree* n_i *over* $\mathfrak{F}$. *Then* $\mathfrak{K}$ *is the domain*

$$(16) \qquad \mathfrak{F}[\xi_1, \ldots, \xi_s]$$

of all polynomials in $\xi_1, \ldots, \xi_s$ *of degrees at most* $n_i - 1$ *in* ξ_i *and coefficients in* $\mathfrak{F}$.

The field $\mathfrak{K}$ is evidently the composite in $\mathfrak{K}$ of the simple algebraic extensions $\mathfrak{F}[\xi_i]$ over $\mathfrak{F}$. We write $\mathfrak{F}_i = \mathfrak{F}(\xi_1, \ldots, \xi_i)$ and see that $\mathfrak{F}_i = \mathfrak{F}_{i-1}(\xi_i)$ has degree at most n_i over $\mathfrak{F}_{i-1}$, $\mathfrak{F}_i = \mathfrak{F}_{i-1}[\xi_i]$. The result (16) for $s = 1$ is given by Theorem 7.6. An evident induction on i gives our desired result.

EXERCISES

1. Let $f(x)$ be an irreducible polynomial of $\mathfrak{F}[x]$ with the factorization (12) in a field $\mathfrak{K}$, and $r(x)$ be given by (14). Use Theorem 7.3 to prove that any quantity $r(\xi)$ of $\mathfrak{F}[\xi]$ is a root of an equation of degree n with coefficients in $\mathfrak{F}$.

2. Let $\mathfrak{K} = \mathfrak{F}[u, v]$ where $u^2 = a$, $v^2 = b$ and a and b are in the field $\mathfrak{F}$. Show that if the characteristic of $\mathfrak{F}$ is not two, the field $\mathfrak{K} = \mathfrak{F}[\xi]$ where $\xi = u(v + 1)$. Hint: Prove that both u and v are in $\mathfrak{F}[\xi]$.

3. Find the inverse of any non-zero element of the field $\mathfrak{K}$ of Ex. 2 using the assumption that $\mathfrak{K} = \mathfrak{F}[\xi]$ is a field and that its degree is 4 over $\mathfrak{F}$. Also find the minimum function of ξ over $\mathfrak{F}$.

4. Algebraic fields over $\mathfrak{F}$. A field $\mathfrak{K}$ over $\mathfrak{F}$ is said to be *algebraic over* $\mathfrak{F}$ if every quantity of $\mathfrak{K}$ is algebraic over $\mathfrak{F}$. The field (16) is one of this type.

Let $\mathfrak{K}$ be a linear set

$$(u_1, \ldots, u_n)$$

of order n over $\mathfrak{F}$. If $\mathfrak{K}$ is a field we say that $\mathfrak{K}$ has *degree n* over $\mathfrak{F}$. Theorem 2.22 states that $1, \xi, \ldots, \xi^n$ are linearly dependent in $\mathfrak{F}$ for every ξ of $\mathfrak{K}$. Thus every ξ of $\mathfrak{K}$ is algebraic of degree at most n over $\mathfrak{F}$. We shall say that $\mathfrak{K}$ is *algebraic of degree n* over $\mathfrak{F}$. Notice that $\mathfrak{K} = \mathfrak{F}[u_1, \ldots, u_n]$.

Theorem 8. *Let $\mathfrak{K} > \mathfrak{L} > \mathfrak{F}$ where $\mathfrak{K}$, $\mathfrak{L}$, $\mathfrak{F}$ are fields and $\mathfrak{K}$ is algebraic of degree* n *over $\mathfrak{F}$. Then $\mathfrak{L}$ is algebraic of degree* m *over $\mathfrak{F}$ and*

$$n = mq ,$$

$\mathfrak{K}$ is algebraic of degree q *over $\mathfrak{L}$.*

For $\mathfrak{L}$ is a proper subset of $\mathfrak{K}$ and Theorem 2.22 states that $\mathfrak{L}$ has finite order m over $\mathfrak{F}$. By Theorem 2.23

$$\mathfrak{L} = (v_1, \ldots, v_m)$$

over $\mathfrak{F}$, and $\mathfrak{L}$ is algebraic of degree m over $\mathfrak{F}$. The quantities $u_1, \ldots, u_n$ which form a basis of $\mathfrak{K}$ over $\mathfrak{F}$ need not be linearly independent over $\mathfrak{L}$ but Theorem 2.21 states that we may so order them that

$$\mathfrak{K} = (u_1, \ldots, u_q)$$

over $\mathfrak{L}$. It remains to prove that $qm = n$.

Every quantity of $\mathfrak{K}$ has the form

$$k = \sum_{i=1}^{q} a_i u_i \tag{17}$$

with unique a_i in $\mathfrak{L}$. Then

$$a_i = \sum_{j=1}^{m} a_{ij} v_j \qquad (a_{ij} \text{ in } \mathfrak{F}) \tag{18}$$

uniquely, and

$$k = \sum_{\substack{i=1,\ldots,q \\ j=1,\ldots,m}} a_{ij} u_i v_j \, . \tag{19}$$

If the mq quantities $u_i v_j$ are linearly dependent in $\mathfrak{F}$ there exist a_{ij} not all zero such that $k = 0$. But when the a_{ij} are not all zero the corresponding a_i of (18) are linear combinations of a basis of $\mathfrak{L}$ over $\mathfrak{F}$ with coefficients not all zero and cannot be all zero. By (17) and the linear independence of the u_i in $\mathfrak{L}$ we have $k \neq 0$, a contradiction. This proves that $\mathfrak{K} = (u_1 v_1, \ldots, u_q v_m)$ over $\mathfrak{F}$ and that $n = mq$.

If $\mathfrak{K}$ is algebraic of degree n over $\mathfrak{F}$, and $\mathfrak{K} \geqq \mathfrak{L}_1 \geqq \mathfrak{F}$ and $\mathfrak{K} \geqq \mathfrak{L}_2 \geqq \mathfrak{F}$, we have seen that $\mathfrak{L}_1 = (v_1, \ldots, v_m)$ over $\mathfrak{F}$, $\mathfrak{L}_2 = (w_1, \ldots, w_t)$ over $\mathfrak{F}$. The field

$$\mathfrak{L} = \mathfrak{F}(v_1, \ldots, v_m, w_1, \ldots, w_t) \tag{20}$$

is the composite in $\mathfrak{K}$ of $\mathfrak{L}_1$ and $\mathfrak{L}_2$ and Theorem 7.7 states that $\mathfrak{L}$ consists of all polynomials in $v_1, \ldots, v_m, w_1, \ldots, w_t$ with coefficients in $\mathfrak{F}$. Every such polynomial is a sum of products of quantities in $\mathfrak{L}_1$ by quantities in $\mathfrak{L}_2$ and is a linear combination of

$$v_1w_1, \ldots, v_mw_t\,. \tag{21}$$

This proves

Theorem 9. *Let $\mathfrak{K}$ be algebraic of degree* n *over $\mathfrak{F}$, and $\mathfrak{K}$ contain $\mathfrak{L}_1$ of degree* m_1 *over $\mathfrak{F}$, $\mathfrak{L}_2$ of degree* m_2 *over $\mathfrak{F}$. Then the composite in $\mathfrak{K}$ of $\mathfrak{L}_1$ and $\mathfrak{L}_2$ is a field $\mathfrak{L}$ of degree* $m \leqq m_1m_2$ *over $\mathfrak{F}$.*

By Theorem 7.8 we see that m_1 and m_2 both divide m. Then m is divisible by the least common multiple of m_1 and m_2.

Definition. *Let the composite $\mathfrak{L}$ in $\mathfrak{K}$ of $\mathfrak{L}_1$ and $\mathfrak{L}_2$ have degree* m_1m_2, *the product of the degrees of $\mathfrak{L}_1$ and $\mathfrak{L}_2$. Then we call $\mathfrak{L}$ the direct product of $\mathfrak{L}_1$ and $\mathfrak{L}_2$ and write*

$$\mathfrak{L} = \mathfrak{L}_1 \times \mathfrak{L}_2\,.$$

Corollary. *Let* m_1 *and* m_2 *be relatively prime in Theorem 9. Then*

$$\mathfrak{L} = \mathfrak{L}_1 \times \mathfrak{L}_2\,.$$

For the least common multiple of m_1 and m_2 is m_1m_2 and $m_1m_2 \geqq m$, m is divisible by m_1m_2, $m = m_1m_2$.

EXERCISES

1. Show that every quadratic field (field of degree two) over $\mathfrak{F}$ of characteristic not two is a linear set $(1, \xi)$ over $\mathfrak{F}$ with $\xi^2 = a$ in $\mathfrak{F}$. Prove also that if $\mathfrak{F}$ has characteristic two we have the additional fields $(1, \eta)$ over $\mathfrak{F}$, $\eta^2 = \eta + b$, b in $\mathfrak{F}$ and no one of these is equivalent to the former $\mathfrak{F}[\xi]$, $\xi^2 = a$. Hint: If $y^2 = ay + b$ and $a \neq 0$ then $(a^{-1}y)^2 = (a^{-1}y) + a^{-2}b$.

2. Let the characteristic of $\mathfrak{F}$ be not two. Prove that a field $\mathfrak{K}$ of degree four over $\mathfrak{F}$ has a quadratic subfield if and only if $\mathfrak{K} = \mathfrak{F}[\xi]$, $\xi^4 + a\xi^2 + b = 0$ with a, b in $\mathfrak{F}$ and $f(x) = x^4 + ax^2 + b$ an irreducible polynomial of $\mathfrak{F}[x]$.

3. Let $f(x) = x^4 + ax^2 + b^2$ have rational coefficients and be an irreducible polynomial of the domain $\mathfrak{R}[x]$, $\mathfrak{R}$ the field of all rational numbers. If ξ is a quantity of a field $\mathfrak{K}$ such that the fourth degree polynomial $f(\xi) = 0$ we call $\mathfrak{F}[\xi]$ a quartic field

over $\mathfrak{F}$. Prove that in our particular case $\mathfrak{F}[\xi]$ contains two quadratic fields and is the direct product of these fields. Hint: Show that ξ, $-\xi$, $b\xi^{-1}$, $-b\xi^{-1}$ are the roots of $f(x) = 0$ and that the two quadratic fields are $\mathfrak{F}[\xi^2]$, $\mathfrak{F}[\xi + b\xi^{-1}]$.

4. Let $\mathfrak{K}$ be a quartic field over $\mathfrak{F}$. Show that if $\mathfrak{K}$ has two distinct quadratic subfields it is the direct product of these subfields.

5. Let $x^2 = x + a$ be irreducible in a field $\mathfrak{F}$ which contains a. Prove that the only quantities of the quadratic field $\mathfrak{F}[u]$ defined by $u^2 = u + a$ which satisfy irreducible equations $y^2 = y + b$ are of the form $c + (1 - 2c)u$ where c ranges over all quantities of $\mathfrak{F}$ such that $2c \neq 1$.

6. Let $x^2 = a$ be irreducible in $\mathfrak{F}$ of characteristic not 2 and $\mathfrak{F}[u]$ be the quadratic field defined by $u^2 = a$. Prove that w in $\mathfrak{F}[u]$ satisfies $w^2 = b$ in $\mathfrak{F}$ if and only if $w = cu$ or $w = c$, c in $\mathfrak{F}$.

7. Let $\mathfrak{K}$ be a quartic field over $\mathfrak{F}$ of characteristic two, and $\mathfrak{K}$ be the direct product of $\mathfrak{F}[u]$, $\mathfrak{F}[v]$, $u^2 = u + a$, $v^2 = v + b$, a and b in $\mathfrak{F}$. Prove that the only quadratic subfields of $\mathfrak{K}$ are the above two fields and the field $\mathfrak{F}[u + v]$. Hint: The field $\mathfrak{K} = (1, v)$ over $\mathfrak{F}[u]$ and we may use Ex. 5 with $2c = 0$.

8. Let $\mathfrak{K}$ be a quartic field over a field $\mathfrak{F}$ of characteristic not two and such that $\mathfrak{K}$ is the direct product of two distinct quadratic subfields $\mathfrak{F}[u]$, $\mathfrak{F}[v]$. Show that we may take $u^2 = a$ in $\mathfrak{F}$, $v^2 = b$ in $\mathfrak{F}$, and that then the only other quadratic subfield of $\mathfrak{K}$ is $\mathfrak{F}[uv]$. (Assume $\mathfrak{K}$ infinite both here and in Ex. 9.)

9. Use the equations of Ex. 8 defining $\mathfrak{K}$ and assume that the characteristic of $\mathfrak{F}$ is two. Prove that in this case $\mathfrak{K}$ has infinitely many distinct quadratic subfields.

10. Let $\mathfrak{K}$ be a quartic field which is not of the types of Exercises 7 and 8 but is still a direct product of two quadratic fields $\mathfrak{F}[u]$, $\mathfrak{F}[v]$ over $\mathfrak{F}$. Prove that $\mathfrak{F}[u]$ and $\mathfrak{F}[v]$ are the only quadratic subfields of $\mathfrak{K}$. Hence show that a quartic field over $\mathfrak{F}$ has exactly three, one, or no quadratic subfields when the characteristic of $\mathfrak{F}$ is not two, but in addition can have two or infinitely many quadratic subfields over $\mathfrak{F}$ of characteristic two.

5. The root field of an equation. The so-called Fundamental Theorem of Algebra states that *every equation with coefficients in the field $\mathfrak{C}$ of all complex numbers has a root in $\mathfrak{C}$.* The known proofs of this theorem are of a complicated nature and the theorem itself is not an adequate one. For example, if ξ is an indeterminate over $\mathfrak{C}$ and $f(x) = 0$ is an equation with coefficients in $\mathfrak{C}(\xi)$ we may wish to find a root of $f(x) = 0$ and cannot use the Fundamental Theorem. We shall not prove the Fundamental Theorem but shall prove instead a more useful and *much simpler* algebraic theorem on the existence of the root fields of equations.

Theorem 10. *Let*

$$f(x) = x^n + a_1x^{n-1} + \dots + a_n \qquad (a_i \textit{ in } \mathfrak{F}). \tag{22}$$

Then there exists a field $\mathfrak{N} = \mathfrak{F}[\xi_1, \ldots, \xi_n]$ *called a* ***root field*** *of the polynomial* f(x) (*and the equation* f(x) = 0) *such that*

(23) $$f(x) = (x - \xi_1) \ldots (x - \xi_n) .$$

The quantities $\xi_1, \ldots, \xi_n$ are in a field $\mathfrak{N}$ and $f(\xi_i) = 0$. Thus the above theorem states that the ξ_i are what we call algebraic quantities over $\mathfrak{F}$. They generate subfields of $\mathfrak{N}$ called the *stem fields* $\mathfrak{F}[\xi_i]$ of $f(x)$, and these fields are simple algebraic extensions of $\mathfrak{F}$. We also call $\mathfrak{N}$ the root field of $\mathfrak{F}[\xi_i]$.

We now prove Theorem 7.10. Consider an irreducible factor $f_1(x)$ of $f(x)$. We then call any two polynomials g and h of $\mathfrak{J} = \mathfrak{F}[x]$ *equivalent modulo* f_1 if $g - h$ is divisible by f_1. The relation so defined is an equivalence relation and classifies the polynomials of $\mathfrak{J}$ into classes $\{g\}$ of equivalent polynomials. Each class $\{g\}$ has a representative given by (14) where $r(x)$ is the unique remainder on division of g by f_1. Then $\{g_1\} = \{g_2\}$ if and only if $r_1 = r_2$. Define

(24) $$\{g\} + \{h\} = \{g + h\} , \quad \{g\}\{h\} = \{gh\} .$$

It can be easily shown that these definitions are independent of the particular representatives g and h of the classes $\{g\}$, $\{h\}$ and uniquely define the sum and product of classes. The set $\mathfrak{A}$ of all classes $\{g\}$ of polynomials in $\mathfrak{J}$ is a ring. We leave the elementary verification of these properties to the reader.

The ring $\mathfrak{A}$ contains the set $\mathfrak{F}_0$ of all classes $\{a\}$ with a in $\mathfrak{F}$ and $\mathfrak{F}_0$ is equivalent to $\mathfrak{F}$ under the correspondence

(25) $$\{a\} \longleftrightarrow a \qquad (a \text{ in } \mathfrak{F}).$$

By Theorem 1.9 we may replace $\mathfrak{A}$ by an equivalent algebra with $\mathfrak{F}$ as subfield. Our definition of multiplication then implies that

$$a\{g\} = \{ag\} \qquad (a \text{ in } \mathfrak{F}).$$

Also if

$$\{x\} = \xi ,$$

we have $\{x\}^i = \{x^i\} = \xi^i$,

(26) $$g(\xi) = \{g(x)\} = \{r(x)\} = r(\xi) ,$$

for every $g(x)$ of $\mathfrak{J}$. In particular

(27) $$f_1(\xi) = \{f_1(x)\} = 0$$

is the zero element of $\mathfrak{A}$. Any polynomial $g(x)$ for which the corresponding class is not the zero class is a polynomial not divisible by the irreducible $f_1(x)$. Then $g(x)$ is prime to $f_1(x)$ by Theorem 2.10 and $g(x)h(x) + f_1(x)b(x) = 1$, $g(\xi)h(\xi) = 1$. This proves that $\mathfrak{A}$ consists of the set of all polynomials $r(\xi)$ with coefficients in $\mathfrak{F}$ and that $\mathfrak{A} = \mathfrak{F}[\xi]$ is a field such that $f_1(\xi) = 0$. Call $\xi = \xi_1$.

Write $f(x) = f_1(x)q_1(x)$ where we have assumed that $f_1(x)$ is irreducible in $\mathfrak{F}$. Then $f(x) = (x - \xi_1)f_2(x)q_2(x)$ where we have taken $\mathfrak{K}_1 = \mathfrak{F}[\xi_1]$, $f_2(x)$ and $q_2(x)$ in $\mathfrak{K}_1[x]$, $f_2(x)$ an irreducible polynomial of $\mathfrak{K}_1[x]$. We apply the above existence proof to $f_2(x)$ and obtain a quantity ξ_2 and a field

$$\mathfrak{K}_2 = \mathfrak{K}_1[\xi_2] = \mathfrak{F}[\xi_1, \xi_2] ,$$

such that $f(x) = (x - \xi_1)(x - \xi_2)f_3(x)q_3(x)$ with f_3 and q_3 in the polynomiall domain $\mathfrak{K}_2[x]$ and f_3 irreducible. After $n - 2$ further extensions of $\mathfrak{F}$ we arrive at the desired root field $\mathfrak{N} = \mathfrak{K}_n = \mathfrak{F}[\xi_1, \ldots, \xi_n]$.

Theorem 7.10 is a statement of the existence of a root field of any equation with coefficients in a field. We now prove its essential uniqueness by first obtaining the almost trivially proved

Theorem 11. *The stem fields of an irreducible polynomial* f(x) *are all equivalent over* $\mathfrak{F}$

For if ξ_1 and ξ_2 are two roots of $f(x) = 0$ we saw in our proof of Theorem 7.10 that every quantity of $\mathfrak{F}[\xi_1]$ has the form $r(\xi_1)$ with $r(x)$ given by (14). It is evident that the correspondence

$$r(\xi_1) \longleftrightarrow r(\xi_2) \tag{28}$$

is preserved under the operations defining the field $\mathfrak{F}[\xi_1]$ and defines an equivalence between $\mathfrak{F}(\xi_1)$ and $\mathfrak{F}(\xi_2)$. For the reductions of any $g(\xi_1)$ to $r(\xi_1)$ and $g(\xi_2)$ to $r(\xi_2)$ are made by the use of the same irreducible $f(x)$.

We now apply the above and obtain

Theorem 12. *The root fields of any equation* f(x) = 0 *with coefficients in a field* $\mathfrak{F}$ *are all equivalent over* $\mathfrak{F}$.

If the degree of $f(x)$ is unity then the root fields of $f(x)$ are all equivalent to $\mathfrak{F}$ and are equivalent. We now make an induction on the degree of $f(x)$ and assume that we have proved our result for polynomials of degree $n - 1$. Suppose that $f(x) = f_1(x)q_1(x)$ has degree n and that $f_1(x)$ is irreducible, $\mathfrak{N}$ and $\mathfrak{H}$ are two root fields of $f(x)$. We let ξ in $\mathfrak{N}$ be a root of $f_1(x) = 0$ and η in $\mathfrak{H}$ be a root of $f_1(x) = 0$. Then $\mathfrak{N}_1 = \mathfrak{F}(\xi)$ is equivalent to $\mathfrak{H}_1 = \mathfrak{F}(\eta)$ and $\mathfrak{N}$ over $\mathfrak{N}_1$ is the root field of $f_2(x)$, $f(x) = (x - \xi)f_2(x)$. Similarly, $\mathfrak{H}$ over $\mathfrak{H}_1$ is the root field of $f_{20}(x)$, $f(x) = (x - \eta)f_{20}(x)$. The field $\mathfrak{H}$ over $\mathfrak{H}_1$ is equivalent to a field $\mathfrak{H}_0$ over $\mathfrak{N}_1$ by Theorem 1.9 and $\mathfrak{H}_0$

is evidently a root field of $f_2(x)$, since we replace η by ξ and f_{20} by f_2 when we replace $\mathfrak{H}_1$ by $\mathfrak{N}_1$. But $f_2(x)$ has degree $n - 1$, $\mathfrak{H}_0$ is equivalent to $\mathfrak{N}$, $\mathfrak{H}$ is then equivalent to $\mathfrak{N}$.

There is evidently no loss in generality if we now speak of *the* root field of $f(x)$. We shall use this terminology in an application of Theorem 7.5.

Theorem 13. *Let α be in the root field $\mathfrak{N}$ of* f(x). *Then α is algebraic over $\mathfrak{F}$ and the root field of its minimum function is in $\mathfrak{N}$.*

For α is a polynomial in the roots $\xi_1, \ldots, \xi_n$ of $f(x) = 0$ and Theorem 7.5 states that α is a root of an equation $g(y) = 0$ with coefficients in $\mathfrak{F}$ and all other roots in $\mathfrak{N}$.

When α lies in a stem field $\mathfrak{F}(\xi)$ of an irreducible polynomial $f(x)$ of degree n the field $\mathfrak{F}[\alpha]$ is a subfield of degree m of $\mathfrak{F}[\xi]$ of degree n over $\mathfrak{F}$. By Theorem 7.8

$$n = mq$$

for an integer q, the degree of $\mathfrak{F}[\xi]$ over $\mathfrak{F}[\alpha]$. We let $f(x) = (x - \xi_1) \ldots (x - \xi_n)$ with $\xi = \xi_1, \ldots, \xi_n$ in a root field $\mathfrak{N}$ of $f(x)$. Then $\alpha = r(\xi)$ where $r(x)$ is given by (14) and define

$$\alpha_i = r(\xi_i) \qquad (i = 1, \ldots, n). \tag{29}$$

Then we obtain a more explicit result in

Theorem 14. *Let* f(x) *be an irreducible polynomial of degree* n *of* $\mathfrak{F}$[x], $\alpha = r(\xi)$ *be in a stem field* $\mathfrak{F}[\xi]$ *of* f(x),

$$\phi(y) = (y - \alpha_1) \ldots (y - \alpha_n) \tag{30}$$

with α_i *as in* (29). *Then if the minimum function of* α *over* $\mathfrak{F}$ *is* ψ(y) *of degree* m *we have*

$$n = mq, \qquad \phi(y) = [\psi(y)]^q. \tag{31}$$

For Theorem 7.3 states that $\psi(y)$ divides $\phi(y)$. The equation $\psi[r(x)] = 0$ is satisfied by ξ and must be divisible by $f(x)$, $\psi[r(\xi_i)] = 0$, and every α_i is a root of $\psi(y) = 0$. Thus if $\phi(y) = [\psi(y)]^q\lambda(y)$ where $\lambda(y)$ is prime to $\psi(y)$ we must have $\lambda(y) = 1$ and (31) holds, $n = mq$.

EXERCISES

1. Apply the fundamental lemma of Sec. 4.4 to give an alternative proof of Theorem 7.10.

2. Let $f(x)$ be a reducible polynomial. What is a necessary and sufficient condition that the stem fields of $f(x)$ be all equivalent over $\mathfrak{F}$?

3. Apply the methods used in the proof of Theorem 7.10 to show the existence of a polynomial algebra $\mathfrak{F}[\xi]$ of order n over $\mathfrak{F}$ defined by a root ξ of (22). Prove that this algebra is not a field if $f(x)$ is a reducible polynomial. We shall henceforth use the notation $\mathfrak{F}[\xi]$ to mean that field generated by a root ξ of an irreducible factor of $f(x)$ and *not* the algebra just defined.

6. Conjugate fields. Let $\mathfrak{K}$, $\mathfrak{H}_1$, $\mathfrak{H}_2$ be fields over $\mathfrak{F}$, that is with $\mathfrak{F}$ as a subfield, and suppose that $\mathfrak{H}_1$ and $\mathfrak{H}_2$ are subfields of $\mathfrak{K}$. Then $\mathfrak{H}_1$ and $\mathfrak{H}_2$ are said to be *conjugate* if they are equivalent over $\mathfrak{F}$.

The stem fields $\mathfrak{F}[\xi_i]$ of an irreducible $f(x)$ are conjugate fields by Theorem 7.11. Conversely, *every conjugate to* $\mathfrak{F}[\xi_i]$ *in the root field* $\mathfrak{K} = \mathfrak{F}[\xi_1, \ldots, \xi_n] = \mathfrak{N}$ *of* $f(x)$ *is one of the* $\mathfrak{F}[\xi_i]$. For if $\mathfrak{H}$ is equivalent to $\mathfrak{F}[\xi_i]$ we may let $\xi \longleftrightarrow \xi_i$, $\mathfrak{H} = \mathfrak{F}[\xi]$, $f(\xi) \longleftrightarrow f(\xi_i)$, $f(\xi) = 0$. Then ξ is in $\mathfrak{K}$ and $(\xi - \xi_1) \ldots (\xi - \xi_n) = 0$, which is true in $\mathfrak{K}$ if and only if $\xi = \xi_i$ for some i.

The quantities $\alpha_i = r(\xi_i)$ of (29) defined by any $\alpha = r(\xi)$ of $\mathfrak{F}[\xi_i]$ are called the *conjugate quantities* of α. The symmetric functions of the α_i which are the coefficients of $\phi(y)$ in (30) are of course important functions of α. There are two of these functions which play an exceedingly important rôle in modern algebra. They are the sum

$$\alpha_1 + \alpha_2 + \ldots + \alpha_n = T(\alpha) , \tag{32}$$

called the *trace* of $\alpha = \alpha_1$, and the product

$$\alpha_1\alpha_2 \ldots \alpha_n = N(\alpha) , \tag{33}$$

called the *norm* of α. Both of these quantities are coefficients of (30) and are in $\mathfrak{F}$. It is easily verified that

$$\begin{cases} T(a\alpha + b\beta) = aT(\alpha) + bT(\beta) , \\ N(a\alpha) = a^nN(\alpha) , \qquad N(\alpha\beta) = N(\alpha)N(\beta) , \end{cases} \tag{34}$$

for every α and β of $\mathfrak{F}[\xi]$ and a and b of $\mathfrak{F}$. We shall obtain some further properties in Section 8.5.

EXERCISES

1. Use Theorem 7.14 to show that if $\mathfrak{F}$ is non-modular and α is in $\mathfrak{F}[\xi]$, the conjugates α_i of α fall into m distinct sets each of q equal conjugates. Show therefore that $\mathfrak{F}[\alpha] = \mathfrak{F}[\xi]$ if and only if the n conjugates of α are all distinct.

2. Let $f(x) = x^4 - (a + b)x^2 + ab$ where a and b are in $\mathfrak{F}$ of characteristic not two. Show that $f(x)$ is a reducible polynomial of $\mathfrak{F}[x]$ but that the root field of $f(x)$ is a quartic field if and only if a, b, ab are each not the square of any quantity of $\mathfrak{F}$.

Hint: Use Exercises 2 and 3 of Section 7.3, 6, 8 of Section 7.4 and show that $\mathfrak{F}[u]$ and $\mathfrak{F}[v]$ defined by $u^2 = a$, $v^2 = b$ are either equivalent or the root field is their direct product.

7. Composites. We have already defined the composite $(\mathfrak{Y}, \mathfrak{Z} : \mathfrak{K})$ of two subfields $\mathfrak{Y}$ and $\mathfrak{Z}$ of $\mathfrak{K}$. We now let $\mathfrak{Y}$ and $\mathfrak{Z}$ be two arbitrary fields over a field $\mathfrak{F}$ and make the

DEFINITION 1. *Let $\mathfrak{K}$ be a field over $\mathfrak{F}$ which has a subfield over $\mathfrak{F}$ equivalent to $\mathfrak{Y}$ and a subfield over $\mathfrak{F}$ equivalent to $\mathfrak{Z}$. Then the $\mathfrak{K}$-composites of $\mathfrak{Y}$ and $\mathfrak{Z}$ are the fields $(\mathfrak{Y}_0, \mathfrak{Z}_0 : \mathfrak{K})$ where $\mathfrak{Y}_0$ ranges over all subfields over $\mathfrak{F}$ of $\mathfrak{K}$ equivalent to $\mathfrak{Y}$ and $\mathfrak{Z}_0$ ranges over all subfields over $\mathfrak{F}$ of $\mathfrak{K}$ equivalent to $\mathfrak{Z}$.*

DEFINITION 2. *The composites of $\mathfrak{Y}$ and $\mathfrak{Z}$ are the $\mathfrak{K}$-composites of $\mathfrak{Y}$ and $\mathfrak{Z}$ where $\mathfrak{K}$ ranges over all fields over $\mathfrak{F}$ with a subfield over $\mathfrak{F}$ equivalent to $\mathfrak{Y}$ and a subfield over $\mathfrak{F}$ equivalent to $\mathfrak{Z}$.*

Composites may be inequivalent. An example is given by the case where $\mathfrak{K} > \mathfrak{Y}$, $\mathfrak{K} > \mathfrak{Z}$ and $\mathfrak{Y}$ and $\mathfrak{Z}$ are distinct equivalent subfields of $\mathfrak{K}$. Then the two $\mathfrak{K}$-composites $(\mathfrak{Y}, \mathfrak{Z} : \mathfrak{K})$ and $(\mathfrak{Y}, \mathfrak{Y} : \mathfrak{K}) = \mathfrak{Y}$ are inequivalent composites of $\mathfrak{Y}$ and $\mathfrak{Z}$. For $(\mathfrak{Y}, \mathfrak{Z} : \mathfrak{K}) > \mathfrak{Y}$.

Every composite of $\mathfrak{Y}$ and $\mathfrak{Z}$ over $\mathfrak{F}$ contains a subfield $\mathfrak{Z}_0$ equivalent to $\mathfrak{Z}$. By Theorem 1.9 we may replace $\mathfrak{Z}_0$ by $\mathfrak{Z}$ and obtain an equivalent composite.

Theorem 15. *Every composite of $\mathfrak{Y}$* and *$\mathfrak{Z}$ over $\mathfrak{F}$ is equivalent to a composite*

$$(\mathfrak{Y}, \mathfrak{Z})$$

over $\mathfrak{Z}$.

We evidently lose no generality in restricting our attention to such composites $(\mathfrak{Y}, \mathfrak{Z})$ and shall do so. Notice also that if $\mathfrak{Y}_1$ and $\mathfrak{Y}$ are equivalent, $\mathfrak{Z}_1$ and $\mathfrak{Z}$ are equivalent, then the composites of $\mathfrak{Y}_1$ and $\mathfrak{Z}_1$ are exactly the same fields as the composites of $\mathfrak{Y}$ and $\mathfrak{Z}$.

Our principal interest is in the composites over $\mathfrak{F}$ of an arbitrary $\mathfrak{Z}$ containing $\mathfrak{F}$ and a field $\mathfrak{Y} = \mathfrak{F}(\xi)$ of finite degree over $\mathfrak{F}$. Then the composites $(\mathfrak{Y}, \mathfrak{Z})$ over $\mathfrak{F}$ are fields over $\mathfrak{Z}$ and over $\mathfrak{F}$. The minimum function $f(x)$ of ξ over $\mathfrak{F}$ has coefficients in $\mathfrak{Z}$ and there exists a root field $\mathfrak{W}$ over $\mathfrak{Z}$ of $f(x)$. If $\mathfrak{K}$ is any field containing $\mathfrak{Z}$ and stem fields $\mathfrak{Y}_i = \mathfrak{F}(\xi_i)$ of $f(x)$ then their $\mathfrak{K}$-composite is contained in a root field $\mathfrak{W}$ over $\mathfrak{Z}$ of $f(x)$ and the $\mathfrak{K}$-composites of $\mathfrak{Y}$ and $\mathfrak{Z}$ are the fields $\mathfrak{Z}(\xi_i)$. Since all root fields $\mathfrak{W}$ over $\mathfrak{Z}$ of $f(x)$ are equivalent we have

Theorem 16. *The composites $(\mathfrak{Y}, \mathfrak{Z})$ over $\mathfrak{F}$ of $\mathfrak{Y} = \mathfrak{F}(\xi)$ and $\mathfrak{Z} \geqq \mathfrak{F}$ are the stem fields $\mathfrak{Z}(\xi_i)$ of the minimum function over $\mathfrak{F}$ of ξ.*

The above result states that the composites over $\mathfrak{F}$ of $\mathfrak{Y}$ and $\mathfrak{Z}$ always exist in the case we are now considering. It also implies that when we study the irreducible $f(x)$ of $\mathfrak{F}[x]$ we find a root field $\mathfrak{Z}(\xi_1, \ldots, \xi_n)$ of $f(x)$, we take $\mathfrak{Y} = \mathfrak{F}(\xi_1)$ with no loss of generality, and that *the composites over* $\mathfrak{F}$ *of* $\mathfrak{Y}$ *and* $\mathfrak{Z}$ *are not only the field* $\mathfrak{Z}(\xi_1)$ *but also all the fields* $\mathfrak{Z}(\xi_i)$. When $f(x)$ is irreducible in $\mathfrak{Z}$ the fields $\mathfrak{Z}(\xi_i)$ are all equivalent. This is evidently the case in the corollary of Theorem 7.9. For in that case $\mathfrak{Z} = \mathfrak{L}_2$, the degree of $\mathfrak{L}_1 \times \mathfrak{L}_2$ over $\mathfrak{L}_2$ is m_1 and $f(x)$ must be irreducible in $\mathfrak{L}_2$.

The fields $\mathfrak{Z}(\xi_i)$ are also equivalent if $f(x)$ factors in $\mathfrak{Z}$ into linear factors. For they are then all equivalent to $\mathfrak{Z}$. But in general

$$f(x) = f_1(x) \ldots f_r(x) \tag{35}$$

where the $f_j(x)$ are irreducible polynomials of $\mathfrak{Z}[x]$. Each $f_i(x)$ defines a field $\mathfrak{Z}[\xi_j]$ and every composite of $\mathfrak{Y}$ and $\mathfrak{Z}$ is equivalent to a field $\mathfrak{Z}[\xi_j]$. But these r fields need not even have the same degrees over $\mathfrak{Z}$.

EXERCISES

1. Let $f(x) = x^3 - 2$, $\mathfrak{F}$ be the field of all rational numbers, $\mathfrak{Z} = \mathfrak{F}(\eta)$, $\mathfrak{Y} = \mathfrak{F}(\xi)$ where $\xi^3 = 2$ and η is a real cube root of two. Prove that the composites over $\mathfrak{Z}$ of $\mathfrak{Z}$ and $\mathfrak{Y}$ are either $\mathfrak{Z}$ or $\mathfrak{Z}(\sqrt{-3})$ where the latter field has degree two over $\mathfrak{F}$. This proves the last statement above the present exercises.

2. Let $\mathfrak{F}$ be a field containing $a \neq 0$ and $\mathfrak{A}$ be the algebra $(1, \xi)$ of order two over $\mathfrak{F}$ defined by the matrices

$$1 = \begin{pmatrix} 1 & 0 \\ 0 & 1 \end{pmatrix}, \qquad \xi = \begin{pmatrix} 0 & a^2 \\ 1 & 0 \end{pmatrix}.$$

Show that if the characteristic of $\mathfrak{F}$ is not two then $\mathfrak{A} = (e_1, e_2)$ over $\mathfrak{F}$ where $e_1^2 = e_1$, $e_2^2 = e_2$, $e_1e_2 = e_2e_1 = 0$, but that if the characteristic of $\mathfrak{F}$ is two then $\mathfrak{A} = (1, \eta)$, $\eta^2 = 0$. We call (η) a *zero algebra* of order one.

3. Prove that the composites of the quadratic fields of Exercise 2 of Section 7.6 are all equivalent to $\mathfrak{F}[u]$ and $\mathfrak{F}[v]$ or all equivalent to their direct product.

4. Let $\mathfrak{K}$ be the composite over $\mathfrak{F}$ of a field $\mathfrak{Z}$ over $\mathfrak{F}$ and a quadratic field $\mathfrak{Q} = \mathfrak{F}[u]$ over $\mathfrak{F}$. Take the field $\mathfrak{K}$ as a field $\mathfrak{Z}[u]$ and prove that $\mathfrak{K}$ has degree one over $\mathfrak{Z}$ if and only if $\mathfrak{Z}$ has a subfield over $\mathfrak{F}$ equivalent to $\mathfrak{Q}$ over $\mathfrak{F}$.

5. Use Ex. 4 to prove that if $\mathfrak{Z}$ has degree n over $\mathfrak{F}$ and if $\mathfrak{Q}$ has degree two over $\mathfrak{F}$ then the composites of $\mathfrak{Z}$ and $\mathfrak{Q}$ are all equivalent to their direct product if and only if $\mathfrak{Z}$ has no quadratic subfield over $\mathfrak{F}$ equivalent to $\mathfrak{Q}$ over $\mathfrak{F}$. Apply this to the case n odd.

8. Fields of characteristic p. There are certain equations with coefficients in a field $\mathfrak{F}$ of finite characteristic p which have rather peculiar properties. We shall discuss these equations in this and the following sections of this chapter.

The properties are, of course, due to the fact that p is a prime such that

$$p \cdot a = a + a + \ldots + a = 0$$

for every a in $\mathfrak{F}$. Then

$$(a + b)^p = a^p + b^p$$

for every a and b of $\mathfrak{F}$. If $p = 2$ then $-1 = 1$ and $(a - b)^2 = (a + b)^2 = a^2 + b^2 = a^2 - b^2$. Otherwise p is odd and $(a - b)^p = [a + (-b)]^p = a^p + (-b)^p = a^p - b^p$. Hence

$$(a \pm b)^p = a^p \pm b^p \tag{36}$$

for every a and b of a field $\mathfrak{F}$ of characteristic p. We now obtain

Theorem 17. *Let $\mathfrak{F}$ be a field of characteristic* p *and*

$$\mathrm{f(x) = x^p - a} \qquad (\mathrm{a} \textit{ in } \mathfrak{F}). \tag{37}$$

Then $\mathrm{f(x)}$ *is a reducible polynomial of* $\mathfrak{F}[\mathrm{x}]$ *if and only if* $\mathrm{a = b^p}$ *for* b *in* $\mathfrak{F}$ *and* $\mathrm{f(x) = (x - b)^p}$. *When* $\mathrm{f(x)}$ *is irreducible in* $\mathfrak{F}$ *any stem field* $\mathfrak{F}(\xi)$ *of* $\mathrm{f(x)}$ *is a root field since*

$$\mathrm{f(x)} = (\mathrm{x} - \xi)^{\mathrm{p}} \tag{38}$$

when $\xi^{\mathrm{p}} = \mathrm{a}$.

For suppose that $f(x) = x^p - a = g(x)h(x)$ where $g(x)$ is a monic irreducible factor of $f(x)$ in $\mathfrak{F}[x]$ and has degree $t > 0$. Then $g(x)$ defines a simple extension $\mathfrak{F}(\xi)$ of $\mathfrak{F}$ such that $\xi^p = a, f(x) = (\mathrm{x} - \xi)^p$. But the factors of $g(x)$ in $\mathfrak{F}(\xi)$ are all $x - \xi$ and $g(x) = (x - \xi)^t$, $\xi^t = c$ is in $\mathfrak{F}$. If $t < p$ there exist integers r and s such that $rp + st = 1$, $\xi = (\xi^p)^r(\xi^t)^s = a^r c^s$ is in $\mathfrak{F}$ as was stated in our theorem. Otherwise $p = t$ and $f(x) = g(x)$ is irreducible in $\mathfrak{F}$.

When $x^p - a$ is irreducible in $\mathfrak{F}$ it defines a field $\mathfrak{K} = \mathfrak{F}(\xi)$ of degree p over $\mathfrak{F}$. Every quantity of $\mathfrak{K}$ has the form $\alpha = c_0 + c_1\xi + \ldots + c_{p-1}\xi^{p-1}$ with c_i in $\mathfrak{F}$ and

$$\alpha^p = c_0^p + c_1^p a + \ldots + c_{p-1}a^{p-1} = c \text{ in } \mathfrak{F}\,. \tag{39}$$

If $c_1 = \ldots = c_{p-1} = 0$ then α is in $\mathfrak{F}$. Otherwise $\mathfrak{F}(\alpha)$ is a subfield of degree $m \neq 1$ over $\mathfrak{F}$ of $\mathfrak{K}$ and $m = p$ by Theorem 7.8, $\alpha^p = c$ in $\mathfrak{F}$. We also see this result from (30) since the conjugates α_i are all equal and $\phi(y) = (y - \alpha)^p = y^p - \alpha^p$, where $N(\alpha) = \alpha^p = c$ must be in $\mathfrak{F}$. Theorem 7.17 states that $y^p - c$ is irreducible in $\mathfrak{F}$ or α is in $\mathfrak{F}$.

An interesting property of the above field $\mathfrak{K} = \mathfrak{F}(\xi)$ is given by

Theorem 18. *Let* g(x) *be in the polynomial domain* $\mathfrak{K}$[x], $\mathfrak{K}$ *as above. Then the* p*th power of* g(x) *is in* $\mathfrak{F}$[x].

For if $g(x) = d_0x^q + d_1x^{q-1} + \ldots + d_q$ then $d_0, \ldots, d_q$ are in $\mathfrak{K}$ and $[g(x)]^p = d_0^px^{pq} + \ldots + d_q^p$ with coefficients d_i^p in $\mathfrak{F}$ by (39).

9. Separable equations. The polynomial

$$f(x) = a_0 + a_1x + \ldots + a_nx^n \qquad (a_i \text{ in } \mathfrak{F}), \tag{40}$$

defines a unique polynomial

$$f'(x) = a_1 + 2a_2x + \ldots + na_nx^{n-1} \tag{41}$$

which is called the *derivative* of $f(x)$. It is easily verified algebraically that

$$(f+g)' = f' + g' , \quad (fg)' = f'g + fg' , \quad (f^t)' = tf^{t-1}f' . \tag{42}$$

The polynomial $f'(x) = 0$ if and only if $ia_i = 0$ $(i = 1, \ldots, n)$. When $\mathfrak{F}$ is non-modular this implies that every $a_i = 0$ and $f = a_0$. But when $\mathfrak{F}$ has characteristic p we have $a_i = 0$ for i not divisible by p,

$$f(x) = \phi(x^p) = b_0 + b_1x^p + \ldots + b_mx^{mp} . \tag{43}$$

Let $\mathfrak{K} = \mathfrak{F}(\beta_0, \ldots, \beta_m)$ be the root field of the polynomial $(x^p - b_0) \ldots (x^p - b_m)$ so that $\beta_i^p = b_i$,

$$f(x) = (\beta_0 + \beta_1x + \ldots + \beta_mx^m)^p$$

in $\mathfrak{K}$. Conversely, if $f = g^p$ where f and g have coefficients in a field $\mathfrak{K}$ of characteristic p we have $f' = pg^{p-1}g' = 0$. We state this result as

Theorem 19. *Let* f(x) *be in the polynomial domain* $\mathfrak{F}$[x]. *Then* f′(x) = 0 *if and only if* f *is in* $\mathfrak{F}$ *or* $\mathfrak{F}$ *has characteristic* p *and there exists a field* $\mathfrak{K} \geq \mathfrak{F}$ *such that* f = g^p, g *in* $\mathfrak{K}$[x].

We now make the

DEFINITION. *A polynomial* f(x) *and the equation* f(x) = 0 *are called separable if* f(x) = 0 *has no multiple roots. Otherwise we call* f(x) *and* f(x) = 0 *inseparable.*

Theorem 20. *A polynomial* f(x) *is separable if and only if the greatest common divisor of* f(x) *and* f′(x) *is unity.*

The greatest common divisor has coefficients in $\mathfrak{F}$ and may be computed by rational methods. This makes the above criterion of great value. For proof we let the g.c.d. of f and f' be unity. If f is inseparable then $f = g^2h$

where $g = x - \xi$, ξ is in the root field of $f(x)$. Also $f' = g^2h' + 2gg'h$ is divisible by the factor g of $f(x)$, a contradiction. Conversely, let $f(x)$ be separable but let the g.c.d. of f and f' be not unity. Then f' is divisible by a factor $x - \xi$ of $f(x) = (x - \xi)\,g(x)$ where $g(x)$ is not divisible by $x - \xi$. Hence $f' = g'(x - \xi) + g$ is not divisible by $x - \xi$, a contradiction.

As a corollary of Theorems 7.19 and 7.20 we have

Theorem 21. *Let* f(x) *be irreducible of degree* n > 1 *over a field* $\mathfrak{F}$ *and* $\mathfrak{F}(\xi)$ *be a corresponding stem field. Then* f(x) *is inseparable if and only if* $\mathfrak{F}$ *has characteristic* p, f *has the form* (43). *When this happens* n = mp *and* $\mathfrak{L} = \mathfrak{F}(\xi^p) = \mathfrak{F}(\eta)$ *has degree* m *over* $\mathfrak{F}$, $\mathfrak{F}(\xi) = \mathfrak{L}(\xi)$, $\xi^p = \eta$ *in* $\mathfrak{L}$.

For $f' \neq 0$ unless $\mathfrak{F}$ has characteristic p and f the form (43). When $f' \neq 0$ it is prime to the irreducible f of degree greater than that of f'. Notice that if $\mathfrak{F}$ is non-modular every irreducible $f(x)$ is separable.

EXERCISES

1. Let $x, a, a_0, \ldots, a_n$ be independent indeterminates over the field $\mathfrak{R}$ of all rational numbers and $f(x) = a_0x + a_1x + \ldots + a_nx^n$. Show that the rth derivative of $f(x)$ is

$$r(r-1)\ldots 2\cdot 1\,a_r + [(r+1)\ldots 2]a_{r+1}x + \ldots$$
$$+ [n(n-1)\ldots(n-r+1)]a_nx^{n-r} \equiv r(r-1)\ldots 1\cdot g_r(x)$$

where $g_r(x)$ has coefficients in the integral domain of all polynomials in $a_r, \ldots, a_n$ *with coefficients rational integers*. Show then that if $g_0(x) = f(x)$, and $g_r(a)$ is obtained by replacing x by a in $g_r(x)$ then the Taylor's expansion $f(x) = g_0(a) + g_1(a)(x - a) + \ldots + g_n(a)(x - a)^n$ is identically true in the indeterminates $a, x, a_0, \ldots, a_n$. Hint: The coefficient $k(k-1)\ldots(k-r+1) = (k!)[(k-r)!]^{-1} = r!C_{n,r}$ where $C_{n,r}$ is a binomial coefficient and is an integer.

2. Let $\mathfrak{F}$ be a field of characteristic p. Prove that our above Taylor's expansion is valid for a, a_i in $\mathfrak{F}$. But it is of course meaningless when we write the usual form $(r!)^{-1}f^{(r)}(a)$ for the coefficients when $r > p$.

3. Restate the results of Exercises 1, 7, 8, 9 of Section 7.4 in terms of separable and inseparable quadratic fields.

4. A field $\mathfrak{F}$ of characteristic p is called *perfect* if every a of $\mathfrak{F}$ is the pth power b^p of b in $\mathfrak{F}$. Define $\mathfrak{F}^{(p)}$ to be the set of all quantities b^p, b in $\mathfrak{F}$ and show that if $\mathfrak{F}$ is not perfect then $\mathfrak{F}^{(p)}$ is a proper subfield of $\mathfrak{F}$ equivalent to $\mathfrak{F}$ under the correspondence $b \longleftrightarrow b^p$, b in $\mathfrak{F}$.

5. Let $\mathfrak{K}$ be a field of degree n over $\mathfrak{F}$ of characteristic p such that every quantity a of $\mathfrak{K}$ is a root of a corresponding equation $x^{p^t} = a$ in $\mathfrak{F}$. Prove that n is a power of p by the use of Theorem 7.8 and that there exists an integer e (called the *exponent* of $\mathfrak{K}$ over $\mathfrak{F}$) such that k^{p^e} is in $\mathfrak{F}$ for every k of $\mathfrak{K}$, and e is the least integer with this property.

6. Let $\mathfrak{Z} > \mathfrak{K} > \mathfrak{F}$ where $\mathfrak{K}$ is as in Ex. 5 and $\mathfrak{Z} = \mathfrak{K}(\xi)$ is a simple separable extension of degree n over $\mathfrak{F}$. Prove that $\mathfrak{Z} = \mathfrak{K} \times \mathfrak{F}(\xi_0)$ where $\xi_0 = \xi^{p^e}$ and e is the exponent of $\mathfrak{K}$ over $\mathfrak{F}$. Prove in fact that $\mathfrak{Z}_0 = \mathfrak{F}_0(\xi_0)$ and $\mathfrak{Z} = \mathfrak{K}(\xi)$ are equivalent (but not equivalent over $\mathfrak{F}$) under the correspondence $a \longleftrightarrow a^{p^e}$ for every a of $\mathfrak{Z}$, where $\mathfrak{F}_0$ is a certain subfield of $\mathfrak{F}$.

10. Finite fields. A field $\mathfrak{F}$ is called a finite field if $\mathfrak{F}$ contains only a finite number of quantities and otherwise is called an infinite field. The most elementary example of a finite field is the field $\mathfrak{C}$ of Theorem 2.3. The elements of $\mathfrak{C}$ may be designated by the integers

$$0, 1, \ldots, p-1, \tag{44}$$

and any one of these integers is understood to represent the class of all integers with the given integer as remainder on division by the prime p. The *Fermat Theorem* on integers is stated here as

$$a^p = a \qquad (a \text{ in } \mathfrak{C}). \tag{45}$$

We refer the reader to Exercise 3 of Section 6.3 for proof.

We designate the algebra $\mathfrak{C}$ by $\mathfrak{P}$ in the case of Theorem 2.3. It is a subfield of every field of characteristic p and has no subfields. Thus we call it the *prime field* of characteristic p. The corresponding field for fields of infinite characteristic is the infinite prime field $\mathfrak{R}$ of all rational numbers.

Every finite field $\mathfrak{K}$ must have finite characteristic p since $\mathfrak{K}$ cannot contain $\mathfrak{R}$. Thus $\mathfrak{K} \geqq \mathfrak{P}$. But $\mathfrak{K}$ must be algebraic of finite degree n over $\mathfrak{P}$ and we may write

$$\mathfrak{K} = (u_1, \ldots, u_n) \text{ over } \mathfrak{P}.$$

The quantities of $\mathfrak{K}$ then have the form $a_1u_1 + \ldots + a_nu_n$ with the a_i assuming the values (44). This proves

Theorem 22. *A finite field $\mathfrak{K}$ has* p^n *elements where* p *is the characteristic of $\mathfrak{K}$ and* n *is the degree of $\mathfrak{K}$ over $\mathfrak{P}$.*

We now write

$$q = p^n \tag{46}$$

and let $a_2, \ldots, a_q$ be the non-zero elements of $\mathfrak{K}$. The non-zero elements of any field form a multiplicative group $\mathfrak{G}$ and now we have $\mathfrak{G} = (a_2, \ldots, a_q)$ of order $q-1$. Every $a \neq 0$ of $\mathfrak{K}$ generates a subgroup of $\mathfrak{G}$ of order a divisor of $q-1$, the identity element of $\mathfrak{G}$ is the unity element 1 of $\mathfrak{K}$, and our theorems on finite groups imply that

$$a^{q-1} = 1 \tag{47}$$

for every non-zero a of $\mathfrak{K}$. Thus

(48) $$a^q = a$$

for every a of $\mathfrak{K}$. The equation $x^q - x = 0$ is divisible by the distinct factors $x, x - a_2, \ldots, x - a_q$ so that

(49) $$x^q - x = \prod_{i=1}^{q}(x - a_i) \qquad (a_1 = 0).$$

Applying Theorem 7.10 we have

Theorem 23. *A finite field with* $q = p^n$ *elements is the root field of the equation* $x^q - x = 0$ *over* $\mathfrak{P}$. *Any two such fields defined for the same* n *are equivalent.*

The derivative of $x^q - x$ is $qx^{q-1} - 1 = -1$ which is prime to $f(x)$. Thus $x^q - x$ is a separable polynomial and its root field $\mathfrak{K}$ contains the distinct roots $a_1 = 0, a_2, \ldots, a_q$ of $x^q - x = 0$. To prove that $\mathfrak{K}$ contains no other elements we notice that every power product of the a_i is in the group $\mathfrak{G}$ and is an a_i. Hence every a of $\mathfrak{K}$ has the form $a = a_2a_2 + \ldots + a_qa_q$ where a_i is an integer (44). We then have $a^p = a_2^pa_2^p + \ldots + a_q^pa_q^p = a_2a_2^p + \ldots + a_qa_q^p$ by (45). Applying this formula repeatedly, we ultimately obtain $a^q = a_2a_2^q + \ldots + a_qa_q^q = a_2a_2 + \ldots + a_qa_q = a$ since $q = p^n$ and $a_i^q = a_i$. But then a is a root of $x^q - x$ in $\mathfrak{K}$ and is one of $a_1, a_2, \ldots, a_q$.

Theorem 24. *The root field* $\mathfrak{K}$ *over* $\mathfrak{P}$ *of the equation* $x^q - x = 0$ *with* $q = p^n$ *is a finite field with* $q = p^n$ *elements and is the only such field. We call* $\mathfrak{K}$ *a* ***Galois field***

(50) $$GF(p^n).$$

The set $\mathfrak{G}$ of all non-zero quantities a of $\mathfrak{K}$ is evidently a *multiplicative* group. Moreover $a^{q-1} = 1$ for every a of $\mathfrak{G}$ and the a's of $\mathfrak{G}$ do not all satisfy any equation $x^t = 1$ with $t < q - 1$ since they are $q - 1$ distinct quantities. By Theorem 6.8 the group $\mathfrak{G}$ is cyclic and $\mathfrak{G} = [\xi]$ where ξ is a root of $x^{q-1} = 1$ in $\mathfrak{K}$.

Theorem 25. *Let* $\mathfrak{K}$ *be a finite field* $GF(p^n)$ *and* $q = p^n$. *Then* $\mathfrak{K} = \mathfrak{P}(\xi)$ *where* ξ *is a root of the separable equation*

$$x^{q-1} = 1,$$

and the quantities of $\mathfrak{K}$ *consist of zero and the roots*

$$\xi, \xi^2, \ldots, \xi^{q-1} = 1$$

of $x^{q-1} = 1$.

The subfields of $\mathfrak{K} = GF(p^n)$ are fields $\mathfrak{L} = GF(p^m)$ where m divides n. The non-zero quantities of $\mathfrak{L}$ form a cyclic subgroup of the group $\mathfrak{G}$ defined by $\mathfrak{K}$. This group is uniquely determined by its order $p^m - 1$ and Theorem 6.3 states that if

$$p^n - 1 = s(p^m - 1)$$

then

$$\mathfrak{L} = \mathfrak{P}(\eta), \qquad \eta = \xi^s,$$

and the non-zero quantities of $\mathfrak{L}$ are the powers η^i with $i = 1, \ldots, p^m - 1$. Hence $\mathfrak{L}$ is unique and every divisor m of n determines a unique subfield $\mathfrak{L}$ of $\mathfrak{K}$.

We have seen that every ξ of $\mathfrak{K}$ is a root of a separable equation over $\mathfrak{F}$. This implies that $x^p - \alpha$ is reducible in $\mathfrak{K}$ for every α of $\mathfrak{K}$. But this is obvious since when α is in $\mathfrak{K}$ then $r = p^{n-1}$ gives $\alpha = (\alpha^r)^p$.

EXERCISES

1. Prove that the pth root of any α of $\mathfrak{K}$ is a unique quantity β of $\mathfrak{K} = GF(p^n)$.

2. Show that our above arguments imply that every finite field $\mathfrak{K}$ is perfect in the sense of Ex. 4 of Section 7.9. Let η be an indeterminate over $\mathfrak{F}$ and show that the field $\mathfrak{F}(\eta)$ of all rational functions of η with coefficients in $\mathfrak{F}$ is an infinite field and is not perfect.

3. Let $\mathfrak{F}$ have characteristic p. Prove that there exists no field over $\mathfrak{F}$ containing a quantity $\zeta \neq 1$ such that $\zeta^p = 1$.

11. Separable fields. We now consider an algebraic field $\mathfrak{K}$ over $\mathfrak{F}$ where $\mathfrak{F}$ is either finite or infinite. Then α in $\mathfrak{K}$ is called *separable over* $\mathfrak{F}$ if the minimum function of α is separable. We also say that the field $\mathfrak{K}$ *is separable over* $\mathfrak{F}$ *if every quantity of* $\mathfrak{K}$ *is separable over* $\mathfrak{F}$. The principal result on such fields is

Theorem 26. *A field* $\mathfrak{K}$ *of degree* n *over* $\mathfrak{F}$ *is separable if and only if*

$$\mathfrak{K} = \mathfrak{F}(\xi),$$

where ξ *is a separable algebraic quantity of degree* n *over* $\mathfrak{F}$.

The above result has been proved true when $\mathfrak{F}$ is finite in Theorem 7.25. We now assume that $\mathfrak{F}$ is infinite, that is, that there are infinitely many distinct quantities in $\mathfrak{F}$.

Let ξ be a root of an irreducible separable equation $f(x) = 0$ of degree n and $\mathfrak{K} = \mathfrak{F}(\xi)$. If any α of $\mathfrak{K}$ is inseparable the field $\mathfrak{F}$ must have finite characteristic p, the minimum equation of α has the form (43), and $\mathfrak{H} = \mathfrak{F}(\alpha)$ has degree $m = pt$ over $\mathfrak{F}$ and degree p over $\mathfrak{L} = \mathfrak{F}(\alpha^p)$. Theorem 7.21

may be applied and $\mathfrak{K}$ has degree $n = mq$ over $\mathfrak{F}$, degree q over $\mathfrak{H}$, degree pq over $\mathfrak{L}$. We factor $f(x)$ in $\mathfrak{H}[x]$ into irreducible factors and have

$$f(x) = g(x) \cdot h(x)$$

where $g(\xi) = 0$, $\mathfrak{K} = \mathfrak{H}(\xi)$, $g(x)$ is an irreducible polynomial of degree q of $\mathfrak{H}[x]$. By Theorem 7.18 the polynomial $[g(x)]^p$ has degree pq, coefficients in $\mathfrak{L}$ and ξ as a root. But $\mathfrak{K} = \mathfrak{L}(\xi)$ has degree pq over $\mathfrak{L}$ and $[g(x)]^p$ must be the minimum function of ξ. Then $f(x)$ is divisible by $[g(x)]^p$ by Theorem 2.10 contrary to our hypothesis that $f(x)$ is a separable equation. Thus α cannot be inseparable and $\mathfrak{K}$ is separable.

We next prove the

LEMMA. *Let $\mathfrak{K} = \mathfrak{F}(\alpha, \beta)$ where α and β are separable algebraic quantities over $\mathfrak{F}$. Then $\mathfrak{K} = \mathfrak{F}(\eta)$ is separable over $\mathfrak{F}$.*

For let $f(x)$ and $g(x)$ be the respective minimum functions of α and β, and let $\mathfrak{N}$ be the root field of $f(x)g(x)$ and contain α and β. Then

$$f(x) = (x - \alpha_1) \dots (x - \alpha_r), \quad g(x) = (x - \beta_1) \dots (x - \beta_s) \qquad (\alpha_i, \beta_j \text{ in } \mathfrak{N}) \tag{51}$$

where $\alpha = \alpha_1$, $\beta = \beta_1$. The α_i are all distinct and the β_j are all distinct since $f(x)$ and $g(x)$ are separable. Then the equations

$$\alpha_i + \beta_j y = \alpha_q + \beta_k y \qquad (i, q = 1, \dots, r; j \neq k) \tag{52}$$

have a unique solution y in $\mathfrak{N}$ for every i. Since $\mathfrak{F}$ is an infinite field we may choose a quantity y_0 in $\mathfrak{F}$ which is distinct from all the solutions y of (52) and have

$$\alpha_i + \beta_j y_0 \neq \alpha_q + \beta_k y_0$$

unless both $i = q$ and $j = k$. Then the polynomial

$$h(x) = \prod_{\substack{i=1,\dots,r \\ j=1,\dots,s}} [x - (\alpha_i + \beta_j y_0)]$$

has coefficients which are symmetric in the roots of *both* $f(x)$ and $g(x)$ and hence are in $\mathfrak{F}$. Moreover, $h(x)$ is separable and has

$$\eta = \alpha + y_0\beta \neq \alpha_i + y_0\beta_j \qquad (i = 1, \dots, r; j = 2, \dots, s)$$

as a root. It remains to prove that α and β are in $\mathfrak{F}(\eta)$ so that $\mathfrak{F}(\eta) \leq \mathfrak{F}(\alpha, \beta)$ implies that $\mathfrak{F}(\eta) = \mathfrak{F}(\alpha, \beta)$.

The polynomial $f(\eta - y_0x)$ in the indeterminate x over $\mathfrak{F}(\eta)$ has β as a root since $f(\eta - y_0\beta) = f(\alpha) = 0$. The greatest common divisor of $g(x)$ and $f(\eta - y_0x)$ also has coefficients in $\mathfrak{F}(\eta)$ and is a non-constant polynomial $d(x)$ whose linear factors are factors $x - \beta_j$ of $g(x)$. If $x - \beta_j$ divides $f(\eta - y_0x)$ then $f(\eta - y_0\beta_j) = 0$, $\eta - y_0\beta_j = \alpha_i$, $\eta = \alpha_i + y_0\beta_j$ which is impossible unless $j = 1$. Hence $d(x) = x - \beta_1$ has the coefficient β_1 in $\mathfrak{F}(\eta)$. Also $\alpha = \eta - y_0\beta_1$ is in $\mathfrak{F}(\eta)$ as desired. This proves our Lemma.

We now let $\mathfrak{K}$ be separable of degree n over $\mathfrak{F}$, $\mathfrak{K} = (u_1, \ldots, u_n)$ over $\mathfrak{F}$. Then u_1 and u_2 are separable over $\mathfrak{F}$ and the above lemma shows that $\mathfrak{F}(u_1, u_2) = \mathfrak{F}(\eta_2)$, $\mathfrak{F}(u_1, u_2, u_3) = \mathfrak{F}(\eta_2, u_3) = \mathfrak{F}(\eta_3)$, $\mathfrak{F}(u_1, \ldots, u_n) = \mathfrak{K} = \mathfrak{F}(\eta_{n-1}, u_n) = \mathfrak{F}(\eta)$ as desired.

COROLLARY. *Let* $\mathfrak{K} = \mathfrak{F}(\xi_1, \ldots, \xi_r)$ *where the* ξ_i *are separable over* $\mathfrak{F}$. *Then* $\mathfrak{K} = \mathfrak{F}(\xi)$ *is separable over* $\mathfrak{F}$.

EXERCISES

1. Show that if $\mathfrak{F}$ has characteristic two the inseparable field of Exercise 9, Section 7.4, is not a simple extension $\mathfrak{F}[\xi]$ of $\mathfrak{F}$. Generalize the proof of this result to obtain the corresponding result for the direct product of two inseparable fields of degree p over $\mathfrak{F}$ of characteristic p.

2. Let $\mathfrak{K}$ be the composite of a separable field of degree n over $\mathfrak{F}$ of characteristic p and an inseparable field of degree p over $\mathfrak{F}$. Prove that $\mathfrak{K}$ is the direct product of these two fields and that $\mathfrak{K} = \mathfrak{F}[\xi]$ is simple of degree np over $\mathfrak{F}$.

12. A lemma on infinite fields. The statement that $ab = 0$ in a field $\mathfrak{F}$ if and only if $a = 0$ or $b = 0$ implies

Theorem 27. *Let* $f(x)$ *be a polynomial of degree* $n > 0$ *and* $\mathfrak{K}$ *be a root field of* $f(x)$. *Then* $f(x) = 0$ *has exactly* n *roots in* $\mathfrak{K}$.

For we have seen that $f(x) = (x - \xi_1) \ldots (x - \xi_n)a_0$ where $a_0 \neq 0$ since the degree of $f(x)$ is $n > 0$. If $f(\xi) = a_0(\xi - \xi_1) \ldots (\xi - \xi_n) = 0$ then at least one $\xi - \xi_i = 0$, $\xi = \xi_i$.

The above result has a consequence

Theorem 28. *Let* $f(x) = a_0x^n + \ldots + a_n$ *with* a_i *in a field* $\mathfrak{F}$ *and suppose that* $f(\xi_i) = 0$ *for* $n + 1$ *distinct* ξ_i. *Then the* $a_i = 0$, *that is* $f(x)$ *is the zero polynomial of* $\mathfrak{F}[x]$.

We may also state this result as

Theorem 29. *Let* $f(x)$ *and* $g(x)$ *be polynomials of* $\mathfrak{F}[x]$ *each of degree at most* n *and suppose that* $f(\xi_i) = g(\xi_i)$ *for* $n + 1$ *distinct* ξ_i *of* $\mathfrak{F}$. *Then* $f(x) = g(x)$ *are the same polynomials.*

Theorem 7.27 may be used to prove a very important tool in algebraic research. We state this lemma as

Theorem 30. *Let* $\mathfrak{F}$ *be an infinite field,* $x_1, \ldots, x_n$ *be independent indeterminates over* $\mathfrak{F}$, $\mathfrak{L}$ *be any infinite subset of* $\mathfrak{F}$ *and*

$$f_i = f_i(x_1, \ldots, x_n) \qquad (i = 1, \ldots, m)$$

be any finite number m *of non-zero polynomials of* $\mathfrak{F}[x_1, \ldots, x_n]$. *Then there exists an infinite subset* $\mathfrak{L}_1$ *of quantities* ξ_1 *of* $\mathfrak{L}$, *an infinite subset* $\mathfrak{L}_2$ *of quantities* ξ_2 *of* $\mathfrak{L}$ *for each* $\xi_1, \ldots,$ *and an infinite subset* $\mathfrak{L}_n$ *of quantities* ξ_n *of* $\mathfrak{L}$ *for each* $\xi_1, \ldots, \xi_{n-1}$ *already chosen such that*

$$f_1(\xi_1, \ldots, \xi_n) \neq 0, \qquad f_2(\xi_1, \ldots, \xi_n) \neq 0, \ldots, \quad f_m(\xi_1, \ldots, \xi_n) \neq 0$$

for every ξ_i *of* $\mathfrak{L}_i$.

It is usually sufficient in the application of the theorem above to know that there exist infinitely many sets $(\xi_1, \ldots, \xi_n)$ of ξ_i in $\mathfrak{F}$ for which the $f_i(\xi_1, \ldots, \xi_n) \neq 0$. In fact most applications require only the existence of a single set. But it may be sometimes important to know that infinitely many sets $(\xi_{1s}, \ldots, \xi_{ns})$ may be chosen with $\xi_{is} \neq \xi_{it}$ for $s \neq t$. This is of course a consequence of our formulation of the above theorem.

Our statement of the theorem indicates its proof. We write

$$f_i = f_i(x_n) = a_{i0}x_n^{e_i} + a_{i1}x_n^{e_i-1} + \ldots + a_{ie_i} \qquad (a_{i0} \neq 0)$$

where the a_{ij} are in $\mathfrak{F}[x_1, \ldots, x_{n-1}]$ and put

$$e = \text{maximum of the } e_i.$$

When $n = 1$ theorem 7.27 states that $f_i(\xi) \neq 0$ for all but at most e_i quantities ξ of $\mathfrak{L}$ so that the $f_i(\xi) \neq 0$ for all but at most *em* quantities of $\mathfrak{L}$. This completes the case $n = 1$ and we assume our theorem true for $n - 1$. Apply this result with specified $\xi_1, \ldots, \xi_{n-1}$ to make $a_{i0}(\xi_1, \ldots, \xi_{n-1}) = a_{i0} \neq 0$ and in $\mathfrak{F}$. Then

$$f_i(\xi_1, \ldots, \xi_{n-1}, x_n) = a_{i0}x_n^{e_i} + \ldots + a_{ie_i}$$

and by Theorem 7.27 we have $f_i(\xi_1, \ldots, \xi_n) \neq 0$ for all but a finite number of ξ_n in $\mathfrak{F}$. This proves our result.

EXERCISES

1. Apply Theorem 7.30 to the case where $\mathfrak{F}$ is the field of all rational numbers and $\mathfrak{L}$ the set of all integers; also to the case $\mathfrak{L}$ the set of all rational numbers whose absolute values are less than 2^{-100}.

2. Let $\mathfrak{K}$ be a field over an infinite field $\mathfrak{F}$, and A, B similar n-rowed square matrices with elements in $\mathfrak{K}$. Prove that if the elements of A and B are in $\mathfrak{F}$, then they are similar in $\mathfrak{F}$ without the use of the invariant factor criteria. Hint: $PAP^{-1} = B$ where P has elements in $\mathfrak{K}$ and we may write $P = P_1\xi_1 + P_2\xi_2 + \ldots + P_t\xi_t$ where the P_i have elements in $\mathfrak{F}$, the ξ_i are in $\mathfrak{K}$ and linearly independent in $\mathfrak{F}$. Show that $P_iA = BP_i$ and that there exists a non-singular $P_0 = P_1\xi_{10} + \ldots + P_t\xi_{t0}$, ξ_{i0} in $\mathfrak{F}$.

CHAPTER VIII

THE GALOIS THEORY

1. Normal fields. A separable field $\mathfrak{K}$ of degree n over $\mathfrak{F}$ is a simple extension $\mathfrak{F}[\xi]$ of $\mathfrak{F}$ by Theorem 7.26. The root field $\mathfrak{N} = \mathfrak{F}[\xi_1, \ldots, \xi_n]$ of the minimum equation $f(x) = 0$ of $\xi = \xi_1$ contains the n stem fields $\mathfrak{F}[\xi_i]$ of $f(x)$ and these stem fields are all equivalent. We make the

DEFINITION. *A simple algebraic extension $\mathfrak{F}[\xi]$ of $\mathfrak{F}$ is called normal (or Galois) over $\mathfrak{F}$ if it is separable and equal to its conjugates over $\mathfrak{F}$.*

The definition above states that $\mathfrak{F}[\xi]$ *is normal if and only if it is separable and its own root field.* Theorems 7.13 and 7.26 also state that *every separable $\mathfrak{F}[\xi]$ is contained in a normal field.* Thus the theory of separable fields is a theory of the subfields of normal fields.

We shall study *the Galois theory of fields.* This is the theory of normal fields $\mathfrak{N} = \mathfrak{F}(\xi)$ of degree n over $\mathfrak{F}$, the automorphisms of $\mathfrak{N}$ over $\mathfrak{F}$, and the subfields of $\mathfrak{N}$ over $\mathfrak{F}$. The theory is of great importance in algebra and has many applications both in algebra and in *Algebraic Number Theory.* Notice that $\mathfrak{N}$ is the root field *of any separable equation.* We shall later apply this result to obtain the *Galois theory of equations.*

2. The automorphisms of $\mathfrak{N}$. The *conjugates* of a generator ξ of a normal field $\mathfrak{N}$ of degree n over $\mathfrak{F}$ are polynomials

$$(1) \qquad \xi_i = \theta_i(\xi) = b_{i0} + b_{i1}\xi + \ldots + b_{i\,n-1}\xi^{n-1} \qquad (b_{ij} \text{ in } \mathfrak{F}).$$

Then the equation $f[\theta_i(x)] = 0$ has a root in common with the irreducible minimum equation $f(x) = 0$ of ξ and is divisible by $f(x)$. Thus $f[\theta_i(\xi_j)] = 0$, that is $\theta_i(\xi_j)$ is a root ξ_k of $f(x) = 0$. We state this in

Theorem 1. *There exist integers* k_{ij} *for every* i *and* j *such that*

$$(2) \qquad \theta_i[\theta_j(\xi)] = \theta_{k_{ij}}(\xi) \qquad (i, j = 1, \ldots, n).$$

The equivalence of the stem fields of $f(x)$ now gives rise to automorphisms

$$S_i: \qquad r = r(\xi) \longleftrightarrow r(\xi_i) = r^{S_i}$$

of $\mathfrak{F}[\xi] = \mathfrak{F}[\xi_i]$ over $\mathfrak{F}$. The product of S_i and

$$S_j: \qquad r(\xi) \longleftrightarrow r(\xi_j) = r^{S_j}$$

is evidently found by writing S_j in the equivalent form

$$S_j\colon \qquad r^{S_i} \longleftrightarrow (r^{S_i})^{S_j},$$

from which

$$S_iS_j\colon \qquad r \longleftrightarrow (r^{S_i})^{S_j} = r^{S_iS_j}.$$

This is in fact given by

$$r^{S_i} = r[\theta_i(\xi)], \qquad r^{S_iS_j} = r[\theta_i(\xi_j)] = r\{\theta_i[\theta_j(\xi)]\}$$

so that

$$(3) \qquad r^{S_iS_j} = r[\theta_{k_{ij}}(\xi)] = r^{S_k}, \qquad k = k_{ij}.$$

Theorem 2. *The automorphisms* S_i *are the only automorphisms of* $\mathfrak{N}$ *over* $\mathfrak{F}$.

For let S be an automorphism of $\mathfrak{N}$ over $\mathfrak{F}$. Then S carries ξ into ξ^S. By definition of an automorphism $(\xi^t)^S = (\xi^S)^t$, $a^S = a$ for every a of $\mathfrak{F}$, $[f(\xi)]^S = f(\xi^S) = 0^S = 0$, so that ξ^S is a root in $\mathfrak{N}$ of $f(x) = 0$. Then ξ^S is one of the ξ_i and S carries ξ into ξ_i. Then S carries every $r(\xi)$ into $r(\xi_i)$ and $S = S_i$.

Notice that Theorem 8.2 provides another proof of Theorem 8.1. For the product of any two automorphisms of $\mathfrak{N}$ over $\mathfrak{F}$ is a product of two transformations of $\mathfrak{N}$ leaving $\mathfrak{F}$ invariant and is an automorphism of $\mathfrak{N}$ over $\mathfrak{F}$. By Theorem 8.2 every automorphism of $\mathfrak{N}$ over $\mathfrak{F}$ is a transformation S_i and hence $S_iS_j = S_{k_{ij}}$. We write $\xi^{S_i} = \xi_i = \theta_i(\xi)$ and (3) holds and yields (2).

The set $\mathfrak{G}$ of all automorphisms of $\mathfrak{N}$ over $\mathfrak{F}$ is a group. We call $\mathfrak{G}$ the *automorphism group or Galois group* of $\mathfrak{N}$ over $\mathfrak{F}$. By Theorem 8.2 we have

Theorem 3. *Let* $\mathfrak{N}$ *be normal of degree* n *over* $\mathfrak{F}$. *Then the automorphism group* $\mathfrak{G}$ *has order* n.

EXERCISES

1. Use Theorem 7.13 to verify that the root field over $\mathfrak{F}$ of any separable polynomial with coefficients in $\mathfrak{F}$ is normal over $\mathfrak{F}$.
2. Let $\mathfrak{N} > \mathfrak{L} > \mathfrak{F}$ and $\mathfrak{N}$ be normal over $\mathfrak{F}$. Show that $\mathfrak{N}$ is normal over $\mathfrak{L}$.

3. The fundamental theorems of the Galois theory. Let $\mathfrak{N}$ be normal of degree n over $\mathfrak{F}$ and $\mathfrak{G}$ be its automorphism group. A quantity λ of $\mathfrak{N}$ is said to be unaltered by an automorphism S of $\mathfrak{G}$ if

$$\lambda = \lambda^S.$$

If also $\lambda = \lambda^T$ then $\lambda^{ST} = (\lambda^S)^T = \lambda^T = \lambda$. Thus the set of all automorphisms leaving λ unaltered forms a subgroup of $\mathfrak{G}$. Every subfield $\mathfrak{L}$ of $\mathfrak{N}$ determines uniquely a subgroup $\mathfrak{H}$ consisting of all automorphisms of $\mathfrak{G}$ leaving the quantities of $\mathfrak{L}$ unaltered. For, $\mathfrak{L} = \mathfrak{F}(\lambda)$. This subfield to subgroup ordering will be designated by

$$\mathfrak{L} \rightarrow \mathfrak{H} .$$

If $\lambda = \lambda^S$ and $\mu = \mu^S$ then $(\lambda \pm \mu)^S = \lambda \pm \mu$, $(\lambda\mu)^S = \lambda\mu$, $(\lambda\mu^{-1})^S = \lambda\mu^{-1}$ for $\mu \neq 0$. Hence the set of all quantities of $\mathfrak{N}$ left unaltered by S is a subfield of $\mathfrak{N}$. We deduce immediately that every subgroup $\mathfrak{H}$ of $\mathfrak{G}$ determines uniquely a subfield $\mathfrak{L}$ over $\mathfrak{F}$ of $\mathfrak{N}$ which consists of all quantities of $\mathfrak{N}$ left unaltered by the automorphisms of $\mathfrak{H}$. This group to field correspondence

$$\mathfrak{H} \rightarrow \mathfrak{L}$$

is a second ordering of the subgroups of $\mathfrak{G}$ and the subfields over $\mathfrak{F}$ of $\mathfrak{N}$. It is not evident that both orderings are the same since if $\mathfrak{L} \rightarrow \mathfrak{H}$ there might be further quantities of $\mathfrak{N}$ unaltered by $\mathfrak{H}$, and we might have $\mathfrak{L} \rightarrow \mathfrak{H} \rightarrow \mathfrak{L}_0$ with $\mathfrak{L}_0 > \mathfrak{L}$. Similarly, if we start with a group $\mathfrak{H}$ and take $\mathfrak{H} \rightarrow \mathfrak{L}$ we may have $\mathfrak{L} \rightarrow \mathfrak{H}_0$ where $\mathfrak{H}_0$ evidently contains $\mathfrak{H}$ and it is possible that $\mathfrak{H}_0 > \mathfrak{H}$. The first fundamental theorem of the Galois theory is exactly the statement that $\mathfrak{L} = \mathfrak{L}_0$, $\mathfrak{H} = \mathfrak{H}_0$ and may be formulated as

Theorem 4. *The subgroup to subfield and subfield to subgroup orderings are the same, that is*

$$\textit{if } \mathfrak{H} \rightarrow \mathfrak{L} \textit{ then } \mathfrak{L} \rightarrow \mathfrak{H} ;$$
$$\textit{if } \mathfrak{L} \rightarrow \mathfrak{H} \textit{ then } \mathfrak{H} \rightarrow \mathfrak{L} .$$

Thus there is a (1–1) *correspondence**

$$\mathfrak{H} \longleftrightarrow \mathfrak{L} , \tag{4}$$

between the subgroups $\mathfrak{H}$ *of* $\mathfrak{G}$ *and the subfields* $\mathfrak{L}$ *over* $\mathfrak{F}$ *of* $\mathfrak{N}$. *Moreover,* $\mathfrak{H}$ *is the automorphism group of* $\mathfrak{N}$ *over* $\mathfrak{L}$ *and its index under* $\mathfrak{G}$ *is the degree of* $\mathfrak{L}$ *over* $\mathfrak{F}$.

It is the final statement in the theorem above that is vital in our proof. It is a rather evident result since if $\mathfrak{N}$ is normal over $\mathfrak{F}$ and $\mathfrak{N} \geqq \mathfrak{L} \geqq \mathfrak{F}$ then $\mathfrak{N}$ over $\mathfrak{F}$ is still the root field of the minimum equation over $\mathfrak{L}$ of its defining element ξ. But the automorphism group of $\mathfrak{N}$ over $\mathfrak{L}$ is the set $\mathfrak{H}$ of all automorphisms of $\mathfrak{N}$ which leave the quantities of $\mathfrak{L}$ unaltered.

* We shall later say that both $\mathfrak{L} = \mathfrak{F}(\lambda)$ and λ *belong to* $\mathfrak{H}$.

These automorphisms then leave the quantities of $\mathfrak{F}$ unaltered and are in $\mathfrak{G}$, $\mathfrak{H} \leqq \mathfrak{G}$, $\mathfrak{H}$ is the group defined by $\mathfrak{L} \rightarrow \mathfrak{H}$. The order of $\mathfrak{H}$ is the degree q of $\mathfrak{N}$ over $\mathfrak{L}$ by Theorem 8.3. Then

$$n = qm \tag{5}$$

where m is the degree of $\mathfrak{L}$ over $\mathfrak{F}$. We now let $\mathfrak{H} \rightarrow \mathfrak{L}_0$, the set of all quantities of $\mathfrak{N}$ unaltered by the automorphisms of $\mathfrak{H}$. The quantities of $\mathfrak{L}$ are of this kind and

$$\mathfrak{L}_0 \geqq \mathfrak{L}, \qquad m_0 \geqq m$$

where m_0 is the degree of $\mathfrak{L}_0$ over $\mathfrak{F}$. Thus $\mathfrak{L}_0 \rightarrow \mathfrak{H}_0$ where $\mathfrak{H}_0$ is the automorphism group of $\mathfrak{N}$ over $\mathfrak{L}_0$ and

$$n = m_0 q_0,$$

$\mathfrak{H}_0$ has order q_0. The crux of our proof is the fact that $\mathfrak{H}_0$ consists of all automorphisms of $\mathfrak{G}$ leaving the quantities of $\mathfrak{L}_0$ unaltered, $\mathfrak{H}$ consists of automorphisms with this property, $\mathfrak{H}_0 \geqq \mathfrak{H}$, $q_0 \geqq q$. Now $n = mq = m_0 q_0$ with $m_0 \geqq m$, $q_0 \geqq q$ so that we must have $m_0 = m$ and $q_0 = q$. Then $\mathfrak{H}_0 = \mathfrak{H}$, $\mathfrak{L}_0 = \mathfrak{L}$ as desired. We also notice that the index of $\mathfrak{H}$ under $\mathfrak{G}$ is m. Conversely, let $\mathfrak{H} = (S_1, \ldots, S_q)$ be any subgroup of $\mathfrak{G}$ and $\mathfrak{H} \rightarrow \mathfrak{L}$. Then $\mathfrak{L} \rightarrow \mathfrak{H}_0$ of order q_0 and $\mathfrak{N} = \mathfrak{L}(\xi)$ of degree q_0 over $\mathfrak{L}$. Clearly $\mathfrak{H}_0 \geq \mathfrak{H}$, $q_0 \geq q$. The polynomial $(x - \xi^{S_1}) \ldots (x - \xi^{S_q})$ has coefficients unaltered by the S_i of $\mathfrak{H}$ and are then in $\mathfrak{L}$. By Theorems 7.6, 2.10 we have $q \geq q_0$, $q = q_0$, $\mathfrak{H} = \mathfrak{H}_0$, as desired.

Our one-to-one correspondence evidently implies that $\mathfrak{F} \longleftrightarrow \mathfrak{G}$, and $\mathfrak{N} \longleftrightarrow (I)$, where I is the identity automorphism and (I) is the identity subgroup of $\mathfrak{G}$. The second of these results is trivial since every S different from I carries ξ_1 to $\xi_i \neq \xi_1$. The first result is non-trivial, however. To see the meaning of this statement we write

$$r = r_0 + r_1\xi + \ldots + r_{n-1}\xi^{n-1} \qquad (r_i \text{ in } \mathfrak{F}) \tag{6}$$

so that

$$r^{S_i} = r_0 + r_1[\theta_i(\xi)] + \ldots + r_{n-1}[\theta_i(\xi)]^{n-1}. \tag{7}$$

It is obvious that r is in $\mathfrak{F}$ if and only if $r_1 = r_2 = \ldots = r_{n-1} = 0$. But our result states that if $r = r^{S_i}$ for all i then r is in $\mathfrak{F}$.

Other fundamental theorems are Theorems 8.5–8.8.

Theorem 5. *Let $\mathfrak{N} \geqq \mathfrak{L}_1 \geqq \mathfrak{L}_2 \geqq \mathfrak{F}$ and $\mathfrak{H}_1$ and $\mathfrak{H}_2$ be the respective corresponding groups. Then $\mathfrak{H}_2 \geqq \mathfrak{H}_1$ and the degree of $\mathfrak{L}_1$ over $\mathfrak{L}_2$ is the index of $\mathfrak{H}_1$ in $\mathfrak{H}_2$.*

For the automorphism group of $\mathfrak{N}$ over $\mathfrak{L}_2$ is $\mathfrak{H}_2$ and $\mathfrak{N}$ over $\mathfrak{L}_2$ contains $\mathfrak{L}_1$. By Theorem 8.4 the group $\mathfrak{H}_1$ is a subgroup of $\mathfrak{H}_2$ whose index is the degree of $\mathfrak{L}_1$ over $\mathfrak{L}_2$.

Theorem 6. *Let $\mathfrak{N} \geqq \mathfrak{L}_1 \geqq \mathfrak{F}$, $\mathfrak{N} \geqq \mathfrak{L}_2 \geqq \mathfrak{F}$ and $\mathfrak{H}_1$ and $\mathfrak{H}_2$ be the respective corresponding subgroups of $\mathfrak{G}$. Then $\mathfrak{L}_1$ and $\mathfrak{L}_2$ are equivalent over $\mathfrak{F}$ if and only if $\mathfrak{H}_1$ and $\mathfrak{H}_2$ are conjugate subgroups of $\mathfrak{G}$.*

For let $\mathfrak{L}_1 = \mathfrak{F}(\eta_1)$ and $\mathfrak{L}_2 = \mathfrak{F}(\eta_2)$ be equivalent such that $\eta_1 \longleftrightarrow \eta_2$. Then η_2 is a root of the minimum equation $g(y) = 0$ of η_1, and since η_2 is in $\mathfrak{N}$, we may write $\eta_1 = \eta(\xi_1)$ and must have $\eta_2 = \eta^S$ by Theorem 7.14. If T is in $\mathfrak{H}_1$ then

$$\eta^T = \eta\,, \qquad (\eta^S)^{S^{-1}TS} = \eta^{TS} = \eta^S\,,$$

so that $S^{-1}TS$ is in $\mathfrak{H}_2$ for every T of $\mathfrak{H}_1$. The order of $\mathfrak{H}_2$ is the degree of $\mathfrak{N}$ over $\mathfrak{L}_2$, and this is evidently the degree of $\mathfrak{N}$ over $\mathfrak{L}_1$, that is the order of $\mathfrak{H}_1$. Thus $\mathfrak{H}_2 \geqq S^{-1}\mathfrak{H}_1S$ which has the same order as $\mathfrak{H}_2$, and $\mathfrak{H}_2 = S^{-1}\mathfrak{H}S$ as desired.

Conversely, write $\eta_1 = \eta(\xi_1)$ and suppose that $\mathfrak{H}_2 = S^{-1}\mathfrak{H}_1S$. Then $\eta_2 = \eta^S$ has the property

$$\eta_2^{S^{-1}TS} = \eta^{SS^{-1}TS} = \eta^{TS} = \eta^S = \eta_2\,,$$

for every $S^{-1}TS$ of $\mathfrak{H}_2$ and η_2 is in $\mathfrak{L}_2$. The subfield $\mathfrak{F}(\eta_2)$ of $\mathfrak{L}_2$ has degree m over $\mathfrak{F}$ where $n = mq$, q is the order of $\mathfrak{H}_1$ and $\mathfrak{H}_2$. But the degree of $\mathfrak{L}_2$ over $\mathfrak{F}$ is m and $\mathfrak{L}_2 = \mathfrak{F}(\eta_2) = \mathfrak{F}(\eta^S)$ is equivalent to $\mathfrak{F}(\eta)$.

We next obtain

Theorem 7. *A subfield $\mathfrak{L}$ of $\mathfrak{N}$ is normal over $\mathfrak{F}$ if and only if the corresponding group $\mathfrak{H}$ is a normal divisor of $\mathfrak{G}$.*

For if $\mathfrak{L}$ is normal over $\mathfrak{F}$, $\mathfrak{L} = \mathfrak{F}(\eta) = \mathfrak{F}(\eta^S)$ for every S of $\mathfrak{G}$. But $\mathfrak{F}(\eta^S)$ corresponds to $S^{-1}\mathfrak{H}S$ by the above proof and also to $\mathfrak{H}$ since $\mathfrak{L} = \mathfrak{F}(\eta^S)$. Thus $S^{-1}\mathfrak{H}S = \mathfrak{H}$, $\mathfrak{H}$ is a normal divisor of $\mathfrak{G}$. Conversely if $S^{-1}\mathfrak{H}S = \mathfrak{H}$ then every η^S is in $\mathfrak{L} = \mathfrak{F}(\eta)$ and $\mathfrak{L}$ is normal over $\mathfrak{F}$.

Theorem 8. *Let $\mathfrak{H}$ be a normal divisor of $\mathfrak{G}$. Then the quotient group $\mathfrak{G}/\mathfrak{H}$ is equivalent to the automorphism group of $\mathfrak{L}$ over $\mathfrak{F}$.*

For we write

$$\mathfrak{G} = S_1\mathfrak{H} + S_2\mathfrak{H} + \ldots + S_m\mathfrak{H}\,, \qquad S_1 = I. \tag{8}$$

Then $\mathfrak{L} = \mathfrak{F}(\eta)$ where $\eta^S = \eta$ for every S of $\mathfrak{H}$. The minimum equation of η over $\mathfrak{F}$ has the form

$$\psi(y) = (y - \eta)(y - \eta^{S_2}) \ldots (y - \eta^{S_m})\,, \tag{9}$$

and the automorphism group of $\mathfrak{L}$ over $\mathfrak{F}$ consists of the automorphisms generated by the correspondences

$$T_i: \qquad \eta \longleftrightarrow \eta^{T_i} = \eta^{S_i} \qquad (i = 1, \ldots, m).$$

When $\mathfrak{H}$ is a normal divisor of $\mathfrak{G}$ we have $\mathfrak{H}S_i = S_i\mathfrak{H}$ and $\mathfrak{G}/\mathfrak{H}$ is the group whose elements are the sets $\mathfrak{H}S_i$ with

$$(S_i\mathfrak{H})(S_j\mathfrak{H}) = S_{k_{ij}}\mathfrak{H}.$$

Then

$$\eta^{T_iT_j} = \eta^{S_iS_j} = \eta^{S_{k_{ij}}} = \eta^{T_{k_{ij}}}$$

and the correspondence

$$T_i \longleftrightarrow \mathfrak{H}S_i$$

is preserved under multiplication and defines an equivalence between the automorphism group of $\mathfrak{L}$ over $\mathfrak{F}$ and $\mathfrak{G}/\mathfrak{H}$.

The form (8) of $\mathfrak{G}$ is the same as $\mathfrak{G} = \mathfrak{H} + \mathfrak{H}S_2 + \ldots + \mathfrak{H}S_m$ when $\mathfrak{H}$ is a normal divisor of $\mathfrak{G}$. This latter form is usually used. But we shall require the former in

Theorem 9. *Let* $\mathfrak{N} = \mathfrak{F}(\xi)$ *be normal of degree* n *over* $\mathfrak{F}$, f(x) *be the minimum function of* ξ *over* $\mathfrak{F}$. *Then if* $\mathfrak{L}$ *is a subfield of* $\mathfrak{N}$ *of degree* m *over* $\mathfrak{F}$ *and*

$$(10) \qquad \mathfrak{H} = (\mathrm{H}_1, \mathrm{H}_2, \ldots, \mathrm{H}_q), \qquad \mathrm{H}_1 = \mathrm{I},$$

is the subgroup of $\mathfrak{G}$ *corresponding to* $\mathfrak{L}$ *then* n = mq, $\mathfrak{G}$ *has the form* (8),

$$(11) \qquad \mathrm{f(x)} = \mathrm{g_1(x)} \cdot \mathrm{g_2(x)} \ldots \mathrm{g_m(x)}$$

where the $\mathrm{g_i(x)}$ *are irreducible polynomials of* $\mathfrak{L}$[x], *and*

$$(12) \qquad \mathrm{g_i(x)} = (\mathrm{x} - \xi^{\mathrm{S_i}})(\mathrm{x} - \xi^{\mathrm{S_iH_2}}) \ldots (\mathrm{x} - \xi^{\mathrm{S_iH_q}}) \qquad (\mathrm{i} = 1, \ldots, \mathrm{m}).$$

The theorem above tells how the minimum function of a generator ξ of the normal field $\mathfrak{N}$ over $\mathfrak{F}$ factors in the subfield $\mathfrak{L}$. Notice that the automorphism S_i which carries the root ξ of $g_1(x) = 0$ to the root ξ^{S_i} of $g_i(x) = 0$ does not carry $g_1(x)$ to $g_i(x)$, but carries it to

$$(x - \xi^{S_i})(x - \xi^{H_2S_i}) \ldots (x - \xi^{H_qS_i})$$

which has coefficients in a field conjugate to $\mathfrak{L}$. This field is $\mathfrak{L}$ and this polynomial is $g_i(x)$ when $\mathfrak{H}$ is a normal divisor of $\mathfrak{G}$. For then $\mathfrak{L}$ is normal over $\mathfrak{F}$ and $\mathfrak{H}S_i = S_i\mathfrak{H}$. We now prove our theorem.

The decomposition (8) of $\mathfrak{G}$ shows that $f(x)$ is the product of the factors $g_i(x)$ of (11), (12). The factors $x - \xi^{S_iH_j}$ of $g_i(x)$ are merely permuted by the automorphisms of $\mathfrak{H}$ and thus the coefficients of $g_i(x)$ are unaltered by $\mathfrak{H}$. Since $\mathfrak{L}$ corresponds to $\mathfrak{H}$ we have shown that $g_i(x)$ has coefficients in $\mathfrak{L}$. The degree of $\mathfrak{N} = \mathfrak{F}(\xi^{S_i})$ over $\mathfrak{L}$ is q by Theorem 7.8, and $g_i(x) = 0$ has degree q, coefficients in $\mathfrak{L}$ and $g_i(\xi^{S_i}) = 0$. Thus $g_i(x)$ is the minimum function of ξ^{S_i} over $\mathfrak{L}$ and is irreducible in $\mathfrak{L}$.

EXERCISES

1. What is the automorphism group of a normal field $\mathfrak{N}$ of prime degree p over $\mathfrak{F}$?

2. Let $f(x)$ be a reducible separable cubic polynomial of $\mathfrak{F}[x]$. Find the possible root fields of $f(x)$ and their automorphism groups.

3. Let $f(x) = (x - \xi_1)(x - \xi_2)(x - \xi_3) = x^3 - bx^2 + cx - d$ where b, c, d are in a field $\mathfrak{F}$, $\mathfrak{N} = \mathfrak{F}(\xi_1, \xi_2, \xi_3)$ is a root field of $f(x)$, $f(x)$ is an irreducible polynomial of $\mathfrak{F}$. Show that if $\mathfrak{F}(\xi_1)$ is separable it is normal over $\mathfrak{F}$ and equal to $\mathfrak{N}$, or $\mathfrak{N}$ has degree six over $\mathfrak{F}$.

4. Let $f(x)$ be a reducible separable quartic polynomial of $\mathfrak{F}[x]$. Find the possible root fields of $f(x)$ and show that one of these has degree four and automorphism group which is the direct product of two cyclic groups of order two. Apply Theorems 8.4–8.9 to the subfields of degree two over $\mathfrak{F}$ of this latter quartic field.

5. Consider the cyclic function $\xi_1^2\xi_2 + \xi_2^2\xi_3 + \xi_3^2\xi_1 = \alpha$ of the roots ξ_i of the cubic $f(x)$ of Ex. 3. Show that $d\xi_2^{-1} = \xi_1\xi_3 = \xi_2^2 - b\xi_2 + c$, whence $\alpha = \xi_1^2\xi_2 + 2\xi_2^2\xi_3 - b\xi_2\xi_3 + c\xi_3 = (\xi_1^2 + 2d\xi_1^{-1} - c)\xi_2 + bc - c\xi_1 - bd\xi_1^{-1}$. Then $\xi_2(3\xi_1^2 - 2b\xi_1 + c) = \alpha + (c - b^2)\xi_1 + b\xi_1^2$. Prove that these results imply that ξ_2 is a polynomial in ξ_1 and α with coefficients in $\mathfrak{F}$ except when the characteristic of $\mathfrak{F}$ is three, $f(x)$ is inseparable, $\alpha = 0$, $\xi_1 = \xi_2 = \xi_3$.

6. In Ex. 7 of Section 7.1 we proved the existence of a quadratic polynomial of $\mathfrak{F}[x]$ with α as a root. Show that the stem field $\mathfrak{F}[\xi_1]$ of an irreducible separable cubic is normal over $\mathfrak{F}$ if and only if this corresponding quadratic has a root α in $\mathfrak{F}$.

7. Write $\xi = \xi_1$, $\xi_2 = A\alpha + B$ where we have shown that if $f(x)$ is separable $A^{-1} = 3\xi^2 - 2b\xi + c \neq 0$ and $B = (b\xi^2 + c\xi - b^2\xi)A$ are in $\mathfrak{F}(\xi)$. Define $\alpha^T = \xi_1^2\xi_3 + \xi_3^2\xi_2 + \xi_2^2\xi_1$ and show that $\xi_3 = A\alpha^T + B$.

8. In Ex. 7 prove that if α is not in $\mathfrak{F}$, so that $\mathfrak{N} = \mathfrak{F}[\xi_1, \xi_2, \xi_3]$ is normal of degree six over $\mathfrak{F}$, the field $\mathfrak{F}[\xi_1]$ corresponds to the subgroup (I, T) of the group $\mathfrak{G}$ of $\mathfrak{N}$ over $\mathfrak{F}$ where $T^2 = I$, T interchanges ξ_2 and ξ_3, leaves ξ_1 unaltered, and carries α to the other root α^T of the irreducible quadratic defined in Ex. 6. Show also that $\xi^S = \xi_2$, $\xi^{S^2} = \xi_3$, $\alpha^S = \alpha$ where $\mathfrak{G} = (I, S, S^2, T, ST, S^2T)$ and $S^3 = I$, $\mathfrak{F}[\alpha]$ corresponds to the cyclic normal divisor (I, S, S^2) of $\mathfrak{G}$. Prove in fact that $\mathfrak{F}[\xi_2]$ corresponds to $(I, S^{-1}TS = ST)$. Hint: $\xi_3 = \xi_2^T = \xi_2^S$, $\xi_2 = \xi_2^{ST}$.

9. Let $f(x) = x^4 + bx^3 + cx^2 + dx + e$ be an irreducible separable polynomial of $\mathfrak{F}[x]$ with roots $\xi_1, \ldots, \xi_4$. Verify (as in Dickson's *First Course in the Theory of*

Equations, p. 51) that the resolvent cubic $g(y) = y^3 - cy^2 + (bd - 4e)y - b^2e + 4ce - d^2 = (y - \eta_1)(y - \eta_2)(y - \eta_3)$ where $\eta_1 = \xi_1\xi_2 + \xi_3\xi_4$, $\eta_2 = \xi_1\xi_3 + \xi_2\xi_4$, $\eta_3 = \xi_1\xi_4 + \xi_2\xi_3$. This holds for an arbitrary field $\mathfrak{F}$ although the Ferrari solution of a quartic is not valid. What is the special form of $g(y)$ over a field $\mathfrak{F}$ of characteristic 2?

10. If the resolvent cubic has a root in $\mathfrak{F}$ we may choose our notation so that $\eta_1 = \xi_1\xi_2 + \xi_3\xi_4$ is in $\mathfrak{F}$. Show that then $\mathfrak{F}(\xi_1)$ and $\mathfrak{F}(\xi_2)$ correspond to the same subgroup of the Galois group of $\mathfrak{N} = \mathfrak{F}[\xi_1, \xi_2, \xi_3, \xi_4]$ over $\mathfrak{F}$ and $\mathfrak{F}[\xi_1] = \mathfrak{F}[\xi_2]$. Hint: Every automorphism of the group $\mathfrak{G}$ carries any root of $f(x)$ into a root. Show that if $\mathfrak{H}$ is the subgroup of $\mathfrak{G}$ to which $\mathfrak{F}(\xi_1)$ corresponds, and if S in $\mathfrak{H}$ alters ξ_2, it carries η_1 into one of the other η_i above, say η_2. But $\eta_1^S = \eta_1$ is in $\mathfrak{F}$ so that $\eta_1 = \eta_2$, $(\xi_1 - \xi_4)(\xi_3 - \xi_2) = 0$. But this is a contradiction of our assumption that $f(x)$ is separable.

11. Prove that the resolvent cubic of a separable quartic is separable.

12. Let $f(x)$ be an irreducible separable quartic with stem field $\mathfrak{Z} = \mathfrak{F}(\xi_1)$. Prove that $\mathfrak{Z}$ is normal over $\mathfrak{F}$ if and only if the roots of the resolvent cubic of $f(x)$ are in $\mathfrak{Z}$. Hint: If S is in the group corresponding to $\mathfrak{Z}$ then $\eta_i^S = \eta_i$ implies that $\xi_i^S = \xi_i$, $\mathfrak{Z}$ is normal. The converse is trivial.

13. Prove that the automorphism group of a normal quartic field is either cyclic or a group $\mathfrak{G}_4 = (I, S, T, ST)$, $TS = ST$, $S^2 = T^2 = I$, according as the roots of the resolvent cubic of a corresponding defining quartic are not or are all in $\mathfrak{F}$.

14. Let the resolvent cubic of an irreducible separable quartic $f(x)$ have the form $(y - \eta_1)h(y)$ with η_1 in $\mathfrak{F}$ and the roots η_2, η_3 not in $\mathfrak{Z} = \mathfrak{F}(\xi_1)$. Show that the root field of $f(x)$ has degree eight over $\mathfrak{F}$.

15. Let the resolvent cubic of a separable irreducible quartic $f(x)$ have a normal cubic stem field. Show that the root field $\mathfrak{N}$ of $f(x)$ has degree twelve over $\mathfrak{F}$. Hint: By Exercise 12 the field $\mathfrak{N}$ has degree four over the root field of the resolvent cubic.

16. Complete the classification of the root fields of irreducible separable quartics by proving that if the root field of the resolvent cubic has degree six over $\mathfrak{F}$ then the root field of the quartic has degree twenty-four over $\mathfrak{F}$.

17. Show that the stem fields of an irreducible quartic $f(x)$ have quadratic subfields if and only if $f(x)$ is inseparable or is separable with a reducible resolvent cubic. Hint: The result is trivial except when $\mathfrak{F}(\xi_1)$ is not normal so that we have the case of Exercise 14. If η_1 is in $\mathfrak{F}$ then we can show that $\mathfrak{F}(\xi_1) = \mathfrak{F}(\xi_2)$, $\mathfrak{F}(\xi_3) = \mathfrak{F}(\xi_4)$ but $\mathfrak{F}(\xi_1) \neq \mathfrak{F}(\xi_3)$. If then $\xi_1^S = \xi_2$ we have $\xi_2^S = \xi_1$ and easily show that $S^2 = I$, the desired quadratic field corresponds to the group $\mathfrak{H} + \mathfrak{H}S$ where $\mathfrak{H} \longleftrightarrow \mathfrak{F}(\xi_1)$.

18. Apply the Sylow theorems on groups (Sec. 6.7) to obtain corresponding theorems on the subfields of a normal field of degree $n = p^e q$, p a prime not dividing q.

4. Composites of $\mathfrak{N}$ and $\mathfrak{Z}$. We considered composites over $\mathfrak{F}$ of a field $\mathfrak{Z}$ over $\mathfrak{F}$ and an algebraic field $\mathfrak{N}$ of finite degree over $\mathfrak{F}$ in Section 7.7, and showed there that they are all equivalent to the stem fields $\mathfrak{Z}(\xi_i)$ of the minimum function $f(x)$ defining $\mathfrak{N} = \mathfrak{F}(\xi)$. Since each ξ_i is in $\mathfrak{F}(\xi)$ when $\mathfrak{N}$ is normal the fields $\mathfrak{Z}(\xi_i)$ are identical. This gives

Theorem 10. *The composites of a normal field $\mathfrak{N}$ over $\mathfrak{F}$ and any $\mathfrak{Z}$ over $\mathfrak{F}$ are all equivalent.*

We have now shown that if $f(x)$ is an irreducible separable polynomial of $\mathfrak{F}[x]$ such that any stem field $\mathfrak{N}$ of $f(x)$ is normal over $\mathfrak{F}$ then the composites $(\mathfrak{N}, \mathfrak{Z})$ may be taken to be an essentially unique field $\mathfrak{Z}(\xi)$ where $\mathfrak{Z}(\xi)$ is any stem field of $f(x)$. This then states that the factors $f_i(x)$ of (7.35) all have the same degree and define the same subfield of the root field over $\mathfrak{Z}$ of $f(x)$. This result is more explicitly given by the following considerations.

We let $\mathfrak{W}$ be any field which contains $\mathfrak{Z}$ as well as a root ξ of $f(x) = 0$. Then we assume without loss of generality that $\mathfrak{N} = \mathfrak{F}(\xi)$ and seek all composites in $\mathfrak{W}$ of $\mathfrak{N}$ and $\mathfrak{Z}$. As we have seen, they are all equal to $\mathfrak{Z}(\xi)$. But now we have

Theorem 11. *Let $\mathfrak{L}$ be the intersection in $\mathfrak{W}$ of $\mathfrak{N}$ and $\mathfrak{Z}$ and $\mathfrak{N} = \mathfrak{L}(\xi)$, where* $g_1(x)$ *is the minimum function of ξ over $\mathfrak{L}$. Then* $g_1(x)$ *is irreducible in $\mathfrak{Z}$ and $(\mathfrak{N}, \mathfrak{Z}) = \mathfrak{Z}(\xi)$ is a stem field of the irreducible* $g_1(x)$. *Moreover, the minimum function of ξ over $\mathfrak{F}$ has the factorization* (11), (12) *in both $\mathfrak{L}$ and $\mathfrak{Z}$.*

For if $g_1(x)$ is not irreducible in $\mathfrak{Z}$ it has an irreducible factor $h(x)$ in $\mathfrak{Z}[x]$. Then

$$h(x) = (x - \xi_1) \dots (x - \xi_r) = x^r + c_1x^{r-1} + \dots + c_r$$

with $\xi_1, \dots, \xi_r$ all in $\mathfrak{W}$ and hence in the unique subfield $\mathfrak{N}$ of $\mathfrak{W}$ which is the root field $\mathfrak{F}(\xi_1, \dots, \xi_r)$ of $f(x)$. The field $\mathfrak{L}(c_1, \dots, c_r) = \mathfrak{L}_1$ is a subfield of both $\mathfrak{N}$ and $\mathfrak{Z}$ since the coefficients c_i of the polynomial $h(x)$ are in $\mathfrak{Z}$ and the elementary symmetric functions $(-1)^ic_i$ of $\xi_1, \dots, \xi_r$ are in $\mathfrak{N}$. Thus $\mathfrak{L}_1 \leqq \mathfrak{L}$, the intersection of $\mathfrak{Z}$ and $\mathfrak{N}$. But $\mathfrak{L}_1 \geqq \mathfrak{L}$ so that $\mathfrak{L}_1 = \mathfrak{L}$, $h(x)$ is in $\mathfrak{L}[x]$. The polynomial $g_1(x)$ is an irreducible polynomial of $\mathfrak{L}[x]$ and is divisible by $h(x)$, a contradiction. Thus $g_1(x)$ is irreducible in $\mathfrak{Z}$. We factor $f(x)$ in $\mathfrak{L}$ and have (11), (12). Each $g_i(x)$ is irreducible in $\mathfrak{L}$ and also in $\mathfrak{Z}$ by our above argument and defines $\mathfrak{Z}(\xi) = (\mathfrak{N}, \mathfrak{Z})$ over $\mathfrak{Z}$.

The automorphism group $\mathfrak{H}$ of $\mathfrak{N}$ over $\mathfrak{L}$ is generated by the correspondences

$$H_j: \qquad \xi \longleftrightarrow \xi^{H_j}$$

with H_j defined by $g_i(x)$ of (10), (11), (12). Since $g_i(x)$ is irreducible in $\mathfrak{Z}$ the correspondences above generate the group $\mathfrak{H}_0$ of $(\mathfrak{N}, \mathfrak{Z})$ over $\mathfrak{Z}$. We cannot say that $\mathfrak{H}_0$ is $\mathfrak{H}$ since they are automorphism groups of different fields but they are certainly equivalent. We now state our results in the useful form

Theorem 12. *Let $\mathfrak{Z} > \mathfrak{F}$, $\mathfrak{N}$ be normal over $\mathfrak{F}$ and $\mathfrak{L}$ be the intersection of $\mathfrak{Z}$ and $\mathfrak{N}$. Then the degree of $\mathfrak{N}$ over $\mathfrak{L}$ is the degree of the normal field $\mathfrak{N}_0 = (\mathfrak{N}, \mathfrak{Z})$ over $\mathfrak{Z}$ and their automorphism groups $\mathfrak{H}$, $\mathfrak{H}_0$ are equivalent.*

We now let $\mathfrak{N}$ and $\mathfrak{C}$ be normal over $\mathfrak{F}$ with respective automorphism groups $\mathfrak{G}, \mathfrak{Q}$. Without loss of generality we may assume that $\mathfrak{N}$ and $\mathfrak{C}$ are subfields of the composite $\mathfrak{K}$ and prove

Theorem 13. *The composite $\mathfrak{K}$ of $\mathfrak{N}$ and $\mathfrak{C}$ is normal over $\mathfrak{F}$. The intersection $\mathfrak{L}$ of $\mathfrak{N}$ and $\mathfrak{C}$ is also normal over $\mathfrak{F}$ and belongs to a normal divisor $\mathfrak{G}_1$ of $\mathfrak{G}$ and $\mathfrak{Q}_1$ of $\mathfrak{Q}$. The group of $\mathfrak{K} = (\mathfrak{N}, \mathfrak{C})$ over $\mathfrak{C}$ is equivalent to $\mathfrak{G}_1$ with a similar result for $\mathfrak{K}$ over $\mathfrak{N}$, and the group of $\mathfrak{L}$ over $\mathfrak{F}$ is equivalent to*

$$\mathfrak{G}/\mathfrak{G}_1 \cong \mathfrak{Q}/\mathfrak{Q}_1 .$$

For if $\mathfrak{N} = \mathfrak{F}(\xi)$ and $\mathfrak{C} = \mathfrak{F}(\eta)$ are distinct then $\mathfrak{K} = \mathfrak{F}(\xi, \eta)$ is the root field of $f(x) \cdot g(x)$ where $f(x)$ is the minimum function of ξ and $g(x)$ is that for η. Thus $\mathfrak{K}$ is normal over $\mathfrak{F}$. The intersection $\mathfrak{L}$ of $\mathfrak{N}$ and $\mathfrak{C}$ is a field defined by a root α of $h(x) = 0$ and the root field $\mathfrak{L}_0$ of $h(x)$ is contained in $\mathfrak{N}$ and $\mathfrak{C}$ by Theorem 7.13. But then $\mathfrak{L}_0 = \mathfrak{L}$ is normal over $\mathfrak{F}$. The group of $\mathfrak{L}$ over $\mathfrak{F}$ is equivalent to $\mathfrak{G}/\mathfrak{G}_1$ and $\mathfrak{Q}/\mathfrak{Q}_1$ by Theorem 8.8 and the group of $\mathfrak{K}$ over $\mathfrak{C}$ is equivalent to $\mathfrak{G}_1$ by Theorem 8.12.

5. Relative traces and norms. Any separable field $\mathfrak{K}$ of degree n over $\mathfrak{F}$ is a simple extension $\mathfrak{F}(\xi)$ defined by a root ξ of a separable irreducible monic polynomial $f(x)$ of the polynomial domain $\mathfrak{F}[x]$. The roots $\xi = \xi_1, \xi_2, \ldots, \xi_n$ of $f(x)$ in a root field $\mathfrak{N}$ are thus all distinct and define n conjugate fields $\mathfrak{K}_i = \mathfrak{F}(\xi_i)$ of $\mathfrak{K}$, and n conjugate quantities

$$\eta_1 = \eta(\xi_1) , \eta_2 = \eta(\xi_2), \ldots , \eta_n = \eta(\xi_n)$$

of any quantity $\eta(\xi)$ of $\mathfrak{K}$. We have already defined (in Section 7.6) the norm and trace of η but shall introduce the more explicit notations

$$T_{\mathfrak{K}|\mathfrak{F}}(\eta) = \eta_1 + \ldots + \eta_n , \qquad N_{\mathfrak{K}|\mathfrak{F}}(\eta) = \eta_1\eta_2 \ldots \eta_n .$$

The former is read "the trace with respect to $\mathfrak{K}$ over $\mathfrak{F}$ of η" and similarly for the norm. If $\mathfrak{K}_0$ is any subfield over $\mathfrak{F}$ of $\mathfrak{K}$ the relative trace and norm

$$(13) \qquad T_{\mathfrak{K}|\mathfrak{K}_0}(\eta) , \qquad N_{\mathfrak{K}|\mathfrak{K}_0}(\eta) ,$$

are of course defined when we replace $\mathfrak{F}$ by the subfield $\mathfrak{K}_0$ of $\mathfrak{K}$. For $\mathfrak{K}$ is a separable algebraic extension of degree q over $\mathfrak{K}_0$ where $n = mq$, $\mathfrak{K}_0$ has degree m over $\mathfrak{F}$. Evidently $T_{\mathfrak{K}_0|\mathfrak{F}}(\beta)$, $N_{\mathfrak{K}_0|\mathfrak{F}}(\beta)$ are defined for any β of $\mathfrak{K}_0$ and the functions (13) are quantities β of $\mathfrak{K}_0$. The relations between our three sets of functions are then given by

Theorem 14. *Let $\mathfrak{K} > \mathfrak{K}_0 > \mathfrak{F}$ where $\mathfrak{K}$ is a separable algebraic field of degree* n *over $\mathfrak{F}$. Then*

$$T_{\mathfrak{K}|\mathfrak{F}}(\eta) = T_{\mathfrak{K}_0|\mathfrak{F}}[T_{\mathfrak{K}|\mathfrak{K}_0}(\eta)], \qquad N_{\mathfrak{K}|\mathfrak{F}}(\eta) = N_{\mathfrak{K}_0|\mathfrak{F}}[N_{\mathfrak{K}|\mathfrak{K}_0}(\eta)]$$

for any η of $\mathfrak{K}$.

For let $\mathfrak{G}$ be the automorphism group of the root field $\mathfrak{N}$ of degree r over $\mathfrak{F}$ of $f(x)$, the minimum function of a generating quantity ξ of $\mathfrak{K}$ over $\mathfrak{F}$. By Theorem 8.5 we have

$$\mathfrak{G} > \mathfrak{H}_0 > \mathfrak{H} \geqq [I], \qquad \mathfrak{F} < \mathfrak{K}_0 < \mathfrak{K} \leqq \mathfrak{N}$$

where $\mathfrak{H}_0$ corresponds to $\mathfrak{K}_0$, $\mathfrak{H}$ to $\mathfrak{K}$. The index of $\mathfrak{H}_0$ under $\mathfrak{G}$ is m, the degree of $\mathfrak{K}_0$ over $\mathfrak{F}$, and the index of $\mathfrak{H}$ under $\mathfrak{H}_0$ is q, the degree of $\mathfrak{K}$ over $\mathfrak{K}_0$. We thus have

$$\mathfrak{G} = \mathfrak{H}_0 + \mathfrak{H}_0 S_2 + \dots + \mathfrak{H}_0 S_m ,$$
$$\mathfrak{H}_0 = \mathfrak{H} + \mathfrak{H} T_2 + \dots + \mathfrak{H} T_q .$$

But then

$$\mathfrak{G} = \mathfrak{H} + \mathfrak{H} T_2 + \dots + \mathfrak{H} T_q + \mathfrak{H} S_2 + \dots + \mathfrak{H} T_q S_m$$

is a decomposition of the group $\mathfrak{G}$ relative to $\mathfrak{H}$ since the index of $\mathfrak{H}$ under $\mathfrak{G}$ is $n = qm$. If $\mathfrak{K}_0 = \mathfrak{F}(\alpha)$ the m conjugates of α are

$$\alpha, \alpha,^{S_2} \dots , \alpha^{S_m} ,$$

and the n conjugates of ξ are

$$\xi, \xi^{T_2}, \dots , \xi^{T_q}, \xi^{S_2}, \dots , \xi^{T_q S_m} .$$

However $\mathfrak{K}$ has degree q over $\mathfrak{K}_0$, the group of $\mathfrak{N}$ over $\mathfrak{K}_0$ is $\mathfrak{H}_0$ and the conjugates of ξ defined by the minimum equation of ξ with respect to $\mathfrak{K}_0$ are

$$\xi, \xi^{T_2}, \dots , \xi^{T_q} .$$

This implies that

$$T_{\mathfrak{K}|\mathfrak{K}_0}(\eta) = \eta + \eta^{T_2} + \dots + \eta^{T_q} , \qquad N_{\mathfrak{K}|\mathfrak{K}_0}(\eta) = \eta\eta^{T_2} \dots \eta^{T_q}$$

for any η of $\mathfrak{K}$, and that obviously

$$T_{\mathfrak{K}_0|\mathfrak{F}}[T_{\mathfrak{K}|\mathfrak{K}_0}(\eta)] = (\eta + \dots + \eta^{T_q}) + (\eta + \dots + \eta^{T_q})^{S_2} + \dots + (\eta + \dots + \eta^{T_q})^{S_m} = \eta + \dots + \eta^{T_q} + \eta^{S_2} + \dots + \eta^{T_q S_m} = T_{\mathfrak{K}|\mathfrak{F}}(\eta)$$

since we have now added all the conjugates of η as a quantity of $\mathfrak{K}$ over $\mathfrak{F}$. Similarly,

$$N_{\mathfrak{K}_0|\mathfrak{F}}[N_{\mathfrak{K}|\mathfrak{K}_0}(\eta)] = (\eta\,\eta^{T_2} \dots \eta^{T_q})(\eta\,\eta^{T_2} \dots \eta^{T_q})^{S_2} \dots (\eta\,\eta^{T_2} \dots \eta^{T_q})^{S_m} = \eta\eta^{T_2} \dots \eta^{T_q S_m} = N_{\mathfrak{K}|\mathfrak{F}}(\eta)\,.$$

6. The Galois group of an equation. Let

(14) $$f(x) = (x - \xi_1) \dots (x - \xi_r)$$

be a separable equation and $\mathfrak{N} = \mathfrak{F}(\xi_1, \dots, \xi_r)$ be the root field of $f(x)$. Then $\mathfrak{N} = \mathfrak{F}(\eta)$ is a normal field of degree n over $\mathfrak{F}$ and has automorphism group $\mathfrak{G}$. The coefficients of $f(x)$ are in $\mathfrak{F}$ and are unaltered by any S of $\mathfrak{G}$. Hence

(15) $$f(x) = (x - \xi_1^S) \dots (x - \xi_r^S)$$

and we have a correspondence

(16) $$S \to P_S\,,$$

where P_S is the permutation (or substitution) on the roots $\xi_1, \dots, \xi_r$ of $f(x)$ given by

(17) $$P_S = \begin{pmatrix} \xi_1 \dots \xi_r \\ \xi_1^S \dots \xi_r^S \end{pmatrix}.$$

If $P_S = P_T$ for $S \neq T$ then $\xi_i^S = \xi_i^T$ $(i = 1, \dots, r)$ and η is in $\mathfrak{F}(\xi_1, \dots, \xi_r)$, $\eta^S = \eta^T$ contrary to Theorem 8.3. Thus the set $\mathfrak{G}_0$ of all permutations P_S is a set of n permutations and (16) is a one-to-one correspondence. The set $\mathfrak{G}_0$ is a group equivalent to $\mathfrak{G}$. For we need only show that (16) is preserved under multiplication. But

$$P_{ST} = \begin{pmatrix} \xi_1 & \dots & \xi_r \\ \xi_1^{ST} & \dots & \xi_r^{ST} \end{pmatrix}, \qquad P_T = \begin{pmatrix} \xi_1^S & \dots & \xi_r^S \\ \xi_1^{ST} & \dots & \xi_r^{ST} \end{pmatrix}$$

since a permutation is unaltered by any permutation of its columns. Then

$$P_S P_T = \begin{pmatrix} \xi_1 & \dots & \xi_r \\ \xi_1^{ST} & \dots & \xi_r^{ST} \end{pmatrix} = P_{ST}\,.$$

We call $\mathfrak{G}_0$ the *Galois group of the equation* (14). It is a group of permutations on the roots of $f(x)$.

The group $\mathfrak{G}_0$ is not merely equivalent to $\mathfrak{G}$ but if $a(\xi_1, \ldots, \xi_r)$ is any quantity of $\mathfrak{N}$ then

$$a^S = a^{P_S}.$$

Thus the automorphisms S of $\mathfrak{N}$ are the correspondences

$$a \longleftrightarrow a^{P_S}. \tag{18}$$

There is therefore no loss of generality when we replace $\mathfrak{G}$ by $\mathfrak{G}_0$, and this is usually done in the *Galois theory of equations*.

7. Some properties of $\mathfrak{G}_0$. All of the theorems which we obtained for $\mathfrak{G}$ of course hold for what is merely a representation $\mathfrak{G}_0$ of $\mathfrak{G}$. But we shall now discuss some special properties of $\mathfrak{G}_0$ which are due to the fact that it is a permutation group.

We write our separable polynomial $f(x)$ as a product $f_1(x) \ldots f_t(x)$ of distinct irreducible polynomials $f_i(x)$ of $\mathfrak{F}[x]$. The coefficients of $f_i(x)$ are unaltered by any automorphism of the group $\mathfrak{G}$ and hence its roots must be merely permuted by $\mathfrak{G}_0$. If $\xi_{i1}, \ldots, \xi_{ir_i}$ are roots of $f_i(x)$ forming a transitive system of $\mathfrak{G}_0$ then the polynomial $(x - \xi_{i1}) \ldots (x - \xi_{ir_i})$ has coefficients in $\mathfrak{F}$, divides the irreducible $f_i(x)$, and must equal $f_i(x)$. Thus the roots of $f_i(x)$ form a transitive system of $\mathfrak{G}_0$ and the number of such systems is the number of irreducible factors of $f(x)$.

COROLLARY. *The equation* $f(x) = 0$ *is irreducible if and only if its Galois group is transitive.*

We next let $f(x)$ be irreducible of degree r. If $r = n$ we call $\mathfrak{G}_0$ a *regular group*. But then $\mathfrak{F}(\xi_1) = \mathfrak{N}$ is a stem field of $f(x)$ and the group $\mathfrak{G}_0$ is determined by

$$\xi_i = \theta_i(\xi), \qquad \xi = \xi_1 \tag{19}$$

and the elements of $\mathfrak{G}_0$ are the permutations

$$P = \begin{pmatrix} \xi & \theta_2(\xi) & \ldots & \theta_r(\xi) \\ \theta_i(\xi) & \theta_2[\theta_i(\xi)] & \ldots & \theta_r[\theta_i(\xi)] \end{pmatrix}. \tag{20}$$

Thus the first column of P determines the whole permutation when $\mathfrak{G}_0$ is regular.

8. Equations with a prescribed group. Let $\mathfrak{Q}$ be a field and $x_1, \ldots, x_n$ independent indeterminates over $\mathfrak{Q}$. The equation

$$f(x) = (x - x_1) \ldots (x - x_n) = x^n - c_1x^{n-1} + \ldots + (-1)^n c_n = 0 \tag{21}$$

is a separable equation with roots in $\mathfrak{Q}[x_1, \ldots, x_n]$ and coefficients in $\mathfrak{F} = \mathfrak{Q}(c_1, \ldots, c_n)$. The Galois group $\mathfrak{G}_0$ of $f(x)$ over $\mathfrak{F}$ is a subgroup of the symmetric $\mathfrak{G}^{(n)}$ of all permutations of $x_1, \ldots, x_n$. But *every* permutation of $x_1, \ldots, x_n$ defines an automorphism of $\mathfrak{N} = \mathfrak{Q}(x_1, \ldots, x_n)$ which leaves the quantities of $\mathfrak{F}$ unaltered. Thus $\mathfrak{G}_0 = \mathfrak{G}^{(n)}$.

As a corollary of this result and our fundamental theorems we have

Theorem 15. *A rational function with coefficients in* $\mathfrak{Q}$ *of* $x_1, \ldots, x_n$ *is unaltered by every permutation of* $x_1, \ldots, x_n$ *if and only if it is in* $\mathfrak{Q}(c_1, \ldots, c_n)$.

Every subfield $\mathfrak{L}$ over $\mathfrak{F}$ of $\mathfrak{N}$ corresponds uniquely to some subgroup $\mathfrak{H}_0$ of $\mathfrak{G}_0$ and the group of $f(x)$ over $\mathfrak{L}$ is the group $\mathfrak{H}_0$. We take $\mathfrak{F} = \mathfrak{L}$ and have proved

Theorem 16. *Let* $\mathfrak{H}_0$ *be any subgroup of the symmetric group. Then there exists a field* $\mathfrak{F}$ *and an equation* $f(x) = 0$ *whose Galois group is* $\mathfrak{H}_0$.

The *general* equation is the equation

$$x^n + \xi_1 x^{n-1} + \ldots + \xi_n = 0, \tag{22}$$

where $\xi_1, \ldots, \xi_n$ are independent indeterminates over some field $\mathfrak{Q}$. We let $\mathfrak{F} = \mathfrak{Q}(\xi_1, \ldots, \xi_n)$ and prove

Theorem 17. *The general equation is separable and its Galois group is the symmetric group.*

For the field $\mathfrak{F}$ is equivalent to $\mathfrak{Q}(c_1, \ldots, c_n)$ where $c_1, \ldots, c_n$ are given by (21) and are independent indeterminates over $\mathfrak{Q}$. We may therefore replace the coefficients ξ_i of (22) by the corresponding new independent indeterminates $(-1)^i c_i$ and have already shown the above properties.

EXERCISES

1. Let $f(x)$ be a cubic separable equation. Use the exercises of Section 8.3 to represent the automorphism groups of the root fields of $f(x)$ as permutation groups on the roots of $f(x)$.

2. Obtain the theory of Ex. 1 for quartic equations.

9. Direct products. We shall prove

Theorem 18. *Let the automorphism group* $\mathfrak{G}$ *of a normal field* $\mathfrak{N}$ *over* $\mathfrak{F}$ *be the direct product of two subgroups* $\mathfrak{H}_1$ *and* $\mathfrak{H}_2$. *Then*

$$\mathfrak{N} = \mathfrak{L}_1 \times \mathfrak{L}_2 \tag{23}$$

where $\mathfrak{L}_1 \longleftrightarrow \mathfrak{H}_1$, $\mathfrak{L}_2 \longleftrightarrow \mathfrak{H}_2$.

For if $\mathfrak{G}$ has order $n = h_1 h_2$ where h_i is the order of $\mathfrak{H}_i$ then $\mathfrak{L}_1$ has degree h_2, $\mathfrak{L}_2$ has degree h_1. If the composite $\mathfrak{L}$ of $\mathfrak{L}_1$ and $\mathfrak{L}_2$ is unaltered by an automorphism S of $\mathfrak{G}$ then $\mathfrak{L}_1$ and $\mathfrak{L}_2$ are separately unaltered by S. Thus S is in both $\mathfrak{H}_1$ and $\mathfrak{H}_2$. Their intersection is the identity group, $S = I$, $\mathfrak{L} \longleftrightarrow (I)$. Then $\mathfrak{L} = \mathfrak{N}$ has degree $n = h_1h_2$ so that $\mathfrak{L} = \mathfrak{L}_1 \times \mathfrak{L}_2 = \mathfrak{N}$.

The result above may be applied to *cyclic fields*, that is normal fields with a cyclic automorphism group. Every cyclic group is a direct product of cyclic groups of prime power orders and a repeated application of Theorem 18 gives the

COROLLARY. *Let $\mathfrak{N}$ be cyclic of degree*

$$n = p_1^{e_1} \dots p_r^{e_r}$$

over $\mathfrak{F}$, where the integers p_i are distinct primes. Then $\mathfrak{N}$ is the direct product

$$\mathfrak{L}_1 \times \dots \times \mathfrak{L}_r, \tag{24}$$

where $\mathfrak{L}_i$ is cyclic of degree $p_i^{e_i}$ over $\mathfrak{F}$.

This theorem essentially reduces all problems on the structure and existence of cyclic fields of degree n over $\mathfrak{F}$ to the case where $n = p^e$, p a prime. There are evidently two distinct cases which arise. These are the cases where p is or is not the characteristic of $\mathfrak{F}$. We shall consider this construction theory in a later chapter.

EXERCISE

State and prove the converses of Theorem 8.18 and its corollary.

10. Metacyclic fields over $\mathfrak{F}$. We consider a normal field $\mathfrak{N}$ of degree n over $\mathfrak{F}$ and prove

Theorem 19. *Let $\mathfrak{L}_1 < \mathfrak{N}$, $\mathfrak{L}_2 < \mathfrak{N}$ with corresponding subgroups $\mathfrak{H}_1$ and $\mathfrak{H}_2$ of the automorphism group $\mathfrak{G}$ of $\mathfrak{N}$. Then $\mathfrak{L}_2$ is cyclic of prime degree* p *over $\mathfrak{L}_1$ if and only if $\mathfrak{H}_2$ is a maximal normal divisor of $\mathfrak{H}_1$ with index the prime* p.

For the group of $\mathfrak{N}$ over its subfield $\mathfrak{L}_1$ is $\mathfrak{H}_1$. By Theorem 8.5 the field $\mathfrak{L}_2$ has degree p over $\mathfrak{L}_1$ if and only if $\mathfrak{H}_2$ is a subgroup of index p of $\mathfrak{H}_1$. Also $\mathfrak{L}_2$ is normal over $\mathfrak{L}_1$ if and only if $\mathfrak{H}_2$ is a normal divisor of $\mathfrak{H}_1$ by Theorem 8.7, and Theorems 6.14 and 6.2 state that $\mathfrak{H}_1/\mathfrak{H}_2$ is simple, $\mathfrak{H}_2$ is maximal.

This argument also implies that if $\mathfrak{L}_2$ has prime degree p over $\mathfrak{L}_1$ there is no subfield $\mathfrak{L}$ of $\mathfrak{L}_2$ such that $\mathfrak{L}_2 > \mathfrak{L} > \mathfrak{L}_1$. But this is an evident consequence of the fact that the degree of $\mathfrak{L}_2$ over $\mathfrak{L}_1$ is the product p of its degree over $\mathfrak{L}$ by the degree of $\mathfrak{L}$ over $\mathfrak{L}_1$.

DEFINITION. *A normal field $\mathfrak{N}$ over $\mathfrak{F}$ is called metacyclic if there exists a chain of subfields*

$$\mathfrak{F} = \mathfrak{N}_0 < \mathfrak{N}_1 < \ldots < \mathfrak{N}_t = \mathfrak{N} \tag{25}$$

where $\mathfrak{N}_i$ is cyclic of prime degree p_i over $\mathfrak{N}_{i-1}$ ($i = 1, \ldots, t$).

Theorem 8.19 states that if $\mathfrak{N}_i \longleftrightarrow \mathfrak{H}_i$ then $\mathfrak{H}_i$ is a maximal normal divisor of index p_i of $\mathfrak{H}_{i-1}$ where $\mathfrak{H}_0 = \mathfrak{G}$. But then $\mathfrak{G}$ is solvable. The converse is also an evident consequence of Theorem 8.19 and we state this result in

Theorem 20. *A normal field $\mathfrak{N}$ over $\mathfrak{F}$ is metacyclic if and only if its automorphism group is solvable.*

EXERCISES

1. Prove that every cyclic field is metacyclic.

2. Let $\mathfrak{N}$ be normal over $\mathfrak{F}$ and $\mathfrak{N} > \mathfrak{K} > \mathfrak{F}$ where $\mathfrak{N}$ is metacyclic over $\mathfrak{K}$ and $\mathfrak{K}$ is metacyclic over $\mathfrak{F}$. Then $\mathfrak{N}$ is metacyclic over $\mathfrak{F}$.

3. Show that the subgroups of the symmetric groups on three and four letters are all solvable. What does this imply about the root fields of separable cubic and quartic equations?

11. Cyclotomic fields. Let p be a prime and

$$f(x) = x^p - a \qquad (a \text{ in } \mathfrak{F}).$$

When p is the characteristic of $\mathfrak{F}$ the equation $f(x) = 0$ is inseparable and when $a = 1$ we have $f(x) = (x - 1)^p$. In this case there is no field $\mathfrak{K} = \mathfrak{F}(\zeta)$, where $\zeta \neq 1$ and $\zeta^p = 1$. We are interested in such cyclotomic fields and shall assume in this section that p is not the characteristic of $\mathfrak{F}$.

The derivative of $f(x) = x^p - 1$ is $px^{p-1} \neq 0$. But $f'(x)$ is then prime to $f(x)$ and

$$f(x) = (x - 1)g(x), \qquad g(x) = x^{p-1} + \ldots + 1$$

is separable. The equation $g(x) = 0$ is called the *cyclotomic equation* and its root field

$$\mathfrak{K} = \mathfrak{F}(\zeta)$$

is called a *cyclotomic field.*

If $\zeta_1^p = 1$ and $\zeta_2^p = 1$ then $(\zeta_1\zeta_2)^p = 1$ and $\zeta_1^{-1} = \zeta_1^{p-1}$ has the property $(\zeta_1^{-1})^p = 1$. Thus the set $\mathfrak{Q}$ of all roots of $f(x) = 0$ forms a multiplicative group. Since $f(x)$ is separable this group has order p and is a cyclic group by Theorem 6.2. We write $\mathfrak{Q} = [\zeta]$ and see that ζ generates $\mathfrak{K}$. Also every root of $f(x) = 0$ is a power of ζ so that $\mathfrak{K}$ is the root field of $f(x)$. Notice that ζ^t also generates $\mathfrak{K}$ for every t not divisible by p.

By Theorem 7.25 with $n = 1$ there exists an integer r such that the remainders on division of

$$(26) \qquad r, r^2, \ldots, r^{p-1}$$

by p are $1, 2, \ldots, p - 1$ in some order. Then the roots of $g(x) = 0$ are

$$(27) \qquad \zeta_q = \zeta^{r^q} \qquad (q = 1, \ldots, p - 1).$$

The field $\mathfrak{K}$ is normal over $\mathfrak{F}$ and every automorphism of $\mathfrak{K}$ over $\mathfrak{F}$ is one of the powers

$$S^q: \qquad \zeta \longleftrightarrow \zeta_q$$

of the* correspondence $\zeta \longleftrightarrow \zeta^r$. Hence the automorphisms of $\mathfrak{K}$ over $\mathfrak{F}$ form a subgroup of the cyclic group $[S]$ and this subgroup is a cyclic group. Thus $\mathfrak{K}$ is a cyclic field of degree n over $\mathfrak{F}$ and its generating automorphism is

$$(28) \qquad \zeta \longleftrightarrow \zeta^t,$$

where t is an integer chosen so that

$$(29) \qquad t^n - 1$$

is divisible by p and $t^m - 1$ is not divisible by p for any $m < n$.

Theorem 21. *The equation* $x^p = a$ *is irreducible in* $\mathfrak{F}$ *if and only if* $a \neq b^p$ *for every* b *of* $\mathfrak{F}$.

For let $x^p - a = h(x) \cdot g(x)$ where $h(x)$ is an irreducible monic polynomial of $\mathfrak{F}[x]$. Then $h(x) = (x - \alpha)(x - \zeta_2\alpha) \ldots (x - \zeta_r\alpha)$, where $\alpha_i^p = a$ and we have written $\alpha_1 = \alpha$, $\alpha_i\alpha^{-1} = \zeta_i$. Also $\zeta_i^p = \alpha_i^p\alpha^{-p} = aa^{-1} = 1$, and ζ_i is a root of $x^p = 1$. The product $b_0 = \alpha\zeta_2\alpha \ldots \zeta_r\alpha = \zeta_0\alpha^r$ is in $\mathfrak{F}$ where $\zeta_0^p = 1$ and $\alpha^{rp} = b_0^p = a^r$. If r is less than p there exist integers s and t such that $rs + tp = 1$, $a^{rs} = a^{1-tp} = b_0^{ps}$, $a = (b_0^s a^t)^p = b^p$ with b in $\mathfrak{F}$. Hence if $a \neq b^p$ with b in $\mathfrak{F}$ we must have $r = p$ and $f(x) = x^p - a = h(x)$ is irreducible in $\mathfrak{F}$. The converse is trivial.

Theorem 22. *Let* $\mathfrak{F}$ *contain a quantity* $\zeta \neq 1$ *such that* $\zeta^p = 1$. *Then the field* $\mathfrak{N}$ *is cyclic of prime degree* p *over* $\mathfrak{F}$ *if and only if* $\mathfrak{N} = \mathfrak{F}(\alpha)$, $\alpha^p = a$ *in* $\mathfrak{F}$, $a \neq b^p$ *with* b *in* $\mathfrak{F}$.

For if $\mathfrak{N} = \mathfrak{F}(\alpha)$ then $\mathfrak{N}$ has degree p over $\mathfrak{F}$ and is cyclic over $\mathfrak{F}$ with generating automorphism $\alpha \longleftrightarrow \zeta\alpha$. Conversely, let $\mathfrak{N} = \mathfrak{F}(\xi)$ be cyclic

* By this notation we mean that any polynomial in ζ corresponds to the same polynomial in ζ_q.

of degree p over $\mathfrak{F}$ and designate the roots of the minimum function of ξ by $\xi_i = \xi^{S^{i-1}}$. These quantities are in $\mathfrak{N}$ and so are

$$a_i = \xi_1 + \zeta^i \xi_2 + \dots + \zeta^{i(p-1)} \xi_p \quad (i = 0, \dots, p-1). \tag{30}$$

The equations (30) may be considered as p equations in p unknowns $\xi_1, \dots, \xi_p$ and the determinant of the coefficients (which are in $\mathfrak{F}$) is

$$d = |\zeta^{(i-1)(j-1)}| \quad (i, j = 1, \dots, p). \tag{31}$$

It is easily shown in the elementary theory of determinants that this *Vandermonde determinant* has the value $d = \prod_{i<j} (\zeta^i - \zeta^j) \neq 0$. But if the a_i are all in $\mathfrak{F}$ the equations (30) state that the ξ_i are all in $\mathfrak{F}$ contrary to our hypothesis that $\mathfrak{N} = \mathfrak{F}(\xi_1) > \mathfrak{F}$. Hence one $a_i = a$ is not in $\mathfrak{F}$ and this is not a_0 which is the trace $\xi_1 + \dots + \xi_p$ of ξ and hence is in $\mathfrak{F}$. It is evident that $a_i^S = \zeta^{-i} a_i \neq a_i$ since $i \neq 0$, $(a^p)^S = \zeta^{ip} a_i^p = a_i^p = a^p = a$ is in $\mathfrak{F}$ and that $\mathfrak{N} = \mathfrak{F}(a)$. By Theorem 8.21 we have $a \neq b^p$ for any b in $\mathfrak{F}$.

12. Solution by radicals. The classical theory of the solution of equations by radicals is a theory over a non-modular field $\mathfrak{F}$. To avoid complications we shall assume in this section that $\mathfrak{F}$ is non-modular.

Definition. *A field $\mathfrak{K}$ over $\mathfrak{F}$ is said to be solvable by radicals relatively to $\mathfrak{F}$ if it is possible to express any quantity of $\mathfrak{K}$ in terms of quantities of $\mathfrak{F}$ by a finite number of rational operations and root extractions.*

This definition implies that if $\mathfrak{K}$ has finite degree over $\mathfrak{F}$ it is solvable by radicals relatively to $\mathfrak{F}$ if and only if there exists a field $\mathfrak{K}_t \geqq \mathfrak{K}$ such that

$$\mathfrak{K}_t \geqq \mathfrak{K}_{t-1} \geqq \dots \geqq \mathfrak{K}_0 = \mathfrak{F}, \tag{32}$$

where either $\mathfrak{K}_i = \mathfrak{K}_{i-1}$ or $\mathfrak{K}_i$ has degree p_i over $\mathfrak{K}_{i-1}$,

$$\mathfrak{K}_i = \mathfrak{K}_{i-1}(a_i) = \mathfrak{F}(a_1, \dots, a_i), \quad a_i^{p_i} = a_i \text{ in } \mathfrak{K}_{i-1} \quad (i = 1, \dots, t), \tag{33}$$

and $p_1, \dots, p_t$ are primes. Thus the field $\mathfrak{K}_t$ is obtained by successive adjunction to $\mathfrak{F}$ of all the radicals occurring in the expression of a generating element ξ of $\mathfrak{K}$ over $\mathfrak{F}$ by radicals in terms of $\mathfrak{F}$.

If $\mathfrak{K} > \mathfrak{K}_0 > \mathfrak{F}$ and $\mathfrak{K}$ is solvable by radicals relatively to $\mathfrak{F}$ then evidently $\mathfrak{K}_0$ is solvable by radicals relatively to $\mathfrak{F}$, and $\mathfrak{K}$ is solvable by radicals relatively to $\mathfrak{K}_0$. The converse states that if $\mathfrak{K}$ is solvable by radicals relatively to $\mathfrak{K}_0$ and $\mathfrak{K}_0$ relatively to $\mathfrak{F}$ then $\mathfrak{K}$ is solvable by radicals relatively to $\mathfrak{F}$. This is also evident. Now we prove

Lemma 1. *Let $\mathfrak{N}$ be metacyclic of degree* $n = p_1 \dots p_r$ *over $\mathfrak{F}$ and let $\mathfrak{F}$*

contain $\zeta_i \neq 1$ *such that* $\zeta_i^{p_i} = 1$ *for primes* p_i ($i = 1, \ldots, r$). *Then* $\mathfrak{N}$ *is solvable by radicals relatively to* $\mathfrak{F}$.

For $\mathfrak{N}$ has the form (25) and thus the form (32) with $\mathfrak{K}_i = \mathfrak{K}_{i-1}(a_i) = \mathfrak{N}_i$ by Theorem 8.22.

LEMMA 2. *Let* $p_1, \ldots, p_r$ *be primes and* $\mathfrak{C}_1, \ldots, \mathfrak{C}_r$ *the corresponding cyclotomic fields over* $\mathfrak{F}$. *Then their composite* $\mathfrak{C}$ *is a metacyclic field over* $\mathfrak{F}$.

This result is evident when $r = 1$ since $\mathfrak{C}_1$ is cyclic over $\mathfrak{F}$. Assume that the composite $\mathfrak{D}$ of $\mathfrak{C}_1, \ldots, \mathfrak{C}_{r-1}$ is metacyclic over $\mathfrak{F}$. Since $\mathfrak{D}$ is normal over $\mathfrak{F}$ and $\mathfrak{C}_r$ is normal over $\mathfrak{F}$ their composite $\mathfrak{C}$ is normal over $\mathfrak{F}$. Theorem 8.12 states that the automorphism group of $\mathfrak{C} = (\mathfrak{C}_r, \mathfrak{D})$ over $\mathfrak{D}$ is equivalent to a subgroup of the cyclic group of $\mathfrak{C}_r$ over $\mathfrak{F}$ and is cyclic and hence solvable. Then $\mathfrak{C}$ is metacyclic over $\mathfrak{D}$ and, since $\mathfrak{D}$ is metacyclic over $\mathfrak{F}$, the field $\mathfrak{C}$ is metacyclic over $\mathfrak{F}$.

LEMMA 3. *A cyclotomic field* $\mathfrak{K} = \mathfrak{F}(\zeta)$, $\zeta^p = 1$, *is solvable by radicals relatively to* $\mathfrak{F}$.

For the theorem is true for primes $p = 1$, $p = 2$ and $\zeta = -1$, $p = 3$ and $\zeta = -\frac{1}{2}(1 + \sqrt{-3})$. Let it be true for all primes less than p. The composite $\mathfrak{C}$ of all the cyclotomic fields $\mathfrak{C}_i$ defined for primes $p_i < p$ is then solvable relatively to $\mathfrak{F}$ and is a normal field over $\mathfrak{F}$ since each $\mathfrak{C}_i$ is normal over $\mathfrak{F}$. The composite $(\mathfrak{K}, \mathfrak{C})$ over $\mathfrak{C}$ is cyclic over $\mathfrak{C}$ by Theorem 8.13 and its degree divides $p - 1$. But then $(\mathfrak{K}, \mathfrak{C}) \geqq \mathfrak{C}$ and is solvable relatively to $\mathfrak{C}$ by Lemma 1. Thus $(\mathfrak{K}, \mathfrak{C})$ is solvable relatively to $\mathfrak{F}$ and so is $\mathfrak{K}$.

An immediate corollary of the lemma above may be stated as

LEMMA 4. *The field* $\mathfrak{C}$ *of Lemma 2 is solvable by radicals relatively to* $\mathfrak{F}$.

We now use these lemmas to prove

Theorem 23. *A normal field* $\mathfrak{N}$ *over* $\mathfrak{F}$ *is solvable by radicals relatively to* $\mathfrak{F}$ *if and only if* $\mathfrak{N}$ *is metacyclic over* $\mathfrak{F}$.

For let $\mathfrak{N}$ be metacyclic of degree n over $\mathfrak{F}$ and let $p_1, \ldots, p_r$ be the distinct prime divisors of n, $\mathfrak{C}$ be the field of Lemma 2. The composite $(\mathfrak{N}, \mathfrak{C})$ is normal over $\mathfrak{C}$ and its automorphism group is equivalent to a subgroup $\mathfrak{H}$ of the group $\mathfrak{G}$ of $\mathfrak{N}$ over $\mathfrak{F}$ by Theorem 8.12. By Theorem 8.20 the group $\mathfrak{G}$ is solvable and so is $\mathfrak{H}$ by Theorem 6.18. We again apply Theorem 8.20 and see that $\mathfrak{N}_0 = (\mathfrak{N}, \mathfrak{C})$ is metacyclic over $\mathfrak{C}$ and its degree m has the form $p_1^{f_1} \ldots p_r^{f_r}$ a divisor of n. By Lemma 1 the field $\mathfrak{N}_0$ is solvable by radicals relatively to $\mathfrak{C}$, and by Lemma 4 the field $\mathfrak{C}$ is solvable by radicals relatively to $\mathfrak{F}$. Thus $\mathfrak{N}_0$ is solvable by radicals relatively to $\mathfrak{F}$ and this is is of course true of its subfield $\mathfrak{N}$.

Conversely, we let $\mathfrak{N}$ be a normal field over $\mathfrak{F}$ with automorphism group $\mathfrak{G}$ and assume that $\mathfrak{N}$ is solvable by radicals relatively to $\mathfrak{F}$. Then $\mathfrak{N}$ is contained in a field $\mathfrak{K}_t$ of (32), (33) and we define $\mathfrak{C}$ as in Lemma 2 for the prime degrees p_i of (33). The group $\mathfrak{G}_1$ of $\mathfrak{C}$ is a solvable group by Lemma 2. If the intersection $\mathfrak{D}$ of $\mathfrak{N}$ and $\mathfrak{C}$ belongs to the subgroup $\mathfrak{H}$ of $\mathfrak{G}$ and $\mathfrak{H}_1$ of $\mathfrak{G}_1$ then we have seen in Theorem 8.13 that $\mathfrak{D}$ is normal over $\mathfrak{F}$ and that its automorphism group is equivalent to both $\mathfrak{G}_1/\mathfrak{H}_1$ and $\mathfrak{G}/\mathfrak{H}$. But $\mathfrak{G}_1/\mathfrak{H}_1$ is a solvable group by Theorem 6.19 and so $\mathfrak{G}/\mathfrak{H}$ is solvable. Also the group $\mathfrak{H}_0$ of $\mathfrak{N}_0 = (\mathfrak{N}, \mathfrak{C})$ over $\mathfrak{C}$ is equivalent to $\mathfrak{H}$. We define $\mathfrak{L}_t = \mathfrak{C}(a_1, \ldots, a_t)$ with a_i as in (33). Then $\mathfrak{L}_i = \mathfrak{L}_{i-1}(a_i)$ contains a quantity $\zeta_i \neq 1$ such that $\zeta_i^{p_i} = 1$, $a_i^{p_i} = a_i$ in $\mathfrak{L}_{i-1}$. Hence either $\mathfrak{L}_i$ is cyclic of prime degree p_i over $\mathfrak{L}_{i-1}$ or $\mathfrak{L}_i = \mathfrak{L}_{i-1}$. The automorphism group $\mathfrak{H}_1$ of the composite $\mathfrak{N}_1 = (\mathfrak{N}_0, \mathfrak{L}_1)$ over $\mathfrak{L}_1$ either is generated by $\mathfrak{H}_0$ or by a normal divisor of index p_1 of $\mathfrak{H}_0$. For, by Theorem 8.19 the index of $\mathfrak{H}_1$ under $\mathfrak{H}_0$ is the degree of the intersection of $\mathfrak{N}_0$ and $\mathfrak{L}_1$ over $\mathfrak{C}$ and since the degree of $\mathfrak{L}_1$ is p_1 or 1 the degree of the intersection is p_1 or 1. We define $\mathfrak{N}_i = (\mathfrak{N}_{i-1}, \mathfrak{L}_i)$ and obtain a sequence of groups (in the sense of equivalence)

$$\mathfrak{H}_0 \geqq \mathfrak{H}_1 \geqq \ldots \geqq \mathfrak{H}_t, \tag{34}$$

where $\mathfrak{H}_t$ is the group of $\mathfrak{N}_t$ over $\mathfrak{L}_t$, and $\mathfrak{H}_i$ is either $\mathfrak{H}_{i-1}$ or a normal divisor of prime index p_i of $\mathfrak{H}_{i-1}$. But $\mathfrak{N}_t = (\mathfrak{N}_0, \mathfrak{L}_t)$ and since $\mathfrak{N} \leqq \mathfrak{K}_t$ we have $\mathfrak{N}_0 \leqq \mathfrak{L}_t$ and $\mathfrak{N}_t = \mathfrak{L}_t$, $\mathfrak{H}_t$ must be the identity group. Each $\mathfrak{H}_i < \mathfrak{H}_{i-1}$ is evidently a maximal normal divisor of $\mathfrak{H}_{i-1}$ and if we delete the repeated groups from (34) we obtain a composition series of $\mathfrak{H}_0$. Then $\mathfrak{H}_0$ is solvable and so is $\mathfrak{H}$. By Theorem 6.19 the group $\mathfrak{G}$ is solvable.

DEFINITION. *An equation* $f(x) = 0$ *is said to be solvable by radicals relatively to its coefficient field* $\mathfrak{F}$ *if its root field* $\mathfrak{N}$ *over* $\mathfrak{F}$ *is solvable by radicals relatively to* $\mathfrak{F}$.

Our Theorems 8.23, 8.20, 8.17, and 6.23 then have as immediate corollaries

Theorem 24. *An equation* $f(x) = 0$ *is solvable by radicals relatively to a non-modular field* $\mathfrak{F}$ *containing its coefficients if and only if its Galois group over* $\mathfrak{F}$ *is a solvable group.*

Theorem 25. *The general equation of degree* $n > 4$ *is not solvable by radicals relatively to the field* $\mathfrak{F}(c_1, \ldots, c_n)$ *of its coefficients.*

We leave the proofs of the theorems above, *where they are valid for modular fields,* as an exercise for the reader. For example, it is easy to show that a separable irreducible equation of degree (=characteristic) $p = 2, 3$ is not solvable by radicals. The reader may also easily prove Theorem 25 for modular fields by first showing that a cyclic field of degree p is not solvable by radicals. But then the general equation of degree $n > p$ is not solvable by radicals. Also equations whose root fields have a separable subfield of degree p are not solvable by radicals. Finally, the proof may be completed by showing that Theorem 24 holds if the degree of the root field of $f(x)$ is less than the characteristic of $\mathfrak{F}$.

CHAPTER IX

CYCLIC FIELDS

1. The structure problem. The simplest finite group is a cyclic group, that is, a group generated by a single element. This evidently implies that a normal field of finite degree n over $\mathfrak{F}$ has the simplest structure when its automorphism group is a cyclic group, so that we call it a *cyclic field.* Cyclic fields are very interesting and we shall study their structure. As the case where $\mathfrak{F}$ is finite has already been studied we shall assume henceforth that $\mathfrak{F}$ is an infinite field.

The corollary to Theorem 8.18 reduced the study of the structure of cyclic fields $\mathfrak{Z}$ of degree n over $\mathfrak{F}$ to the case where

$$n = p^e \tag{1}$$

and p is a prime. Then $\mathfrak{Z}$ is metacyclic and

$$\mathfrak{Z} = \mathfrak{Z}_e > \mathfrak{Y} = \mathfrak{Z}_{e-1} > \mathfrak{Z}_{e-2} > \dots > \mathfrak{Z}_1 > \mathfrak{Z}_0 = \mathfrak{F}, \tag{2}$$

where $\mathfrak{Z}_i$ is cyclic of degree p^i over $\mathfrak{F}$, cyclic of degree p over $\mathfrak{Z}_{i-1}$ for $i = 1, \dots, e$. Moreover, each $\mathfrak{Z}_i$ is the only subfield of degree p^i over $\mathfrak{F}$ of $\mathfrak{Z}$.

For, the automorphism group $\mathfrak{G}$ of $\mathfrak{Z}$ over $\mathfrak{F}$ consists of the powers $S^a(a = 0, 1, \dots, n-1)$ of a generating automorphism S. The group $\mathfrak{G}$ has a unique subgroup $\mathfrak{H}_i$ of order p^{e-i} by Theorem 6.3 and $\mathfrak{Z}_i$ is the subfield of $\mathfrak{Z}$ corresponding to $\mathfrak{H}_i$. Since $\mathfrak{G}$ is evidently a solvable group we obtain (2). But the group of $\mathfrak{Z}_i$ over $\mathfrak{F}$ is $\mathfrak{G}/\mathfrak{H}_i$ which is cyclic of order p^i and generating automorphism $S_i = \mathfrak{H}_i S$. This proves our result above. Notice that $\mathfrak{Z}$ has precisely $e - 1$ proper subfields over $\mathfrak{F}$ distinct from $\mathfrak{F}$.

We shall study the structure of a cyclic field $\mathfrak{Z}$ of degree p^e over $\mathfrak{F}$ by an induction on e. Our first step is thus a consideration of the case where $e = 1$, $\mathfrak{Z}$ is cyclic of degree p over $\mathfrak{F}$. We then write

$$\mathfrak{Z} > \mathfrak{Y} > \mathfrak{F}, \tag{3}$$

where $\mathfrak{Y}$ is a cyclic field of degree p^{e-1} over $\mathfrak{F}$, and determine necessary and sufficient conditions on the field $\mathfrak{Y}$ of known structure that $\mathfrak{Z}$ shall be cyclic of degree p^e over $\mathfrak{F}$ with $\mathfrak{Y}$ as its unique subfield $\mathfrak{Z}_{e-1}$. Evidently $\mathfrak{Z}$ is cyclic of degree p over $\mathfrak{Y}$ and the case $e = 1$ determines the structure of $\mathfrak{Z}$ relative to $\mathfrak{Y}$. But the assumptions that $\mathfrak{Z}$ is cyclic over $\mathfrak{Y}$, $\mathfrak{Y}$ is cyclic over $\mathfrak{F}$, do not imply that $\mathfrak{Z}$ is cyclic over $\mathfrak{F}$. For, $\mathfrak{Z}$ need not even be normal over $\mathfrak{F}$.

EXERCISE

Prove the existence of a field of degree four over the field $\mathfrak{R}$ of all rational numbers which has a quadratic subfield but which is not normal over $\mathfrak{R}$.

2. Generating automorphisms and elements of $\mathfrak{Z}$ over $\mathfrak{F}$. We write

$$n = p^e , \qquad m = p^{e-1} , \qquad U = S^m , \tag{4}$$

so that

$$U^p = I , \qquad \mathfrak{H} = (I, U, \dots , U^{p-1}) \tag{5}$$

is the group of $\mathfrak{Z}$ over $\mathfrak{Y}$. The group of $\mathfrak{Y}$ over $\mathfrak{F}$ is the quotient group

$$\mathfrak{L} = \mathfrak{G}/\mathfrak{H} = [\Sigma] , \qquad \Sigma = \mathfrak{H}S, \qquad \Sigma^m = \mathfrak{H} . \tag{6}$$

But

$$y^\Sigma = y^S$$

for every y of $\mathfrak{Y}$. It follows that the automorphism S of $\mathfrak{G}$ *induces* a generating automorphism of $\mathfrak{Y}$ over $\mathfrak{F}$. Thus there is no loss of generality if we identify these automorphisms and write

$$\mathfrak{L} = [S] , \qquad S^m = U , \tag{7}$$

where U is the identity automorphism for $\mathfrak{Y}$ over $\mathfrak{F}$.

We now prove the useful

LEMMA 1. *Let $\mathfrak{Z} = \mathfrak{Y}(\xi)$ be cyclic of degree p^e over $\mathfrak{F}$, $\mathfrak{Y}$ be cyclic of degree p^{e-1} over $\mathfrak{F}$, ξ be a quantity of $\mathfrak{Z}$ not in $\mathfrak{Y}$. Then $\mathfrak{Z} = \mathfrak{F}(\xi)$.*

For $\mathfrak{F}(\xi)$ is contained in $\mathfrak{Z}$ and is either $\mathfrak{Z}$ or a proper subfield $\mathfrak{Z}_i$ of $\mathfrak{Z}$. But every proper subfield of $\mathfrak{Z}$ was seen above to be contained in $\mathfrak{Z}_{e-1} = \mathfrak{Y}$ and ξ is not in $\mathfrak{Y}$. Hence $\mathfrak{F}(\xi) = \mathfrak{Z}$.

The lemma above states that $\mathfrak{Z}$ is generated as a field over $\mathfrak{F}$ by any ξ which generates $\mathfrak{Z}$ over $\mathfrak{Y}$. Thus if we show that $\mathfrak{Z} = \mathfrak{Y}(\xi)$ is cyclic of degree p^e over $\mathfrak{F}$ the minimum equation of ξ over $\mathfrak{F}$ will be an irreducible equation of degree p^e over $\mathfrak{F}$ with the cyclic group. We call such equations *cyclic equations* (of degree p^e) and have reduced the problem of finding a set of cyclic equations defining all cyclic fields of degree p^e over $\mathfrak{F}$ to an inductive construction with the induction on e. It is important to notice that we originally obtain only a ξ satisfying our equation with coefficients in $\mathfrak{Y}$ and that for non-cyclic fields it is possible that $\mathfrak{F}(\xi) < \mathfrak{Y}(\xi)$ so that ξ does not generate $\mathfrak{Z} = \mathfrak{Y}(\xi)$ over $\mathfrak{F}$. But in our case $\mathfrak{Z}$ *is generated as a field over* $\mathfrak{F}$ *by any* ξ *which generates it over* $\mathfrak{Y}$.

Lemma 1 is also useful in another way. We may now replace ξ by any η not in $\mathfrak{Y}$ and still have $\mathfrak{Z} = \mathfrak{F}(\eta)$. The general quantity of $\mathfrak{Z}$ over $\mathfrak{Y}$ is

$$\eta = b_0 + b_1\xi + \dots + b_{p-1}\xi^{p-1} \qquad (b_i \text{ in } \mathfrak{Y})$$

and η is not in $\mathfrak{Y}$, $\mathfrak{Z} = \mathfrak{F}(\eta)$ if and only if some $b_i \neq 0$ for $i > 0$.

3. Necessary conditions for fields of characteristic *p*. The structure of $\mathfrak{Z}$ of degree p^e over $\mathfrak{F}$ depends essentially on the characteristic of $\mathfrak{F}$. There are two evident cases to consider. In the first case we assume that p is the characteristic of $\mathfrak{F}$.

The unity element 1 of $\mathfrak{F}$ generates the additive cyclic group

$$1, 2, \dots, p - 1, p = 0, \tag{8}$$

so that any correspondence defined by a transformation

$$x \longleftrightarrow x + 1$$

on x in a field over $\mathfrak{F}$ has order p. The *normalized* equation,

$$x^p - x - a = 0 \qquad (a \text{ in } \mathfrak{F}) \tag{9}$$

with a root ξ, has the property that

$$(\xi + h)^p = \xi^p + h^p = (\xi + a) + h,$$

for every integer h, since $h^p = h$ when $p = 0$. Then

$$\xi, \xi + 1, \dots, \xi + p - 1 \tag{10}$$

are the roots of (9) in a root field containing ξ. We may prove

Theorem 1. *The equation* $x^p = x + a$ *is either irreducible in* $\mathfrak{F}$ *or has all of its roots* (10) *in* $\mathfrak{F}$. *In the former case* $\mathfrak{F}(\xi)$ *is cyclic of degree* p *over* $\mathfrak{F}$ *and every cyclic field* $\mathfrak{Z}$ *of degree* p *over* $\mathfrak{F}$ *is a field* $\mathfrak{F}(\xi)$, ξ *a root of a normalized equation* (9).

The result above is analogous to that of Theorem 8.22. In that case we assumed that $\mathfrak{F}$ contained a primitive pth root of unity ζ and obtained $\mathfrak{Z} = \mathfrak{F}(\xi)$, $\xi^p = a$ in $\mathfrak{F}$, with generating automorphism defined by

$$\xi \longleftrightarrow \xi^S = \zeta\xi.$$

In our present case $\xi^p = \xi + a$, $\xi^S = \xi + 1$. Our results are parallel. In the former case we had a multiplicative group consisting of the powers of ζ and in the latter we use an additive group. This parallelism will run through our whole theory.

To prove our theorem we let ξ be a root of $x^p - x - a = 0$. Then the minimum function of ξ over $\mathfrak{F}$ is an irreducible factor $g(x)$ of $f(x)$. If $g(x)$ has degree $q > 1$ and thus the root ξ is not in $\mathfrak{F}$ its other roots all have the form $\xi + h$, h an integer. The correspondence generated by

$$\xi \longleftrightarrow \xi^S = \xi + h \tag{11}$$

is an automorphism of $\mathfrak{Z} = \mathfrak{F}(\xi)$ over $\mathfrak{F}$, and S^q is the identity automorphism. Hence

$$\xi^{S^q} = \xi + qh = \xi\,,$$

so that qh is divisible by p. But $0 < h < p$, $1 < q \leqq p$, so that $q = p$ and $\mathfrak{Z} = \mathfrak{F}(\xi)$ is cyclic of degree p over $\mathfrak{F}$. Then obviously $\xi \longleftrightarrow \xi + 1$ defines a generating automorphism of $\mathfrak{Z}$ over $\mathfrak{F}$. When the degree of $g(x)$ is unity we have ξ in $\mathfrak{F}$, and the roots (10) are all in $\mathfrak{F}$.

Conversely, let $\mathfrak{Z}$ be cyclic of degree p over $\mathfrak{F}$ and write $\xi_i = \xi^{S^i}$, S a generating automorphism of the group of $\mathfrak{Z}$ over $\mathfrak{F}$. As in Section 8.11 the square of the Vandermonde determinant

$$V = |\xi_{i-1}^{j-1}| \qquad (i, j = 1, \ldots, p) \tag{12}$$

is the discriminant of the separable minimum equation of ξ over $\mathfrak{F}$ and is not zero. Then $V \neq 0$. The first column of V consists of p ones and if we add all the last $p - 1$ rows of V to its first row we obtain the determinant

$$\begin{vmatrix} 0 & s_1 & s_2 & \ldots & s_{p-1} \\ 1 & \xi_1 & \xi_1^2 & \ldots & \xi_1^{p-1} \\ \cdot & \cdot & \cdot & \ldots & \cdot \\ 1 & \xi_{p-1} & \xi_{p-1}^2 & \ldots & \xi_{p-1}^{p-1} \end{vmatrix} = V \neq 0\,, \tag{13}$$

where s_k is the sum of the kth powers of the ξ_i. If all the $s_k = 0$ we would have $V = 0$, a contradiction. Hence at least one $s_k \neq 0$ and for this k we write

$$s = \sum_{i=0}^{p-1} \xi_i^k \neq 0\,. \tag{14}$$

The quantity s is symmetric in the ξ_i and is in $\mathfrak{F}$.

Define

$$\eta = -s^{-1}\sum_{i=0}^{p-1} i\xi_i^k \tag{15}$$

for k defined by s of (14). Then $\xi_i^S = \xi_{i+1}$,

$$\begin{cases} \eta^S = -s^{-1}\sum_{i=0}^{p-1} i\xi_{i+1}^k = -s^{-1}\sum_{i=0}^{p-1}[(i+1)\xi_{i+1}^k - \xi_{i+1}^k] \\ \qquad = -s^{-1}\left[\sum_{i=0}^{p-1} i\xi_i^k - s\right] = \eta + 1\,, \end{cases} \tag{16}$$

where we have used the fact that $p = 0$, $\xi_p = \xi_0$ in replacing $\sum_{i=0}^{p-1}(i+1)\xi_{i+1}^k$ by $\sum_{i=0}^{p-1} i\xi_i^k$ and $\sum_{i=0}^{p-1}\xi_{i+1}^k$ by s. The quantity η is not in $\mathfrak{F}$ since $\eta^S \neq \eta$. But any quantity in $\mathfrak{Z}$ of degree p over $\mathfrak{F}$ and not in $\mathfrak{F}$ generates $\mathfrak{Z}$ over $\mathfrak{F}$. Thus $\mathfrak{Z} = \mathfrak{F}(\eta)$. Moreover, $(\eta^p - \eta)^S = (\eta + 1)^p - (\eta + 1) = \eta^p - \eta$ is in $\mathfrak{F}$ since it is unaltered by the generating automorphism of the group of $\mathfrak{Z}$ over $\mathfrak{F}$. This proves our theorem.

We shall use

Theorem 2. *Let $\mathfrak{Z} = \mathfrak{F}(\xi)$ where ξ is a root of the irreducible normalized equation $\mathrm{x}^p = \mathrm{x} + \mathrm{a}$. Then η in $\mathfrak{Z}$ is a root of a normalized equation $\eta^p = \eta + \mathrm{c}$ over $\mathfrak{F}$ if and only if*

$$\eta = \mathrm{k}\xi + \mathrm{b} \qquad (\mathrm{k}\ \textit{an integer},\ \mathrm{b}\ \textit{in}\ \mathfrak{F}). \tag{17}$$

For when $\eta^p = \eta + c$ we also have $(\eta^S)^p = \eta^S + c$ and η^S must have the form $\eta + k$, k an integer. Then $(\eta - k\xi)^S = \eta + k - k(\xi + 1) = \eta - k\xi = b$ is in $\mathfrak{F}$, $\eta = k\xi + b$. Conversely $(k\xi + b)^p = k^p\xi^p + b^p = k^p(\xi + a) + b^p = (k\xi + b) + c$ where $c = b^p - b + ka$ and we have used the property that $k^p = k$ (modulo p) for every integer k of our field $\mathfrak{F}$.

The quantity η of (17) generates $\mathfrak{Z}$ over $\mathfrak{F}$ if and only if k is not divisible by p. This is of course due to the fact that η is in $\mathfrak{F}$ if and only if $k = 0$.

Theorem 9.1 is the case $e = 1$ of our structure theory. We apply it together with our auxiliary Theorem 9.2 to cyclic fields

$$\mathfrak{Z} = \mathfrak{Y}(\xi) \tag{18}$$

of degree p^e over $\mathfrak{F}$, with $\mathfrak{Y}$ defined as usual. Since $\mathfrak{Z}$ is cyclic of degree p over $\mathfrak{Y}$ we may write

$$\xi^p = \xi + \alpha \tag{19}$$

where α is in $\mathfrak{Y}$. This is of course the application of Theorem 9.1. We now obtain the following necessary condition

Theorem 3. *Let $\mathfrak{Z}$ be cyclic of degree* p^e *over $\mathfrak{F}$ of characteristic* p *and $\mathfrak{Y}$ the unique subfield of $\mathfrak{Z}$ over $\mathfrak{F}$ defined so that $\mathfrak{Y}$ is cyclic of degree* p^{e-1} *over $\mathfrak{F}$ with generating automorphism Σ. Then there exists a quantity β in $\mathfrak{Y}$ such that the relative trace*

$$T_{\mathfrak{Y}|\mathfrak{F}}(\beta) = 1\,, \tag{20}$$

and

$$\mathfrak{Z} = \mathfrak{F}(\xi)\,, \qquad \xi^p = \xi + \alpha\,, \tag{21}$$

where α in $\mathfrak{Y}$ has the property

$$\alpha^\Sigma - \alpha = \beta^p - \beta\,, \tag{22}$$

so that

$$T_{\mathfrak{Y}|\mathfrak{F}}(\beta^p) = T_{\mathfrak{Y}|\mathfrak{F}}(\beta)\,. \tag{23}$$

Moreover, the generating automorphism S *of $\mathfrak{Z}$ over $\mathfrak{F}$ is uniquely determined by*

$$\xi^S = \xi + \beta\,, \qquad \gamma^S = \gamma^\Sigma\,, \tag{24}$$

for every γ of $\mathfrak{Y}$.

The automorphism S has order p^e and by Theorem 6.4 we may take a power S^u of S as a generator of the group of $\mathfrak{Z}$ over $\mathfrak{F}$ if and only if u is prime to p. Now

$$S^m = U\,, \qquad \xi^U = \xi + h \neq \xi$$

where h is an integer. Since $0 < h < p$ we may choose an integer u so that $hu = 1$ in the finite subfield of $\mathfrak{F}$ and $S^{um} = U^u = U_0$, $\xi^{U_0} = \xi + uh = \xi + 1$. Hence we may assume with no loss of generality that

$$\xi^U = \xi + 1\,,$$

where we have already seen that $\xi^p = \xi + \alpha$. We now apply S and have

$$(\xi^S)^p = \xi^S + \alpha^S = \xi^S + \alpha^\Sigma\,,$$

where α is of course in $\mathfrak{Y}$. Then by Theorem 9.2 we have $\xi^S = k\xi + \beta$, and β in $\mathfrak{Y}$. Hence $\xi^{S^2} = k\xi^S + \beta^S = k^2\xi + \beta_2$, and ultimately $\xi^{S^n} = k^n\xi + \beta_n = \xi$ where $n = p^e$. This shows that $\beta_n = 0$, $k^n = 1$. But we know that $k^p = k$ for every integer k of $\mathfrak{F}$ and therefore that $k^{p^2} = k^p = k = k^{p^3} = \ldots = k^n = 1$ and have proved the first part of (24). Since every quantity of $\mathfrak{Z}$ over $\mathfrak{F}$ has the form $g = \sum_{i=0}^{p-1} \gamma_i\xi^i$ with γ_i in $\mathfrak{Y}$ it is evident that

$$g^S = \sum_{i=0}^{p-1} \gamma_i^\Sigma(\xi + \beta)^i \tag{25}$$

uniquely determines a generating automorphism S of $\mathfrak{Z}$ over $\mathfrak{F}$ if and only if $(\xi^S)^p = \xi^S + \alpha^\Sigma$. This is the same as $(\xi + \beta)^p = \xi + \beta + \alpha^\Sigma = \beta^p + \xi + \alpha$ and we have (22). Finally, $\xi^{S^2} = (\xi + \beta) + \beta^S = \xi + \beta + \beta^\Sigma$, and $\xi^{S^m} = \xi + \beta + \beta^\Sigma + \ldots + \beta^{\Sigma^{m-1}} = \xi^U = \xi + 1$. This gives (20). The trace of any two conjugate elements α, α^Σ of $\mathfrak{Y}$ over $\mathfrak{F}$ is the same and $T_{\mathfrak{Y}|\mathfrak{F}}(\alpha^\Sigma - \alpha) = 0 = T_{\mathfrak{Y}|\mathfrak{F}}(\beta^p - \beta)$ so that we have (23) and our theorem.

The conditions of our theorem will actually be sufficient as well as necessary. This will partly be due to the following properties of the trace function in a cyclic field $\mathfrak{Z}$ of degree p^e over $\mathfrak{F}$ of characteristic p. We showed in Section 8.5 that

$$T_{\mathfrak{Z}|\mathfrak{F}}(\delta) = T_{\mathfrak{Y}|\mathfrak{F}}[T_{\mathfrak{Z}|\mathfrak{Y}}(\delta)]\,, \tag{26}$$

for every δ of $\mathfrak{Z}$. When δ is in $\mathfrak{Y}$ the trace of δ in $\mathfrak{Z}$ over $\mathfrak{Y}$ is the sum of $\delta + \delta + \ldots + \delta = p\delta = 0$ in our bizarre environment. Hence

$$T_{\mathfrak{Z}|\mathfrak{F}}(\delta) = 0 \text{ for every } \delta \text{ of } \mathfrak{Y}\,. \tag{27}$$

We now take an arbitrary δ of $\mathfrak{Z}$ and write

$$\delta = \delta_0 + \delta_1\xi + \ldots + \delta_{p-1}\xi^{p-1} \qquad (\delta_i \text{ in } \mathfrak{Y}) \tag{28}$$

where ξ satisfies $\xi^p = \xi + \alpha$. Then

$$T_{\mathfrak{Z}|\mathfrak{Y}}(\delta) = \sum_{i=1}^{p-1} \delta_i T_{\mathfrak{Z}|\mathfrak{Y}}(\xi^i)\,. \tag{29}$$

In the elementary theory of symmetric functions* it was shown that if

* Cf. Ex. 5 of Sec. 7.1.

s_k is the sum of the kth powers of the roots of an equation $x^p + c_1 x^{p-1} + \ldots + c_p = 0$ then we have the identities

$$s_k + c_1 s_{k-1} + \ldots + c_{k-1} s_1 + k c_k = 0 \qquad (k = 1, 2, \ldots, p). \tag{30}$$

In our case $c_{p-1} = -1$, $c_1 = c_2 = \ldots = c_{p-2} = 0$. Since $s_k = T_{\mathfrak{Z}|\mathfrak{Y}}(\xi^k)$ we have $(p-1)c_{p-1} = 1 - p = 1$,

$$T_{\mathfrak{Z}|\mathfrak{Y}}(\xi^{p-1}) = -1, \qquad T_{\mathfrak{Z}|\mathfrak{Y}}(\xi^i) = 0 \qquad (i = 1, \ldots, p-2). \tag{31}$$

We apply this result in (29) and obtain $-\delta_{p-1}$ for the right member. Applying (26) we have the

Lemma 2. *Let $\mathfrak{Z} = \mathfrak{F}(\xi) = \mathfrak{Y}(\xi)$ where $\xi^p = \xi + a$ and thus every δ of $\mathfrak{Z}$ has the form* (28). *Then*

$$T_{\mathfrak{Z}|\mathfrak{F}}(\delta) = -T_{\mathfrak{Y}|\mathfrak{F}}(\delta_{p-1}). \tag{32}$$

The above lemma implies the following existence theorem:

Theorem 4. *Let $\mathfrak{Z}$ be cyclic of degree p^e over $\mathfrak{F}$ of characteristic p. Then there exists a quantity β in $\mathfrak{Z}$ such that*

$$T_{\mathfrak{Z}|\mathfrak{F}}(\beta) = T_{\mathfrak{Z}|\mathfrak{F}}(\beta^p) = 1. \tag{33}$$

For if $e = 1$ we have $\mathfrak{Z} = \mathfrak{F}(\xi)$, $\xi^p = \xi + a$, and have (32), $T_{\mathfrak{Z}|\mathfrak{F}}(-\xi^{p-1}) = -(-1) = 1$. Also $\beta = -\xi^{p-1}$, $\beta^p = (-\xi^{p-1})^p = (-1)^p(\xi^p)^{p-1} = (-1)^p \cdot (\xi + a)^{p-1}$. By our lemma above $T_{\mathfrak{Z}|\mathfrak{F}}[(\xi + a)^{p-1}] = T_{\mathfrak{Z}|\mathfrak{F}}(\xi^{p-1}) = -1$, $T_{\mathfrak{Z}|\mathfrak{F}}(\beta^p) = (-1)^{p+1}$. If $p = 2$ then $-1 = 1$, $(-1)^{p+1} = 1$. When p is odd $(-1)^{p+1} = 1$. This completes the case where $e = 1$. We make an induction on e and assume that $T_{\mathfrak{Y}|\mathfrak{F}}(\beta_{e-1}) = T_{\mathfrak{Y}|\mathfrak{F}}({\beta_{e-1}}^p) = 1$ with β_{e-1} in the subfield $\mathfrak{Y}$ of degree p^{e-1} of the cyclic field $\mathfrak{Z}$. Put $\beta = -\beta_{e-1}\xi^{p-1}$, $\mathfrak{Z} = \mathfrak{Y}(\xi)$, $\xi^p = \xi + a$, a in $\mathfrak{Y}$ and use (32). Then $T_{\mathfrak{Z}|\mathfrak{F}}(\beta) = -T_{\mathfrak{Y}|\mathfrak{F}}(-\beta_{e-1}) = T_{\mathfrak{Y}|\mathfrak{F}}(\beta_{e-1}) = 1$ and $T_{\mathfrak{Z}|\mathfrak{F}}(\beta^p) = T_{\mathfrak{Z}|\mathfrak{F}}[(-\beta_{e-1})^p(\xi + a)^{p-1}] = (-1)^{p+1}T_{\mathfrak{Y}|\mathfrak{F}}(\beta_{e-1}^p) = 1$ by (32) and $(-1)^{p+1} = 1$.

We shall see in a later exercise that the above theorem does not have a parallel result for the case where $\mathfrak{F}$ does not have characteristic p. It is this property as well as the fact that a field $\mathfrak{F}$ does not in general contain a primitive pth root of unity that makes the case where the characteristic of $\mathfrak{F}$ is p so much simpler than the case of characteristic not p.

4. The norm and trace theorems. There are two analogous theorems about norms and traces in cyclic fields. These existence theorems will be essential for our sufficiency proofs and we shall study them. They are theorems on cyclic fields $\mathfrak{Z}$ of arbitrary degree n over an infinite field $\mathfrak{F}$. The first is given as

Theorem 5. *Let* S *be a generating automorphism of the cyclic group of* $\mathfrak{Z}$ *over* $\mathfrak{F}$. *Then a quantity* δ *of* $\mathfrak{Z}$ *has the property*

$$N_{\mathfrak{Z}|\mathfrak{F}}(\delta) = 1 \tag{34}$$

if and only if there exists a $\gamma \neq 0$ *in* $\mathfrak{Z}$ *such that*

$$\delta = \gamma^{S}\gamma^{-1}. \tag{35}$$

The solution γ *is unique in the sense that if also* $\delta = \gamma_1^{S}\gamma_1^{-1}$ *then* $\gamma_1 = c\gamma$ *where* c *is in* $\mathfrak{F}$.

For if $\gamma^{S}\gamma^{-1} = \delta$ we have $N_{\mathfrak{Z}|\mathfrak{F}}(\delta) = N_{\mathfrak{Z}|\mathfrak{F}}(\gamma^{S})\, N_{\mathfrak{Z}|\mathfrak{F}}(\gamma^{-1}) = \nu\nu^{-1} = 1$ where $\nu = N_{\mathfrak{Z}|\mathfrak{F}}(\gamma) = N_{\mathfrak{Z}|\mathfrak{F}}(\gamma^{S})$ is in $\mathfrak{F}$. Conversely, let the norm of δ over $\mathfrak{F}$ be unity, x be an indeterminate over $\mathfrak{Z} = \mathfrak{F}(\xi)$ and write

$$x + \xi = y^{-1}, \qquad \gamma(x) = 1 + y^{-1}[y^{S}\delta + y^{S^2}\delta\delta^{S} + \ldots + y^{S^{n-1}}\delta\delta^{S} \ldots \delta^{S^{n-2}}].$$

Here, of course, we have used the abbreviation $y^{-1}y^{S^q}$ for the rational function of x

$$\frac{x + \xi}{x + \xi^{S^q}}.$$

When we replace x by $-\xi \neq -\xi^{S^q}$ for $q = 1, 2, \ldots, n-1$ we obtain $\gamma(-\xi) = 1$. It follows that

$$\gamma(x) = \frac{A(x)}{B(x)}, \qquad B(x) = (x + \xi^{S}) \ldots (x + \xi^{S^{n-1}})$$

where $A(x)$ is a polynomial of $\mathfrak{Z}[x]$ and is not zero. Since $\mathfrak{F}$ is an infinite field we may apply Theorem 7.30 and obtain an s in $\mathfrak{F}$ such that $\gamma(s) \neq 0$. But $s \neq -\xi^{S^i}$ and if

$$y_0 = (s + \xi)^{-1} \neq 0$$

then

$$\gamma_0 = \gamma(s) = 1 + y_0^{-1}(\delta y_0^{S} + \delta\delta^{S}y_0^{S^2} + \ldots + \delta\delta^{S} \ldots \delta^{S^{n-2}}y_0^{S^{n-1}}) \neq 0 \tag{36}$$

is in $\mathfrak{Z}$. The norm $\delta\delta^{S} \ldots \delta^{S^{n-1}}$ of δ is unity and

$$\begin{aligned}\delta y_0^{-1}y_0^{S}\gamma_0^{S} &= \delta y_0^{-1}(y_0^{S} + \delta^{S}y_0^{S^2} + \ldots + \delta^{S} \cdot \delta^{S^2} \ldots \delta^{S^{n-2}}y_0^{S^{n-1}}) + y_0^{-1}y_0N_{\mathfrak{Z}|\mathfrak{F}}(\delta)\\ &= 1 + y_0^{-1}(\delta y_0^{S} + \ldots + \delta \cdot \delta^{S} \ldots \delta^{S^{n-2}}y_0^{S^{n-1}}) = \gamma_0.\end{aligned}$$

Hence

$$\delta = (\gamma_0 y_0)(\gamma_0^S y_0^S)^{-1} = \gamma^S \gamma^{-1}\,, \tag{37}$$

where $\gamma = (\gamma_0 y_0)^{-1}$. If also $\delta = \gamma_1^S \gamma_1^{-1}$ we have $\gamma_1^S \gamma_1^{-1} = \gamma^S \gamma^{-1}$, $\gamma_1 = \gamma c$, where $c = (\gamma_1 \gamma^{-1})^S = \gamma_1 \gamma^{-1}$ is in $\mathfrak{F}$.

The analogous theorem on traces is proved by an application of the important

LEMMA. *Let $\mathfrak{Z} = \mathfrak{F}(x)$ be cyclic of degree n over $\mathfrak{F}$ with generating automorphism S. Then there exists a quantity u of $\mathfrak{Z}$ which has non-zero trace and is such that*

$$u,\ u^S,\ u^{S^2},\ \ldots,\ u^{S^{n-1}} \tag{38}$$

form a basis of $\mathfrak{Z}$ over $\mathfrak{F}$.

We write

$$u^{S^i} = \lambda_1 + \lambda_2 x^{S^i} + \ldots + \lambda_n (x^{n-1})^{S^i} \qquad (i = 0, \ldots, n-1) \tag{39}$$

with λ_i in $\mathfrak{F}$ and replace x^{S^i} by the corresponding polynomial of $\mathfrak{F}[x]$. Then $u^{S^i} = \Sigma \lambda_{ij} x^j$ with λ_{ij} in $\mathfrak{F}$ and (38) form a basis of $\mathfrak{Z}$ over $\mathfrak{F}$ if and only if the determinant $|\lambda_{ij}| \neq 0$. This determinant is a polynomial

$$\Delta(\lambda_1, \ldots, \lambda_n) \tag{40}$$

with coefficients in $\mathfrak{F}$. Similarly,

$$T(u) = u + \ldots + u^{S^{n-1}} = T(\lambda_1, \ldots, \lambda_n) \tag{41}$$

is a polynomial in $\lambda_1, \ldots, \lambda_n$ with coefficients in $\mathfrak{F}$.

The defining element x of $\mathfrak{Z}$ over $\mathfrak{F}$ may be taken to be an n-rowed matrix. We use the diagonal matrix representation

$$x = \xi_1 e_1 + \ldots + \xi_n e_n \tag{42}$$

of x and our notation means that e_i is the diagonal matrix with unity in the ith place and zeros elsewhere. The quantities ξ_i are in a scalar extension $\mathfrak{K} = \mathfrak{F}(\xi_1)$ over $\mathfrak{F}$ and this field is equivalent to $\mathfrak{F}(x)$. Moreover, if $x^S = \theta(x)$ so that $x^{S^i} = \theta^i(x)$ is the ith iterative of the polynomial $\theta(x)$ then

$$\xi_i = \theta^{i-1}(\xi_1) \qquad (i = 2, \ldots, n). \tag{43}$$

We form the polynomial $\theta(x)$ in the diagonal matrix x and see that since $\theta(\xi_i) = \xi_{i+1}$, $\theta(\xi_n) = \xi_1$,

$$(44) \qquad x^S = \theta(x) = \xi_1 e_n + \xi_2 e_1 + \dots + \xi_n e_{n-1} .$$

In a similar fashion

$$(45) \qquad x^{S^i} = \xi_1 e_{n-(i-1)} + \xi_2 e_{n-i+2} + \dots + \xi_n e_{n-i} .$$

Then

$$(46) \quad e_{n-(i-1)} = (x^{S^i} - \xi_2)(x^{S^i} - \xi_3) \dots (x^{S^i} - \xi_n)[(\xi_1 - \xi_2) \dots (\xi_1 - \xi_n)]^{-1} ,$$

so that $e_{n-i+1} = e_1^{S^i}$ is obtained from e_1 by replacing x by x^{S^i}.

The algebra $\mathfrak{Z}_{\mathfrak{K}}$ of all polynomials in the matrix x with coefficients in $\mathfrak{K}$ has order n over $\mathfrak{K}$ and a basis $1, x, \dots , x^{n-1}$. By (46) the e_i are in $\mathfrak{Z}_{\mathfrak{K}}$. They are evidently linearly independent in $\mathfrak{K}$ and also form a basis of $\mathfrak{Z}_{\mathfrak{K}}$ over $\mathfrak{K}$. If $e_1 = \lambda_{10} + \lambda_{20}x + \dots + \lambda_{n0}x^{n-1}$ with λ_{i0} in $\mathfrak{K}$ then $\Delta(\lambda_{10}, \dots , \lambda_{n0}) \neq 0$ since $e_1, e_1^S, \dots , e_1^{S^{n-1}}$ form a basis as we have said. Hence $\Delta(\lambda_1, \dots , \lambda_n)$ is not identically zero in $\lambda_1, \dots , \lambda_n$. Also $T(e_1) = e_1 + \dots + e_n = 1 \neq 0$ so that $T(\lambda_1, \dots , \lambda_n)$ is not identically zero. By Theorem 7.30 there exist quantities $\bar{\lambda}_1, \dots , \bar{\lambda}_n$ in $\mathfrak{F}$ such that $\Delta(\bar{\lambda}_1, \dots , \bar{\lambda}_n) \neq 0$, $T(\bar{\lambda}_1, \dots , \bar{\lambda}_n) \neq 0$ and $u = \bar{\lambda}_1 + \bar{\lambda}_2 x + \dots + \bar{\lambda}_n x^{n-1}$ has the properties of our lemma.

We now prove

Theorem 6. *Let* S *be a generating automorphism of the cyclic group of* $\mathfrak{Z}$ *over* $\mathfrak{F}$. *Then a quantity* δ *of* $\mathfrak{Z}$ *has the property*

$$(47) \qquad T_{\mathfrak{Z}|\mathfrak{F}}(\delta) = 0$$

if and only if there exists a γ *in* $\mathfrak{Z}$ *such that*

$$(48) \qquad \delta = \gamma^S - \gamma .$$

Any other solution γ_1 *of* (48) *has the form* $\gamma_1 = \gamma + c$ *with* c *in* $\mathfrak{F}$.

For choose u as in our lemma and see that any δ of $\mathfrak{Z}$ has the form $a_1u + a_2u^S + \dots + a_nu^{S^{n-1}}$ with a_i in $\mathfrak{F}$. Then $\gamma = b_1u + b_2u^S + \dots + b_nu^{S^{n-1}}$ and

$$(49) \quad \gamma^S - \gamma = (b_n - b_1)u + (b_1 - b_2)u^S + \dots + (b_{n-1} - b_n)u^{S^{n-1}} .$$

The equation $\delta = \gamma^S - \gamma$ gives

$$b_{n-1} = b_n + a_n , \qquad b_{n-2} = b_{n-1} + a_{n-1} = b_n + a_{n-1} + a_n$$

and by an evident induction

$$b_i = b_{i+1} + a_{i+1} = (a_{i+1} + \dots + a_n) + b_n \quad (i = 1, \dots, n-1). \tag{50}$$

These equations uniquely determine $b_1, \dots, b_{n-1}$ in terms of b_n and we now see that $\delta = \gamma^S - \gamma$ if and only if

$$b_n = b_1 + a_1 = (a_1 + \dots + a_n) + b_n, \tag{51}$$

that is $a_1 + \dots + a_n = 0$. But $T(\delta) = (a_1 + \dots + a_n)T(u) = 0$ if and only if $a_1 + \dots + a_n = 0$, so that $\delta = \gamma^S - \gamma$ if and only if $T(\delta) = 0$. If also $\gamma_1^S - \gamma_1 = \delta$ then $(\gamma_1 - \gamma)^S = \gamma_1 - \gamma = c$ is in $\mathfrak{F}$. This proves our theorem.

The basis $u, u^S, \dots, u^{S^{n-1}}$ of $\mathfrak{Z}$ over $\mathfrak{F}$ has been called a *normal basis* of $\mathfrak{Z}$ over $\mathfrak{F}$ and is a basis all of whose elements are conjugates. It is known* that such a basis exists for any field $\mathfrak{Z}$ which is normal over $\mathfrak{F}$ but we shall not prove this result.

5. Sufficiency theorems for fields of characteristic p. The fundamental theorem on the existence of cyclic fields whose degree over $\mathfrak{F}$ is a power of the prime characteristic of $\mathfrak{F}$ is given by

Theorem 7. *There exist cyclic fields $\mathfrak{Z}$ of degree p^{e} over $\mathfrak{F}$ of characteristic* p *for every exponent* e *if and only if there exist quantities* a *in $\mathfrak{F}$ for which*

$$\mathrm{x}^{\mathrm{p}} - \mathrm{x} - \mathrm{a} \tag{52}$$

is an irreducible polynomial of $\mathfrak{F}[\mathrm{x}]$.

Thus the existence of a cyclic field of degree p over $\mathfrak{F}$ of characteristic p implies the existence of cyclic fields $\mathfrak{Z}$ of degree p^e over $\mathfrak{F}$ *with the given field as subfield* and for any e.

For proof let $\mathfrak{Z}$ be cyclic of degree p^e over $\mathfrak{F}$ so that $\mathfrak{Z}$ contains a subfield $\mathfrak{W} = \mathfrak{F}(\xi)$, $\xi^p = \xi + a$, a in $\mathfrak{F}$ by Theorem 9.1, and the polynomial $x^p - x - a$ is irreducible. Conversely, we make an induction on $e \geqq 2$ and assume that we have constructed a cyclic field $\mathfrak{Y}$ of degree p^{e-1} over $\mathfrak{F}$ which contains $\mathfrak{W} = \mathfrak{F}(\xi)$, $\xi^p = \xi + a$, a in $\mathfrak{F}$, as its unique cyclic subfield of degree p over $\mathfrak{F}$. Let Σ be a generating automorphism of $\mathfrak{Y}$ over $\mathfrak{F}$.

Theorem 9.4 states that there exists a quantity β in $\mathfrak{Y}$ for which $T_{\mathfrak{Y}|\mathfrak{F}}(\beta) = T_{\mathfrak{Y}|\mathfrak{F}}(\beta^p) = 1$. Then

$$\delta = \beta^p - \beta, \qquad T_{\mathfrak{Y}|\mathfrak{F}}(\delta) = 0 \tag{53}$$

* Cf. M. Deuring, *Mathematische Annalen*, CVII (1933), 140–44.

and we apply Theorem 9.6 to obtain a quantity a in $\mathfrak{Y}$ such that

$$\beta^p - \beta = a^\Sigma - a . \tag{54}$$

By Theorem 9.1 the root field $\mathfrak{Z}$ of the normalized equation

$$x^p = x + a \tag{55}$$

is either $\mathfrak{Y}$ or a field $\mathfrak{Z} = \mathfrak{Y}(X)$ of degree p over $\mathfrak{Y}$. The equation $x^p = x + a^\Sigma$ has $X + \beta$ as a root since direct computation gives

$$(X + \beta)^p = X^p + \beta^p = X + a + \beta^p = X + \beta + a^\Sigma . \tag{56}$$

If X is in $\mathfrak{Y}$ the roots of $x^p = x + a^\Sigma$ are $X^\Sigma, X^\Sigma + 1, \ldots , X^\Sigma + p - 1$ and $X + \beta = X^\Sigma + k$, k an integer. Then $\beta = X^\Sigma - X + k$ and $1 = T_{\mathfrak{Y}|\mathfrak{F}}(\beta) = T_{\mathfrak{Y}|\mathfrak{F}}(X^\Sigma - X) + T_{\mathfrak{Y}|\mathfrak{F}}(k) = 0 + p^{e-1}k = 0$, a contradiction. Hence X is not in $\mathfrak{Y}$, $\mathfrak{Z}$ has degree p over $\mathfrak{Y}$ and degree p^e over $\mathfrak{F}$ by Theorem 7.8.

Every quantity of $\mathfrak{Z}$ is uniquely expressible as a polynomial in X with coefficients in $\mathfrak{Y}$ and degree at most $p - 1$. Thus the equations

$$(X^t)^S = (X + \beta)^t , \quad \eta^S = \eta^\Sigma \qquad (t = 1, \ldots , p - 1) \tag{57}$$

for every η of $\mathfrak{Y}$, define a correspondence of $\mathfrak{Z}$ and itself in which $\phi^S = \phi^\Sigma = \phi$ for every ϕ of $\mathfrak{F}$. The property $(X^S)^p = X^S + a^S$ then implies that S is an automorphism of $\mathfrak{Z}$ over $\mathfrak{F}$. We leave the details of the proof to the reader.

We easily compute

$$X^{S^r} = X + \beta + \beta^S + \ldots + \beta^{S^{r-1}} \qquad (r = 1, 2, \ldots) ,$$

and see that

$$X^{S^m} = X + T_{\mathfrak{Y}|\mathfrak{F}}(\beta) = X + 1 , \quad X^{S^n} = X^{S^{mp}} = X + p = X ,$$

where $n = mp = p^e$. Hence S^n is the identity automorphism I of $\mathfrak{Z}$ over $\mathfrak{F}$. It follows that the order of S divides n and must be n or a divisor of $m = p^{e-1}$. But in the latter case $S^m = I$ contrary to our proof that $X^{S^m} \neq X$. Hence S has order n, the degree of $\mathfrak{Z}$ over $\mathfrak{F}$, and $\mathfrak{Z}$ is cyclic with generating automorphism S. We have proved our theorem and have given an inductive construction of $\mathfrak{Z}$.

We have seen that every a in $\mathfrak{F}$ for which $x^p - x - a$ is irreducible in $\mathfrak{F}$ defines a cyclic field $\mathfrak{Z}$ of degree p^e over $\mathfrak{F}$ for any e with $\mathfrak{Z}_1 = \mathfrak{F}(\xi_1)$,

$\xi_1^p = \xi_1 + a$, as the unique subfield of degree p over $\mathfrak{F}$. Conversely, every $\mathfrak{Z}$ defines an a. But we wish to determine all fields $\mathfrak{Z}$ defined by a given a (and integer e).

The determination above will be accomplished by induction on e. Assume that all the cyclic fields $\mathfrak{Y}$ of degree p^{e-1} over $\mathfrak{F}$ determined by a have been found and let $\mathfrak{Y}$ be one such field. Then we fix $\mathfrak{Y}$ and prove

Theorem 8. *Let β, α be determined in $\mathfrak{Y}$ as in* (54) *and Theorem* 9.4. *Then every cyclic field $\mathfrak{Z}$ of degree p^e over $\mathfrak{F}$ with $\mathfrak{Y}$ as a subfield is a field $\mathfrak{Z} = \mathfrak{F}(\xi)$,*

$$\xi^p = \xi + \alpha + w \,, \tag{58}$$

where w ranges over all quantities of $\mathfrak{F}$. Moreover the generation (58) *of $\mathfrak{Z}$ implies that*

$$\xi^S = \xi + \beta \,. \tag{59}$$

For in Theorem 9.3 we proved the existence of β_0, α_0 such that $\mathfrak{Z} = \mathfrak{Y}(\xi_0) = \mathfrak{F}(\xi_0)$ with

$$\xi_0^p = \xi_0 + \alpha_0 \,, \qquad \xi_0^S = \xi_0 + \beta_0 \,, \qquad T_{\mathfrak{Y}|\mathfrak{F}}(\beta_0) = 1 \,, \qquad \alpha_0^S - \alpha_0 = \beta_0^p - \beta_0 \,.$$

Then $T_{\mathfrak{Y}|\mathfrak{F}}(\beta_0) = T_{\mathfrak{Y}|\mathfrak{F}}(\beta)$ for our fixed β, and $T_{\mathfrak{Y}|\mathfrak{F}}(\beta_0 - \beta) = 0$. Apply Theorem 9.6 and obtain $\beta = \beta_0 + \gamma^S - \gamma$ with γ in $\mathfrak{Y}$. We define $\xi = \xi_0 + \gamma$ and Lemma 1 states that $\mathfrak{Z} = \mathfrak{F}(\xi)$. Now $\xi^p = \xi_0^p + \gamma^p = \xi_0 + \alpha_0 + \gamma^p - \gamma + \gamma = \xi + \alpha_1$ with α_1 in $\mathfrak{Y}$, $\xi^S = \xi_0^S + \gamma^S = \xi_0 + \beta_0 + \gamma^S = \xi_0 + \beta + \gamma = \xi + \beta$ as desired. By Theorem 9.3 we have $\alpha_1^S - \alpha_1 = \beta^p - \beta = \alpha^S - \alpha$, so that $\alpha_1 = \alpha + w$ where $w = w^S$ is in $\mathfrak{F}$.

The theory above completes the structure theory for cyclic fields of degree p^e over $\mathfrak{F}$ of characteristic p. There remains only the case where the characteristic of $\mathfrak{F}$ is not p and we shall now treat this more complicated case.

EXERCISES

1. Find the minimum equation of a generating element of any cyclic field of degree four over $\mathfrak{F}$ of characteristic two with a given quadratic subfield defined by $x^2 = x + a$.

2. Let $\mathfrak{Y} = \mathfrak{F}(\xi)$ be cyclic of degree four over $\mathfrak{F}$ of characteristic two. Find a quantity β in $\mathfrak{Y}$ whose trace and that of β^2 are both unity, find the corresponding α, and use these quantities to construct all cyclic fields of degree eight over $\mathfrak{F}$ of characteristic two.

3. Find all cyclic fields of degree nine over $\mathfrak{F}$ of characteristic three with a given cubic subfield defined by $x^3 = x + a$.

6. Cyclic fields of degree p^e over a field containing a primitive pth root of unity. The theory of cyclic fields of degree p^e over $\mathfrak{F}$ of characteristic not p is parallel for a great part to the theory we have already treated in case we assume that $\mathfrak{F}$ contains a primitive pth root of unity ζ, that is, a quantity $\zeta \neq 1$ such that $\zeta^p = 1$. We shall make this assumption and shall later treat the case where ζ is not in $\mathfrak{F}$.

Theorem 8.22 was the strict parallel of Theorem 9.1, as we have already seen. The parallel to Theorem 9.2 is given by

Theorem 9. *Let $\mathfrak{Z} = \mathfrak{F}(\xi)$ where ξ is a root of the cyclic irreducible pure equation $\xi^p = a$ in $\mathfrak{F}$. Then η in $\mathfrak{Z}$ is a root of a pure equation $\eta^p = c$ in $\mathfrak{F}$ if and only if*

$$\eta = b\xi^k \qquad (k \textit{ an integer}, b \textit{ in } \mathfrak{F}). \tag{60}$$

Thus $\mathfrak{Z} = \mathfrak{F}(\eta)$ if and only if k *is not divisible by* p.

For, as in the proof of Theorem 9.2, we notice that $(\eta^S)^p = c$ and η^S must have the form $\zeta^k\eta$. Then $(\eta\xi^{-k})^S = (\zeta^k\eta\zeta^{-k}\xi^{-k}) = \eta\xi^{-k} = b$ is in $\mathfrak{F}$ since $\xi^S = \zeta\xi$ and $\eta\xi^{-k}$ is unaltered by the generating automorphism S of $\mathfrak{Z}$ over $\mathfrak{F}$. Hence $\eta = b\xi^k$. Conversely, $\eta^p = b^p\xi^{kp} = b^pa^k = c$ is in $\mathfrak{F}$. Every η of $\mathfrak{Z}$ which is not in $\mathfrak{F}$ generates $\mathfrak{Z}$ over $\mathfrak{F}$ and η is in $\mathfrak{F}$ if and only if k is divisible by p.

We have already seen that $k^{p^e} \equiv k \pmod p$ for every integer k. This is a trivial consequence of the so-called "little" **Fermat** theorem which states that $k^p \equiv k \pmod p$ for every integer k. We use this result to obtain

$$\xi^{k^n} = \xi^k a^q \qquad (n = p^e) \tag{61}$$

for every integer k, where, of course, $k^n - k = qp$.

We now let $\mathfrak{Z}$ be cyclic of degree p^e over $\mathfrak{F}$ and prove the parallel of Theorem 9.3.

Theorem 10. *Let $\mathfrak{Z}$ contain $\mathfrak{F}$, $\mathfrak{Y}$ be the unique subfield of $\mathfrak{Z}$ over $\mathfrak{F}$ defined so that $\mathfrak{Y}$ is cyclic of degree* p^{e-1} *over $\mathfrak{F}$ with generating automorphism Σ. Then there exists a quantity β in $\mathfrak{Y}$ such that the relative norm*

$$N_{\mathfrak{Y}|\mathfrak{F}}(\beta) = \zeta. \tag{62}$$

The field

$$\mathfrak{Z} = \mathfrak{F}(\xi), \qquad \xi^p = \alpha, \tag{63}$$

where α in $\mathfrak{Y}$ has the property

$$\frac{\alpha^\Sigma}{\alpha} = \beta^p \tag{64}$$

and thus $N_{\mathfrak{Y}|\mathfrak{F}}(\beta^p) = 1$. *Moreover, the generating automorphism* S *of* $\mathfrak{Z}$ *over* $\mathfrak{F}$ *is uniquely determined by*

$$\gamma^S = \gamma^\Sigma, \qquad \xi^S = \beta\xi \tag{65}$$

for every γ *of* $\mathfrak{Y}$.

As in the proof of Theorem 9.3 we have (63), apply Theorem 9.9 and see that $\xi^S = \beta\xi^k$ for $0 < k < p$. By (61) we have

$$\xi^{S^n} = \beta_n\xi^{k^n} = \beta_n\alpha^q\xi^k = \xi .$$

Then Theorem 9.9 states that $k - 1$ is divisible by p so that since $0 < k < p$ we have $k = 1$. As in the proof of Theorem 9.3 we may choose S so that if $U = S^m$, $m = p^{e-1}$, then $\xi^U = \zeta\xi$. This gives $\xi^S = \beta\xi$, $\xi^{S^2} = \beta\beta^\Sigma\xi$, $\ldots$, and we finally obtain $\xi^{S^m} = \beta\beta^\Sigma \ldots \beta^{\Sigma^{m-1}}\xi = \zeta\xi$ and (62). Since $(\xi^S)^p = \alpha^S = \alpha^\Sigma$, we have (64). The remainder of the argument is as in the proof of Theorem 9.3.

There is no parallel to Theorem 9.4. In fact, it is true that there may exist a cyclic field $\mathfrak{Z}$ of degree p over $\mathfrak{F}$ and such that $N_{\mathfrak{Z}|\mathfrak{F}}(\beta) \neq \zeta$ for any β of $\mathfrak{Z}$. For let $\mathfrak{R}$ be the field of all rational numbers, $p = 2$, $\zeta = -1$, $\mathfrak{Z} = \mathfrak{R}(\xi)$ where $\xi^2 = -1$. Then $\mathfrak{Z}$ is cyclic of degree two over $\mathfrak{R}$ and $\beta = a_1 + a_2\xi$ with rational a_1, a_2,

$$N_{\mathfrak{Z}|\mathfrak{F}}(\beta) = (a_1 + a_2\xi)(a_1 - a_2\xi) = a_1^2 + a_2^2 \neq -1 .$$

for any rational a_1 and a_2. This implies that there exist cyclic fields $\mathfrak{Y}$ of degree p^{e-1} over $\mathfrak{F}$ with no over field $\mathfrak{Z}$ which is cyclic of degree p^e over $\mathfrak{F}$. The parallel to Theorem 9.7 is therefore given by

Theorem 11. *Let* $\mathfrak{Y}$ *be cyclic of degree* p^{e-1} *over* $\mathfrak{F}$ *and let* $\mathfrak{F}$ *contain a quantity* $\zeta \neq 1$ *such that* $\zeta^p = 1$. *Then there exists a field* $\mathfrak{Z}$ *which is cyclic of degree* p^e *over* $\mathfrak{F}$ *and contains* $\mathfrak{Y}$ *if and only if there exists a quantity* β *in* $\mathfrak{Y}$ *such that*

$$N_{\mathfrak{Y}|\mathfrak{F}}(\beta) = \zeta . \tag{66}$$

We have already proved the condition necessary in Theorem 9.10. Conversely, let β be in $\mathfrak{Y}$ and Σ be a generating automorphism of $\mathfrak{Y}$ over $\mathfrak{F}$. Then $N_{\mathfrak{Y}|\mathfrak{F}}(\beta^p) = \zeta^p = 1$ so that by Theorem 9.5

$$\beta^p = \frac{\alpha^\Sigma}{\alpha} \qquad (\alpha \text{ in } \mathfrak{Y}). \tag{67}$$

Let $X^p = \alpha$. Then if X were in $\mathfrak{Y}$, we should have $(X^\Sigma)^p = \alpha^\Sigma$ and $(\beta X)^p = \alpha^\Sigma\alpha\alpha^{-1} = \alpha^\Sigma$, so that we must have $\beta X = \zeta^r X^\Sigma$, $\beta = \zeta^r X^\Sigma X^{-1}$. Then

$N_{\mathfrak{Y}|\mathfrak{F}}(\zeta) = 1 = N_{\mathfrak{Y}|\mathfrak{F}}(\zeta^r)$ and $N_{\mathfrak{Y}|\mathfrak{F}}(\beta) = N_{\mathfrak{Y}|\mathfrak{F}}(X^{\Sigma}X^{-1}) = 1$, contrary to hypothesis. The field $\mathfrak{Z} = \mathfrak{Y}(X)$ thus has degree p over $\mathfrak{Y}$ by Theorem 8.22. By Theorem 7.8 the field $\mathfrak{Z}$ has degree p^e over $\mathfrak{F}$. The transformation of $\mathfrak{Z}$ generated by the partial correspondences

$$X^S = \beta X\,, \qquad \gamma^S = \gamma^{\Sigma} \qquad (\gamma \text{ in } \mathfrak{Y}), \tag{68}$$

defines an automorphism of $\mathfrak{Z}$ over $\mathfrak{F}$, as is evident by a trivial argument. Then

$$X^{S^r} = \beta\beta^S \ldots \beta^{S^{r-1}}X\,,$$

and

$$X^{S^m} = N_{\mathfrak{Y}|\mathfrak{F}}(\beta)X = \zeta X\,, \qquad X^{S^n} = X\,.$$

As in the proof of Theorem 9.7, this implies that S has order $n = p^e$ over $\mathfrak{F}$, $\mathfrak{Z}$ is cyclic over $\mathfrak{F}$, $\mathfrak{Z} = \mathfrak{F}(X)$.

We finally obtain a result occupying the same role in our present theory as that of Theorem 9.8 for the earlier case.

Theorem 12. *Let* a *be determined as in* (64) *for any given* β *in* $\mathfrak{Y}$ *satisfying* (62). *Then every cyclic field* $\mathfrak{Z}$ *of degree* p^e *over* $\mathfrak{F}$ *with* $\mathfrak{Y}$ *as a subfield is a field* $\mathfrak{Z} = \mathfrak{F}(X)$,

$$X^p = ac\,, \tag{69}$$

where c *ranges over all non-zero quantities of* $\mathfrak{F}$ *and* $\mathfrak{Z}$ *has the property*

$$X^S = \beta X\,. \tag{70}$$

For $\mathfrak{Z} = \mathfrak{Y}(X_0)$, $X_0^p = a_0$, $X_0^S = \beta_0 X_0$, $N_{\mathfrak{Y}|\mathfrak{F}}(\beta_0) = \zeta$ by Theorem 9.10. Then $\beta\beta_0^{-1}$ has unity as its norm and $\beta = \gamma^S\gamma^{-1}\beta_0$ by Theorem 9.5. We take $X = \gamma X_0$, γ in $\mathfrak{Y}$ and have $\mathfrak{Z} = \mathfrak{Y}(X)$. By Lemma 1 we have $\mathfrak{Z} = \mathfrak{F}(X)$. Now $X^p = \gamma^p a_0 = a_1$ in $\mathfrak{Y}$, $X^S = \gamma^{\Sigma}\beta_0 X_0 = (\gamma^{\Sigma}\gamma^{-1}\beta_0)\gamma X_0 = \beta X$ as desired. By Theorem 9.5 we have $\beta^p = a_1^{\Sigma}a_1^{-1} = a^{\Sigma}a^{-1}$, $a_1 = ac$, c in $\mathfrak{F}$.

EXERCISES

1. Let $\mathfrak{F}(\xi)$ be a quadratic field over $\mathfrak{F}$ of characteristic not two. Show that we may take $\xi^2 = \tau$ in $\mathfrak{F}$ and prove that there exists a cyclic quartic field over $\mathfrak{F}$ containing $\mathfrak{F}(\xi)$ if and only if $\tau = \delta^2 + \epsilon^2$ with δ and ϵ in $\mathfrak{F}$.

2. Use Theorem 9.12 and Ex. 1 to find all canonical cyclic quartic equations, that is, a set of irreducible quartics whose root (stem) fields are cyclic of degree four over $\mathfrak{F}$.

3. Let $\mathfrak{F}$ be any ordered field, $\xi^2 = -a$, where $a > 0$ in $\mathfrak{F}$. Show that there is no cyclic quartic field with $\mathfrak{F}(\xi)$ as subfield.

7. Adjunction of ζ. We have studied the field $\mathfrak{K} = \mathfrak{F}(\zeta)$, where $\zeta \neq 1$, $\zeta^p = 1$, in Section 8.11. When $\mathfrak{K} = \mathfrak{F}$, our theory of cyclic fields over $\mathfrak{F}$ is complete. Let $\mathfrak{K} > \mathfrak{F}$ so that $\mathfrak{K}$ has degree ν over $\mathfrak{F}$. Then, as we saw in Section 8.11, the field $\mathfrak{K}$ is cyclic of degree ν over $\mathfrak{F}$ with generating automorphism determined by

$$\zeta^T = \zeta^t \neq \zeta, \tag{71}$$

where the integer ν is chosen so that

$$t^\nu \equiv 1 \pmod{p}, \qquad t^k \not\equiv 1 \pmod{p} \tag{72}$$

for any positive integer $k < \nu$. The integer t is prime to p and there exists an integer s such that

$$st \equiv 1 \pmod{p}. \tag{73}$$

This means that ζ^s is a primitive pth root of unity such that

$$(\zeta^s)^T = \zeta. \tag{74}$$

We shall introduce an equivalence relation in $\mathfrak{K}$ for later simplicity. A quantity λ of $\mathfrak{K}$ is said to be *p-equal to* μ in $\mathfrak{K}$, and we write

$$\lambda \underset{(p)}{=} \mu, \tag{75}$$

if $\mu = \delta^p\lambda$ where δ is in $\mathfrak{K}$. Every cyclic field $\mathfrak{Z}$ of degree p over $\mathfrak{K}$ has been shown to be a field $\mathfrak{Z} = \mathfrak{K}(\xi)$, $\xi^p = \lambda$ in $\mathfrak{K}$. Then $\mathfrak{Z} = \mathfrak{K}(\delta\xi)$, $(\delta\xi)^p = \mu$. This is the main reason for calling λ and $\mu = \delta^p\lambda$ p-equal. The relation of p-equality is trivially an equivalence relation as the reader may easily verify.

The integer ν of (72) is prime to p since it was shown in Section 8.11 to be a divisor of $p - 1$. Hence there exists an integer ν_0 such that

$$\lambda^{\nu\nu_0} \underset{(p)}{=} \lambda \tag{76}$$

for every λ of $\mathfrak{K}$, that is, $\nu_0\nu - 1$ is divisible by p. We define

$$s_k = \nu_0 s^k = s_{k-1}s \qquad (k = 0, 1, \ldots, \nu), \tag{77}$$

and have

$$\sigma = \sum_{k=1}^{\nu} t^k s_k = \nu_0 \sum_{k=1}^{\nu} (ts)^k \equiv \nu_0\nu \equiv 1 \pmod{p}, \tag{78}$$

since $st \equiv 1 \pmod p$, $(st)^k \equiv 1 \pmod p$. Consider now the function

$$M(\lambda) = \prod_{k=1}^{\nu} (\lambda^{T^k})^{s_k}.$$

The function $M(\lambda)$ is defined for every λ of $\mathfrak{K}$. We now prove

Theorem 13. *A quantity μ of $\mathfrak{K}$ has the property*

$$\mu^T \underset{(p)}{=} \mu^t \tag{79}$$

if and only if $\mu \underset{(p)}{=} M(\lambda)$ for some λ of $\mathfrak{K}$.

For if $\mu \underset{(p)}{=} M(\lambda)$, we have

$$\mu^T \underset{(p)}{=} \prod_{k=1}^{\nu} (\lambda^{T^{k+1}})^{s_k}, \qquad \mu^t \underset{(p)}{=} \prod_{k=1}^{\nu} (\lambda^{T^k})^{s_k t}.$$

But $\lambda^{s_k t} = \lambda^{s_{k-1} s t} \underset{(p)}{=} \lambda^{s_{k-1}}$, and

$$\mu^t \underset{(p)}{=} \prod_{k=1}^{\nu} (\lambda^{T^{k+1}})^{s_k} \underset{(p)}{=} \mu^T,$$

since $T^{\nu+1} = T$, $s_0 \equiv s_\nu \pmod p$. Conversely, if $\mu^T \underset{(p)}{=} \mu^t$, we form $M(\mu)$. Then

$$M(\mu) \underset{(p)}{=} \prod_{k=1}^{\nu} \mu^{t^k s_k} \underset{(p)}{=} \mu^\sigma \underset{(p)}{=} \mu$$

by (78).

As a consequence of Theorem 9.13, we may prove

Theorem 14. *Let $\mu \underset{(p)}{=} \lambda^T \lambda^{-t}$ where $\lambda \neq 0$ is in $\mathfrak{K}$. Then*

$$\mu^T \underset{(p)}{=} \mu^t \tag{80}$$

if and only if $\lambda^T \underset{(p)}{=} \lambda^t$, that is $\mu \underset{(p)}{=} 1$.

For if $\lambda^T \underset{(p)}{=} \lambda^t$, we have $\mu \underset{(p)}{=} 1$, and $\mu^T \underset{(p)}{=} 1 \underset{(p)}{=} \mu^t$. Conversely, let $\mu^T \underset{(p)}{=} \mu^t$. Then $\lambda^T \underset{(p)}{=} \lambda^t \mu$ implies that

$$\lambda^{T^2} \underset{(p)}{=} (\lambda^T)^t \mu^T \underset{(p)}{=} (\lambda^t \mu)^t \mu^t \underset{(p)}{=} \lambda^{t^2} \mu^{2t},$$

and

$$\lambda^{T^3} \underset{(p)}{=} (\lambda^T)^{t^2} (\mu^T)^{2t} \underset{(p)}{=} (\lambda^t \mu)^{t^2} (\mu^t)^{2t} \underset{(p)}{=} \lambda^{t^3} \mu^{3t^2},$$

so that ultimately by induction

$$\lambda^{T^r} \underset{(p)}{=} (\lambda^T)^{t^{r-1}}[\mu^{(r-1)t^{r-2}}]^T \underset{(p)}{=} \lambda^{t^r}\mu^{rt^{r-1}}$$

for every integer r. Take $r = \nu$ and have $T^\nu = I$,

$$\lambda \underset{(p)}{=} \lambda^{t^\nu}\mu^{\nu t^{\nu-1}} \underset{(p)}{=} \lambda\mu^{\nu t^{\nu-1}}, \qquad \mu^{\nu t^{\nu-1}} \underset{(p)}{=} 1 .$$

Since $\nu t^{\nu-1}$ is prime to p, we must have $\mu \underset{(p)}{=} 1$ and $\lambda^T \underset{(p)}{=} \lambda^t$ as desired.

Theorem 9.13 gives a construction of all quantities a in $\mathfrak{K}$ such that $a^T \underset{(p)}{=} a^t$. The additional condition stating that $a \underset{(p)}{=} M(\lambda)$ is not the pth power of any quantity of $\mathfrak{K}$ is an irreducibility condition on the equation $x^p = a$ and depends fundamentally on the structure of the field $\mathfrak{F}$ itself. We cannot give any more precise form of this condition for a general field $\mathfrak{F}$ and shall leave it in the present form. We now determine uniquely* all cyclic fields $\mathfrak{Z}$ of degree p over $\mathfrak{F}$ in

Theorem 15. *A field $\mathfrak{Z}$ is cyclic of degree* p *over $\mathfrak{F}$ if and only if $\mathfrak{Z}$ is the unique subfield of degree* p *over $\mathfrak{F}$ of the cyclic field $\mathfrak{Z}_0$ over $\mathfrak{F}$ given by*

$$\mathfrak{Z}_0 = \mathfrak{K} \times \mathfrak{Z} = \mathfrak{K}(\xi) ,$$

where this latter field has degree pν *over $\mathfrak{F}$, ν as in* (72), *a is not* p*-equal to one,*

(81) $$\xi^p = a , \qquad a^T \underset{(p)}{=} a^t \qquad (a \text{ in } \mathfrak{K}) .$$

For the degree over $\mathfrak{K}$ of the composite of a cyclic field $\mathfrak{Z}$ of degree p over $\mathfrak{F}$ and $\mathfrak{K}$ of degree ν prime to p over $\mathfrak{F}$ was shown in Section 8.4 to be p. Hence this composite has degree $p\nu$ over $\mathfrak{F}$. By the converse to Theorem 8.18 this composite field is cyclic of degree νp over $\mathfrak{F}$, and the products

(82) $$S^iT^j \qquad (i = 0, \ldots, p-1; j = 0, \ldots, \nu - 1)$$

define the automorphisms of $\mathfrak{Z}_0$ over $\mathfrak{F}$. Here S is a generating automorphism of $\mathfrak{Z}$ over $\mathfrak{F}$ and

$$ST = TS .$$

* Notice that in the construction of a satisfying (81) we need not choose a as the μ of Theorem 13 but can take $a \underset{(p)}{=} M(\lambda)$ since any $\mu \underset{(p)}{=} a$ defines the same $\mathfrak{Z}_0$ as a.

By Theorem 8.22 we have $\mathfrak{Z}_0 = \mathfrak{K}(\xi)$, $\xi^p = \alpha$ in $\mathfrak{K}$, and we may choose S so that $\xi^S = \zeta\xi$. Then α is not p-equal to one, $(\xi^T)^p = \alpha^T$, so that by Theorem 9.9,

$$\xi^T = b\xi^k \qquad (b \text{ in } \mathfrak{F}),$$

and k is an integer. The quantity ξ^T cannot be in $\mathfrak{K}$ since $(\xi^T)^{T^{-1}} = \xi$ is not in $\mathfrak{K}$, and k must be one of the integers $1, 2, \ldots, p-1$. Then $b^S = b$ in $\mathfrak{K}$,

$$(\xi^T)^S = b(\xi^S)^k = b\zeta^k\xi^k = (\xi^S)^T = (\zeta\xi)^T = \zeta^t b\xi^k\,,$$

and $k = t$, $\xi^T = b\xi^t$,

$$\alpha^T = (\xi^T)^p = b^p\alpha^t \underset{(p)}{=} \alpha^t\,,$$

as desired. Conversely, if $\alpha^T \underset{(p)}{=} \alpha^t$ and α in $\mathfrak{K}$ is not p-equal to unity, the field $\mathfrak{Z}_0 = \mathfrak{K}(\xi)$, $\xi^p = \alpha$, is cyclic of degree p over $\mathfrak{K}$ and has degree $p\nu$ over $\mathfrak{F}$. Then $\alpha^T = b^p\alpha^t$ and the computation above shows that

$$\xi^T = b\xi^t \tag{83}$$

is a root of $x^p = \alpha^T$. The correspondence T of $\mathfrak{Z}_0$ defined by (83) and the automorphism T in $\mathfrak{K}$ is an automorphism of $\mathfrak{Z}_0$ over $\mathfrak{F}$ and our computation shows that $\xi^{ST} = \xi^{TS}$. Every quantity of $\mathfrak{Z}_0$ has the form $\delta_0 + \delta_1\xi + \ldots + \delta_{p-1}\xi^{p-1}$ with δ_i in $\mathfrak{K}$ and $\delta_i^{ST} = \delta_i^T = \delta_i^{TS}$. Hence $ST = TS$ in $\mathfrak{Z}_0$. The operations imposed on ξ and ζ by S and T show that the automorphisms (82) are all distinct and, since they are $p\nu$ in number, that $\mathfrak{Z}_0$ is normal over $\mathfrak{F}$. But (82) is a direct product of $[S]$ and $[T]$ which are cyclic of relatively prime orders. Thus (82) is a cyclic group of order $p\nu$ and has a unique cyclic subgroup $[T]$ of order ν. The field $\mathfrak{Z}$ is the unique field corresponding to $[T]$ in Theorem 8.4.

We have completely determined all cyclic fields of prime degree p over a field $\mathfrak{F}$. We now pass to fields $\mathfrak{Z}$ of degree p^e over $\mathfrak{F}$. Assume that $\mathfrak{Y}$ is given cyclic of degree p^{e-1} over $\mathfrak{F}$ and that $\mathfrak{Y}_0 = \mathfrak{Y} \times \mathfrak{K}$. Then $\mathfrak{Z}_0 = \mathfrak{Z} \times \mathfrak{K} = \mathfrak{Y}_0(\xi)$ exists as a cyclic field over $\mathfrak{K}$ if and only if there is a quantity β in $\mathfrak{Y}_0$ such that

$$N_{\mathfrak{Y}_0|\mathfrak{K}}(\beta) = \zeta\,. \tag{84}$$

We must of course assume the existence of β.

The field $\mathfrak{Z}_0$ must be cyclic of degree $p^e\nu$ over $\mathfrak{F}$ and $\mathfrak{Z}$ is uniquely determined by Theorem 8.4 and Theorem 9.12. For $\mathfrak{Z}_0$ is uniquely determined by β in Theorem 9.12, and $\mathfrak{Z}$ corresponds to a subgroup of the cyclic group of $\mathfrak{Z}_0$ over $\mathfrak{F}$. Hence it is sufficient for our construction of $\mathfrak{Z}$ over $\mathfrak{F}$ to construct our field $\mathfrak{Z}_0$ so that it is cyclic over $\mathfrak{F}$. We first prove

Theorem 16. *The quantity β assumed to exist in $\mathfrak{Y}_0 = \mathfrak{Y} \times \mathfrak{K}$ may be so chosen that there exists a solution* a *of*

$$\beta^p = a^S a^{-1} \tag{85}$$

with

$$a^T = d^p a^t, \qquad \beta^T = \zeta^\tau d^S d^{-1} \beta^t \qquad (d \text{ in } \mathfrak{Y}_0), \tag{86}$$

and $0 \leqq \tau < p$, τ is an integer uniquely determined by $\mathfrak{Y}_0$.

For $\beta = \beta_1$ was any quantity of $\mathfrak{Y}_0$ with the property that $N_{\mathfrak{Y}_0|\mathfrak{K}}(\beta_1) = \zeta$. Suppose that $a_1^S a_1^{-1} = \beta_1^p$. We know that such an a_1 exists by Theorem 9.5. We then form $a = M(a_1)$ and have $a^T = d^p a^t$ by Theorem 9.13, where now d is in $\mathfrak{Y}_0$ which is evidently the field $\mathfrak{Y}(\zeta)$ of degree ν over $\mathfrak{Y}$. But then $ST = TS$ in $\mathfrak{Y}_0$ implies that if

$$\beta = M(\beta_1),$$

then

$$\beta^p = M(\beta_1^p) = M(a_1^S a_1^{-1}) = a^S a^{-1}$$

since $M(\lambda)$ is obviously a multiplicative function. The property $ST = TS$ in $\mathfrak{Y}_0$ implies that

$$N_{\mathfrak{Y}_0|\mathfrak{K}}(\beta) = M[N_{\mathfrak{Y}_0|\mathfrak{K}}(\beta_1)] = M(\zeta) = \prod_{k=1}^{\nu} \zeta^{t^k e_k} = \zeta^\sigma = \zeta .$$

Thus we replace β_1 by β. Now use $ST = TS$ in

$$\begin{aligned}(a^S a^{-1})^T &= (\beta^T)^p = a^{TS}(a^T)^{-1} = (d^p a^t)^S (d^p a^t)^{-1} \\ &= (d^S d^{-1})^p \beta^{pt} = (\beta^p)^T = (\beta^T)^p ,\end{aligned}$$

which is true since $a^S a^{-1} = \beta^p$. It follows that the pth power of $\rho = \beta^T \beta^{-t} (d^S d^{-1})^{-1}$ is unity and ρ must be a power ζ^τ of the primitive pth root of unity ζ. This gives the existence of the integer τ in (86).

To prove τ unique we let β_0 and a_0 be given so that

$$\beta_0^p = a_0^S a_0^{-1}, \qquad a_0^T = d_0^p a_0^t, \qquad \beta_0^T = \zeta^{\tau_0} d_0^S d_0^{-1} \beta_0^t \qquad (0 \leqq \tau_0 < p).$$

Then $N_{\mathfrak{Y}_0|\mathfrak{K}}(\beta_0\beta^{-1}) = 1$ so that $\beta_0 = \gamma^S\gamma^{-1}\beta$ with γ in $\mathfrak{Y}_0$ by Theorem 9.5. It follows that $\beta_0^p = (\gamma^S\gamma^{-1})^p a^S a^{-1} = a_0^S a_0^{-1}$, so that $(a_0 a^{-1}\gamma^{-p})^S = a_0 a^{-1}\gamma^{-p}$ is in $\mathfrak{K}$. Hence

$$a_0 = \gamma^p a \sigma \qquad (\sigma \text{ in } \mathfrak{K}),$$

and

$$d_0^p = a_0^T a_0^{-t} = (\gamma^T)^p d^p a^t \sigma^T \gamma^{-tp} a^{-t} \sigma^{-t} = d^p (\gamma^T \gamma^{-t})^p \sigma^T \sigma^{-t}\,.$$

We have shown that

$$(87) \qquad \sigma^T = \sigma^t \Delta^p\,, \qquad \Delta = d_0 d^{-1} (\gamma^T \gamma^{-t})^{-1}\,.$$

If Δ is not in $\mathfrak{K}$, its pth power is in $\mathfrak{K}$ and it lies in a subfield of degree p over $\mathfrak{K}$ of $\mathfrak{Y}_0$. This subfield is of course cyclic of degree p over $\mathfrak{K}$. By our hypothesis on $\mathfrak{Y}_0$ this subfield $\mathfrak{K}(\Delta)$ is the direct product of a cyclic field $\mathfrak{Y}_1$ of degree p over $\mathfrak{F}$ and $\mathfrak{K}$ since $\mathfrak{Y}_0 = \mathfrak{Y} \times \mathfrak{K}$, $\mathfrak{Y} \geqq \mathfrak{Y}_1$ of degree p over $\mathfrak{F}$. Hence $\Delta^p = \delta$ in $\mathfrak{K}$ must have the property $\delta^T = \delta^t$ where this now means $\delta^T = \mu^p \delta^t$, μ in $\mathfrak{K}$. But $\delta = \sigma^T \sigma^{-t}$ with σ in $\mathfrak{K}$. By the proof of Theorem 9.14 we have $\sigma^T = \sigma^t \epsilon^p$, ϵ in $\mathfrak{K}$. Since $\epsilon^p = \Delta^p$ we have Δ in $\mathfrak{K}$, a contradiction. Thus in all cases Δ is in $\mathfrak{K}$ and

$$(88) \qquad d_0 = \Delta \gamma^T \gamma^{-t} d\,.$$

We now compute

$$(89) \quad \beta_0^T = \gamma^{ST}(\gamma^{-1})^T \zeta^\tau d^S d^{-1} \beta^t = \zeta^{\tau_0} (\Delta\gamma^T\gamma^{-t}d)^S (\Delta\gamma^T\gamma^{-t}d)^{-1} (\gamma^S\gamma^{-1})^t \beta^t$$

and obtain $\zeta^\tau = \zeta^{\tau_0}$, $\tau = \tau_0$ as desired.

The integer τ is uniquely determined by the field $\mathfrak{Y}_0$ and of course is then determined by $\mathfrak{Y}$. It is an invariant of $\mathfrak{Y}$ and we now completely determine all cyclic fields $\mathfrak{Z}$ by proving

Theorem 17. *The field $\mathfrak{Y}$ possesses cyclic overfields $\mathfrak{Z}$ of degree* p^e *over $\mathfrak{F}$ if and only if the corresponding integer τ of Theorem* 9.16 *is zero. Every such cyclic field $\mathfrak{Z}$ is the unique subfield of $\mathfrak{Z}_0 = \mathfrak{Y}_0(\xi) = \mathfrak{Z} \times \mathfrak{K}$, $\xi^p = \alpha$ in $\mathfrak{Y}_0$, such that*

$$(90) \qquad \alpha = \lambda a, \qquad \lambda^T = \sigma^p \lambda^t \qquad (\lambda, \sigma \textit{ in } \mathfrak{K}),$$

with the quantity a *given as in* (85) *and* (86).

For we have already shown in Theorem 9.12 that if $\mathfrak{Z}$ is cyclic of degree p

over $\mathfrak{Y}$, then $\mathfrak{Z}_0 = \mathfrak{Z} \times \mathfrak{K}$ has the form $\mathfrak{Z}_0 = \mathfrak{Y}_0(\xi)$, $\xi^p = \lambda a = \alpha$, λ in $\mathfrak{K}$. Our hypothesis $\mathfrak{Z}_0 = \mathfrak{Z} \times \mathfrak{K}$ implies that

$$\alpha^T \underset{(p)}{=} \alpha^t$$

by the proof of Theorem 9.15. Hence $\alpha^T = d_0^p\alpha^t$, d_0 in $\mathfrak{Y}_0$. We choose a as in Theorem 9.16 and have $a^T = d^p a^t$. By Theorem 9.12 we know that $\alpha^S\alpha^{-1} = \beta^p$, and

$$\alpha^T = \lambda^T a^T = d_0^p\alpha^t = d^p a^t \lambda^T = d_0^p \lambda^t a^t .$$

Hence $\lambda^T\lambda^{-t} = (d_0d^{-1})^p$. As in the proof of Theorem 9.16, this implies that $\sigma = d_0d^{-1}$ is in $\mathfrak{K}$ and we have proved (90). We use $d_0 = \sigma d$ with $\sigma = \sigma^S$ in $\mathfrak{K}$, and (86) may be replaced by

$$\beta^T = \zeta^\tau \frac{d_0^S}{d_0} \beta^t .$$

We now apply the condition that automorphisms S and T of $\mathfrak{Z}_0$ be commutative. Compute

$$(\xi^T)^p = \alpha^T = d_0^p\alpha^t = d_0^p\xi^{tp} ,$$

and obtain

$$\xi^T = \zeta^g d_0 \xi^t , \tag{91}$$

where g is an integer and $\zeta \neq 1$ is our pth root of unity. Then $\xi^S = \beta\xi$, $\zeta^S = \zeta$,

$$\xi^{TS} = \zeta^g d_0^S \beta^t \xi^t , \tag{92}$$

while

$$\xi^{ST} = (\beta\xi)^T = \zeta^\tau d_0^S d_0^{-1} \beta^t \zeta^g d_0 \xi^t = \zeta^\tau (\zeta^g d_0^S \beta^t \xi^t) \tag{93}$$

is equal to ξ^{TS} if and only if $\tau = 0$.

Conversely, let $\mathfrak{Y}$ be cyclic of degree p^{e-1} over $\mathfrak{F}$, $\mathfrak{Y}_0 = \mathfrak{Y} \times \mathfrak{F}(\zeta)$ contain β such that $N_{\mathfrak{Y}_0|\mathfrak{K}}(\beta) = \zeta$, and choose a, β in $\mathfrak{Y}_0$ satisfying (84), (85), (86). We let λ range over all quantities of $\mathfrak{K} = \mathfrak{F}(\zeta)$ satisfying (90) and have already proved that $\mathfrak{Z}_0 = \mathfrak{Y}_0(\xi) = \mathfrak{K}(\xi)$ is cyclic of degree p^e over $\mathfrak{K}$. It remains to show that the condition $\tau = 0$ implies that $\mathfrak{Z}_0 = \mathfrak{F}(\xi, \zeta)$ is cyclic of degree $p^e\nu$ over $\mathfrak{F}$. This is immediately true when we define an automorphism T of $\mathfrak{Z}_0$ over $\mathfrak{F}$ by

$$\gamma(\zeta)^T = \gamma(\zeta^t) , \qquad \eta^T = \eta , \qquad \xi^T = \sigma d\xi^t$$

for every $\gamma(\zeta)$ of $\mathfrak{K}$ and every η of $\mathfrak{Y}$, and we of course have the equation $a^T = d^p a^t$ defining d. For, our above computation (91), (92), and (93) with $g = 1$ shows that $ST = TS$ and $\mathfrak{Z}_0$ over $\mathfrak{F}$ has the automorphisms (S^iT^j) with $i = 0, 1, \ldots, p^e - 1$, and $j = 0, 1, \ldots, \nu - 1$. The order of S is p^e, the order of T is ν prime to p, and (S^iT^j) is evidently the direct product of the cyclic groups $[S]$, $[T]$, and has order $p^e\nu$. This is the degree of $\mathfrak{Z}_0$ over $\mathfrak{F}$ and $\mathfrak{Z}_0$ is cyclic over $\mathfrak{F}$.

CHAPTER X

ALGEBRAS OF MATRICES

1. Algebras of order *n* over $\mathfrak{F}$. The theory of what are called *linear associative algebras* or simply *algebras over a field* $\mathfrak{F}$ is the theory of linear sets

$$(1) \qquad \mathfrak{A} = (u_1, \ldots, u_n) ,$$

of order n over a field $\mathfrak{F}$, which have been made into (associative)* rings over $\mathfrak{F}$ by defining multiplication in $\mathfrak{A}$ in a suitable manner. This theory is one of the most important branches of modern algebra. It arises very naturally from a consideration of square matrices forming a linear set $\mathfrak{A}$ with the property that the product of any two matrices of $\mathfrak{A}$ is in $\mathfrak{A}$. The set $\mathfrak{M}_m$ of all m-rowed square matrices with elements in $\mathfrak{F}$ is an example of such an algebra of order $n = m^2$ over $\mathfrak{F}$. Another elementary example is the algebra $\mathfrak{F}[A]$ of all polynomials, with coefficients in $\mathfrak{F}$, in an m-rowed square matrix A and the m-rowed identity matrix. In this case the order is the degree of the minimum function of A. There are, of course, many other less simple examples of algebras of matrices.

The interest and importance of the theory of algebras leads us to give an introduction to the theory here by a consideration of some of the simpler properties of *matric* algebras resulting from certain of our theorems on matrices. It is not only natural to do this but we shall increase the importance of these results by showing that every algebra $\mathfrak{A}$ of order n over $\mathfrak{F}$ is equivalent to an algebra of square matrices with elements in $\mathfrak{F}$.

2. The regular representation of an algebra. Consider a linear set $\mathfrak{A}$ of (1) and define n^3 quantities γ_i^{kj} in $\mathfrak{F}$ called the multiplication constants† of $\mathfrak{A}$. We recall that

$$(2) \qquad u_k u_j = \sum_{i=1}^{n} u_i \gamma_i^{kj} \qquad (j, k = 1, \ldots, n)$$

$$(3) \qquad \left(\sum_{k=1}^{n} \xi_k u_k\right)\left(\sum_{j=1}^{n} \eta_j u_j\right) = \sum_{k,j=1}^{n} \xi_k \eta_j \, u_k u_j \qquad (\xi_k, \eta_j \text{ in } \mathfrak{F})$$

* Sets satisfying all of our postulates for a ring except the postulate $a(bc) = (ab)c$ are called "non-associative rings." We shall not consider such rings here. Notice that our definition of $\mathfrak{A}$ over $\mathfrak{F}$ of Chap. II does not imply that $\mathfrak{A}$ contains $\mathfrak{F}$ unless $\mathfrak{A}$ has a unity element and may be made to contain $\mathfrak{F}$ by Theorem 1.9.

† We wrote γ_{kji} in an earlier chapter, but our present notation is more satisfactory.

define multiplication in $\mathfrak{A}$ such that $\mathfrak{A}$ is an (associative) algebra if and only if

$$u_k(u_j u_i) = (u_k u_j)u_i \qquad (i, j, k = 1, \ldots, n). \tag{4}$$

The equations (4) are equivalent to certain conditions on the γ_i^{kj} and we shall consider only algebras $\mathfrak{A}$ in which they are satisfied. Matrix multiplication is associative and when the u_j are matrices satisfying (2) and (3) the condition (4) is automatically satisfied.

Every quantity x of $\mathfrak{A}$ has the form

$$x = \xi_1 u_1 + \ldots + \xi_n u_n = \sum_{k=1}^{n} \xi_k u_k \tag{5}$$

with the ξ_k in $\mathfrak{F}$. We compute

$$x u_j = \sum_{i=1}^{n} u_i \xi_{ij} \qquad (j = 1, \ldots, n), \tag{6}$$

where the

$$\xi_{ij} = \sum_{k=1}^{n} \xi_k \gamma_i^{kj} \tag{7}$$

are in $\mathfrak{F}$. If U is the one-rowed matrix whose elements are $u_1, \ldots, u_n$ and

$$X = (\xi_{ij}) \qquad (i, j = 1, \ldots, n), \tag{8}$$

we may write equations (6) in matrix form as

$$xU = UX. \tag{9}$$

Then we have defined a correspondence

$$x \to X \tag{10}$$

from the algebra $\mathfrak{A}$ to the set $\mathfrak{B}$ of all matrices X of (8).

If y is in $\mathfrak{A}$ and $yU = UY$, we have

$$(x + y)U = U(X + Y), \quad (xy)U = x(yU) = x(UY) = (xU)Y = (UX)Y = U(XY),$$

since all our products are associative. Also

$$axU = xUa = UaX \qquad (a \text{ in } \mathfrak{F}).$$

Thus the correspondence (10) is preserved under addition, multiplication and the scalar multiplication of our quantities by elements a of $\mathfrak{F}$.

DEFINITION. *The algebra* $\mathfrak{B}$ *of matrices* (8), (7) *is called the regular representation of* $\mathfrak{A}$ *with respect to the basis* $u_1, \ldots, u_n$ *of* $\mathfrak{A}$.

We shall usually omit the reference to the basis and shall speak of the regular representation. The basis will of course be given in each case.

Theorem 1. *The algebra* $\mathfrak{B}$ *is an algebra of order* n *over* $\mathfrak{F}$ *equivalent to* $\mathfrak{A}$ *under the correspondence* (10) *if and only if there is no quantity* $x \neq 0$ *in* $\mathfrak{A}$ *such that*

$$\text{(11)} \qquad xa = 0$$

for every a *of* $\mathfrak{A}$.

For if (10) is a one-to-one correspondence the algebras $\mathfrak{A}$ and $\mathfrak{B}$ are equivalent. Thus $\mathfrak{B}$ and $\mathfrak{A}$ are not equivalent if and only if $Z = Y$ in $\mathfrak{B}$ for correspondents z and y in $\mathfrak{A}$ such that $z \neq y$. Then $x = z - y \neq 0$ and $x \to 0$. But $xU = 0$, which means that $xu_i = 0$, $xa = 0$ for every a of $\mathfrak{A}$. Conversely, when $xa = 0$ for every a of $\mathfrak{A}$ and $x \neq 0$ in $\mathfrak{A}$, we have $x \to 0$ and $x \neq 0$ so that (10) is not a one-to-one correspondence.

Theorem 2. *Let* $\mathfrak{A}$ *be an algebra of order* n *with a unity element. Then the regular representation of* $\mathfrak{A}$ *is an algebra* $\mathfrak{B}$ *of* n*-rowed square matrices with elements in* $\mathfrak{F}$ *equivalent to* $\mathfrak{A}$ *under a correspondence in which the unity element of* $\mathfrak{A}$ *corresponds to the* n*-rowed identity matrix.*

For if e is the unity element we have $xe = x = 0$ if and only if $x = 0$. Thus $\mathfrak{B}$ and $\mathfrak{A}$ are equivalent by Theorem 10.1. Also $eu_i = u_i$ and $eU = UI$ where I is the n-rowed identity matrix.

Let $\mathfrak{B}$ and $\mathfrak{C}$ be algebras of m-rowed square matrices over $\mathfrak{F}$ which are equivalent under a correspondence

$$B_i \longleftrightarrow C_i\,,$$

where $B_1, \ldots, B_n$ is a basis of $\mathfrak{B}$ over $\mathfrak{F}$, and $C_1, \ldots, C_n$ a corresponding basis of $\mathfrak{C}$ over $\mathfrak{F}$. We call $\mathfrak{B}$ and $\mathfrak{C}$ *similar* if there exists a non-singular m-rowed square matrix T with elements in $\mathfrak{F}$ such that

$$\text{(12)} \qquad C_i = TB_iT^{-1} \qquad (i = 1, \ldots, n).$$

We now prove

Theorem 3. *Any two regular representations of an algebra with a unity element are similar.*

This result states that our regular representations of algebras are essentially independent of the particular bases used in their definition. For proof let $\mathfrak{B}$ be a regular representation defined by a basis $u_1, \ldots, u_n$ of $\mathfrak{A}$ and similarly $\mathfrak{C}$ be defined by $v_1, \ldots, v_n$. Then

$$v_j = \sum u_i \tau_{ij} \qquad (\tau_{ij} \text{ in } \mathfrak{F}), \tag{13}$$

where the matrix

$$T = (\tau_{ij}) \qquad (i, j = 1, \ldots, n)$$

is necessarily non-singular by Theorem 3.16. The regular representation $\mathfrak{C}$ is obtained from

$$xV = VX_0 ,$$

where (13) is equivalent to

$$V = UT .$$

But $xU = UX$ and

$$x(UT) = UTX_0 , \qquad xU = UTX_0T^{-1}$$

since T is non-singular. Hence $X = TX_0T^{-1}$ for the general x of $\mathfrak{A}$,

$$X_0 = T^{-1}XT . \tag{14}$$

This means that (12) holds for the correspondence

$$x \longleftrightarrow X \longleftrightarrow X_0 .$$

If $\mathfrak{A} = (u_1, \ldots, u_n)$ over $\mathfrak{F}$ is an algebra without a unity element, we consider a linear set

$$\mathfrak{A}_0 = (u_0, u_1, \ldots, u_n)$$

with an additional basal element u_0, and make $\mathfrak{A}_0$ into an algebra of order $n + 1$ over $\mathfrak{F}$ with u_0 as unity element by defining multiplication by (3) for $j, k = 0, \ldots, n$, (2) for $j, k = 1, \ldots, n$ as in $\mathfrak{A}$, and

$$u_ju_0 = u_0u_j = u_j \qquad (j = 0, \ldots, n).$$

The regular representation $\mathfrak{B}_0$ of $\mathfrak{A}_0$ is equivalent to $\mathfrak{A}_0$ and has a subalgebra $\mathfrak{B}$ equivalent to $\mathfrak{A}$. The matrices representing the elements of $\mathfrak{A}$ are easily seen to have the form

$$\begin{pmatrix} 0 & 0_{1n} \\ \xi & X \end{pmatrix} \longleftrightarrow x = \sum_{k=1}^{n} \xi_k u_k$$

where X is given by (8), (7) and is the regular representation of x, 0_{1n} is the $1 \times n$ zero matrix, ξ is the $n \times 1$ column of elements $\xi_1, \ldots, \xi_n$. For,

$$u_k u_0 = u_k,\ u_k u_j = 0u_0 + \sum_{k=1}^{n} u_i \gamma_i^{kj}.$$

We have now seen that every algebra $\mathfrak{A}$ of order n over $\mathfrak{F}$ *is equivalent* to an algebra of n- or $(n + 1)$-rowed square matrices with elements in $\mathfrak{F}$ and *is a subalgebra* of an algebra $\mathfrak{A}_0$ over $\mathfrak{F}$ with a unity element. The regular representation of $\mathfrak{A}_0$ provides a representation of $\mathfrak{A}$ by an equivalent algebra of matrices. We are not interested here in the problem of finding all representations of $\mathfrak{A}$ by algebras of matrices. We shall therefore restrict our further attention to the study of algebras with a unity element and the properties resulting from their regular matrix representations. Notice that (14) implies that it is the properties of matrices invariant under similarity transformations that are of greatest importance here. We shall therefore apply the theorems of Chapter IV.

EXERCISES

1. Find the regular representation of the algebra $\mathfrak{F}[A]$ with basis $I, A, \ldots, A^{n-1}$ over $\mathfrak{F}$, where A is an m-rowed square matrix with minimum function $\lambda^n + a_1\lambda^{n-1} + \ldots + a_n$, a_i in $\mathfrak{F}$. Hint: Show that A is represented by the transpose of the matrix of the fundamental lemma of Section 4.4.

2. Find the regular representation of the algebra $\mathfrak{M}_m = (e_{ij})$ of all m-rowed square matrices with elements in $\mathfrak{F}$.

3. Let $\mathfrak{G} = (S_1, S_2, \ldots, S_n)$ be a finite group where $S_1 = I$ is the identity element of $\mathfrak{G}$, $S_iS_j = S_{k_{ij}}$. Then we make $\mathfrak{G}$ into an algebra called the *group algebra* $\mathfrak{A}(\mathfrak{G})$ defined by $\mathfrak{G}$ by assuming that $S_i = u_i$ are the basal elements of a linear set. Find the regular representation of this algebra.

4. The multiplication constants of a group algebra are either 0 or 1, so that the reference field $\mathfrak{F}$ may be taken at our choice. Take $\mathfrak{F}$ to be the field of all rational numbers and prove that there is no matrix Y in the regular representation of a group algebra such that $T(XY) = 0$ for every X of the representation. Here we define $T(B)$ to be the trace (sum of the diagonal elements) of the matrix B. Hint: Write $Y = \sum_{i=1}^{n} \eta_i U_i$ where U_i is the matrix representation of S_i, and show that there

exists an i such that $U_i^{-1}Y = \lambda_1 U_1 + \ldots + \lambda_n U_n$ where $\lambda_1 \neq 0$. Complete the proof by showing that $T(U_1) = n$, $T(U_j) = 0$ $(j = 2, \ldots, n)$, where of course U_1 is now the identity matrix.

5. Find the regular representation of the algebra (u_1, u_2, u_3, u_4) over $\mathfrak{F}$ where u_1 is the unity element, $u_2^2 = -1$, $u_3^2 = 3$, $u_2u_3 = -u_3u_2 = u_4$, $\mathfrak{F}$ is the field of all rational numbers.

6. Two algebras $\mathfrak{A}$ and $\mathfrak{A}'$ are said to be *reciprocal* over $\mathfrak{F}$ if the postulates for equivalence are satisfied except that $(ab)' = b'a'$. State this definition in full. Show that if we form u_jx in (6) we obtain an algebra like the $\mathfrak{B}$ of Theorem 10.1 but now reciprocal over $\mathfrak{F}$ to $\mathfrak{A}$.

3. The characteristic and minimum functions, scalar extension. Let $\mathfrak{A}$ be an algebra of order n over $\mathfrak{F}$ so that every quantity of $\mathfrak{A}$ is uniquely expressible in the form

$$a = a_1u_1 + \ldots + a_nu_n \qquad (a_i \text{ in } \mathfrak{F}). \tag{15}$$

The u_i are in $\mathfrak{A}$ and $\mathfrak{A}$ is a linear set $(u_1, \ldots, u_n)$ over $\mathfrak{F}$ such that $u_ku_j = \sum_{i=1}^{n} u_i\gamma_i^{kj}$ with γ_i^{kj} in $\mathfrak{F}$. We assume that $\mathfrak{A}$ is an algebra with a unity element and have shown that $\mathfrak{A}$ is equivalent to its regular representation $\mathfrak{A}_0$ of n-rowed square matrices A with elements in $\mathfrak{F}$.

Every A of $\mathfrak{A}_0$ is a matrix and has a unique characteristic function and a unique minimum function as in Chapter IV. These polynomials

$$f(\lambda) = |\lambda I - A| \cdot I\,, \qquad \phi(\lambda)$$

are two scalar matrices in the set of all n-rowed square matrices whose elements are in the polynomial integral domain $\mathfrak{F}[\lambda]$, λ an indeterminate over $\mathfrak{F}$. The ring $\mathfrak{F}[\lambda]$ is equivalent to the ring of all scalar matrices with elements in $\mathfrak{F}[\lambda]$, that is, the ring of all matrices $g(\lambda) \cdot I$ where $g(\lambda)$ is in $\mathfrak{F}[\lambda]$ and I is the n-rowed identity matrix. We have already agreed in Chapter IV to identify these two rings and therefore say that $f(\lambda)$ and $\phi(\lambda)$ are in $\mathfrak{F}[\lambda]$. These latter polynomials have the properties

$$f(A) = \phi(A) = 0\,.$$

If also $g(\lambda)$ in $\mathfrak{F}[\lambda]$ has the property $g(A) = 0$ the polynomial $g(\lambda)$ is divisible by $\phi(\lambda)$.

The equivalence $a \longleftrightarrow A$ between the algebras $\mathfrak{A}$ and $\mathfrak{A}_0$ implies that when we replace the identity matrix of $\mathfrak{A}_0$ by the unity element of $\mathfrak{A}$ we have

$$f(a) = \phi(a) = 0\,.$$

Thus every quantity a of an algebra $\mathfrak{A}$ over $\mathfrak{F}$ satisfies some equation with coefficients in $\mathfrak{F}$, and the degree of one such equation (e.g., $f(a) = 0$) is at most* n.

When we pass to a new regular representation of $\mathfrak{A}$ we replace $A \longleftrightarrow a$ by $T^{-1}AT$ similar to A and do not change either $\phi(\lambda)$ or $f(\lambda)$. Thus we may make the

Definition. *Let an algebra $\mathfrak{A}$ with a unity element be equivalent to its regular representation $\mathfrak{A}_0$ under the correspondence*

$$a \longleftrightarrow A \qquad (a \textit{ in } \mathfrak{A},\ A \textit{ in } \mathfrak{A}_0).$$

Then the characteristic and minimum functions (equations) of the matrix A *are called the characteristic and minimum functions (equations) of the quantity* a *of* $\mathfrak{A}$.

Consider any algebra $\mathfrak{B}$ of matrices such that $\mathfrak{B}$ is equivalent to $\mathfrak{A}$ under a correspondence $a \longleftrightarrow B$ in $\mathfrak{B}$. Then the characteristic function of the quantity a may be different from that of B and these polynomials may evidently have quite different degrees. But $g(B) = 0$ if and only if $g(a) = 0$ and this implies that the minimum functions of B and a are identical. For let them be $\phi_1(\lambda)$, $\phi(\lambda)$ respectively so that $\phi(B) = \phi_1(B) = 0$, $\phi_1(a) = \phi(a) = 0$. Thus $\phi(\lambda)$ is divisible by $\phi_1(\lambda)$, $\phi_1(\lambda)$ is divisible by $\phi(\lambda)$. Since both are monic polynomials they are identical.

The argument above shows that the minimum function of any quantity of an algebra is independent of the particular equivalent algebra of matrices defining it. We shall characterize this important function further.

Let $\mathfrak{K}$ be any field containing $\mathfrak{F}$ and define $\mathfrak{A}_{\mathfrak{K}}$ to be the algebra

$$(u_1, \ldots, u_n) \text{ over } \mathfrak{K},$$

defined by our original $\mathfrak{A} = (u_1, \ldots, u_n)$ over $\mathfrak{F}$. The quantites of $\mathfrak{A}_{\mathfrak{K}}$ are all quantities of the form $\alpha_1 u_1 + \ldots + \alpha_n u_n$ with α_i now in $\mathfrak{K}$ and $\mathfrak{A}_{\mathfrak{K}}$ contains $\mathfrak{A}$ as a subalgebra (but not as a subalgebra over $\mathfrak{K}$). Moreover, $\mathfrak{A}_{\mathfrak{K}}$ is not in general an algebra of finite order over $\mathfrak{F}$. We call $\mathfrak{A}_{\mathfrak{K}}$ a *scalar*† *extension* of $\mathfrak{A}$ and say that $\mathfrak{A}_{\mathfrak{K}}$ is obtained from $\mathfrak{A}$ by a scalar extension of its coefficient field.

* This property is true of course for the subalgebras of $\mathfrak{A}$ and hence true whether or not $\mathfrak{A}$ has a unity element.

† An algebra $\mathfrak{A}$ over $\mathfrak{F}$ may contain a field $\mathfrak{K}$ over $\mathfrak{F}$ and $\mathfrak{A}$ may be an algebra over $\mathfrak{K}$. This is evidently not the same case as when $\mathfrak{K}$ is a scalar extension over $\mathfrak{F}$. E.g., in matric algebras $\mathfrak{K}$ may be an algebra of non-scalar matrices and thus be not a scalar extension field of $\mathfrak{F}$.

The matrices of a regular representation $\mathfrak{A}_0$ of $\mathfrak{A}$ consist of the linear combinations with coefficients in $\mathfrak{F}$ of n basal matrices $U_i \longleftrightarrow u_i$. These are n matrices linearly independent with respect to $\mathfrak{F}$ of the algebra $\mathfrak{M}_n$ of all n-rowed square matrices with elements in $\mathfrak{F}$. *They are then linearly independent with respect to any scalar extension* $\mathfrak{K}$ of $\mathfrak{F}$. Hence the regular representation $(\mathfrak{A}_{\mathfrak{K}})_0$ of $\mathfrak{A}_{\mathfrak{K}}$ is the algebra $(\mathfrak{A}_0)_{\mathfrak{K}}$ consisting of all linear combinations of the U_i with coefficients in $\mathfrak{K}$.

We have seen in Chapter IV that the minimum function (equation) of a matrix A with elements in a field is independent of this field and depends only on the elements of the matrix itself. Thus the minimum function (equation) of any a of an algebra $\mathfrak{A}$ over $\mathfrak{F}$ is the same as its minimum function (equation) in any scalar extension $\mathfrak{A}_{\mathfrak{K}}$ over $\mathfrak{K}$. We state our results as

Theorem 4. *Let $\mathfrak{A}$ be an algebra of order* n *over $\mathfrak{F}$ and $\mathfrak{K}$ be any scalar extension of $\mathfrak{F}$. Then the minimum function of any* a *of $\mathfrak{A}$ is the monic polynomial $\phi(\lambda)$ of least degree with coefficients in any such $\mathfrak{K}$ such that* $\phi(a) = 0$. *It has coefficients in $\mathfrak{F}$ and divides any* $g(\lambda)$ *of* $\mathfrak{K}[\lambda]$ *such that* $g(a) = 0$. *Moreover, $\phi(\lambda)$ is the minimum function of the matrix* $A \longleftrightarrow a$ *in a regular representation of the algebra $\mathfrak{A}$. We call* $\phi(\lambda) = 0$ *the* **minimum equation** *of* a.

Let $x_1, \ldots, x_n$ be independent indeterminates over the algebra $\mathfrak{A}$ so that the function field $\mathfrak{K} = \mathfrak{F}(x_1, \ldots, x_n)$ of all rational functions of $x_1, \ldots, x_n$ with coefficients in $\mathfrak{F}$ is a scalar extension of $\mathfrak{F}$. Then $\mathfrak{A}_{\mathfrak{K}}$ is an algebra over $\mathfrak{K}$ and contains the quantity

$$x = x_1u_1 + \ldots + x_nu_n , \tag{16}$$

called the *general quantity* of $\mathfrak{A}$. It is of course not in $\mathfrak{A}$ but in $\mathfrak{A}_{\mathfrak{K}}$. However, we replace x by any a of $\mathfrak{A}$ when we replace the indeterminates x_i by a_i in $\mathfrak{F}$.

DEFINITION. *The minimum and characteristic functions (equations) of the general quantity of $\mathfrak{A}$ are called* the minimum and characteristic functions (equations) respectively of $\mathfrak{A}$, and the degree of the minimum function is called the degree of $\mathfrak{A}$.*

The minimum function of x divides its characteristic function. Replace algebra $\mathfrak{A}$ by its regular representation, and thus x of (16) by the corresponding matrix

$$X = x_1U_1 + \ldots + x_nU_n .$$

* The minimum function has been called the *rank function* and its degree the *rank* of $\mathfrak{A}$. We shall prefer the terminology of the definition above which has been used in recent literature.

The elements of X are homogeneous polynomials of degree one (linear forms) of $\mathfrak{F}[x_1, \ldots, x_n]$ and thus the characteristic function of X and consequently of x (by definition) is

$$f(\lambda; x_1, \ldots, x_n) = |\lambda I - X| = \lambda^n + f_1(x_1, \ldots, x_n)\lambda^{n-1} + \ldots + f_n(x_1, \ldots, x_n) ,$$

where a simple computation shows that $f_i(x_1, \ldots, x_n)$ is a homogeneous polynomial of total degree at most i. The minimum function

$$(17) \qquad \phi(\lambda; x_1, \ldots, x_n) = \lambda^r + \phi_1(x_1, \ldots, x_n)\lambda^{r-1} + \ldots + \phi_r(x_1, \ldots, x_n)$$

divides f and has coefficients in $\mathfrak{K} = \mathfrak{F}(x_1, \ldots, x_n)$. We apply Theorem 2.16 and have proved

Theorem 5. *The coefficients* $\phi_i(x_1, \ldots, x_n)$ *of the minimum function* (17) *of an algebra* $\mathfrak{A}$ *over* $\mathfrak{F}$ *are polynomials in* $x_1, \ldots, x_n$ *with coefficients in* $\mathfrak{F}$.

The equation

$$\phi(x; x_1, \ldots, x_n) = \phi(x_1u_1 + \ldots + x_nu_n; x_1, \ldots, x_n) = 0$$

is the statement that a certain polynomial with coefficients in the algebra $\mathfrak{A}$ of n independent indeterminates x_i over $\mathfrak{A}$ is zero. Then these coefficients are the zero elements of $\mathfrak{A}$ and the equation remains true when $x_1, \ldots x_n$ are replaced by any quantities α_i of $\mathfrak{F}$. This replacement substitutes a quantity a of $\mathfrak{A}$ for x. Hence every

$$a = \alpha_1u_1 + \ldots + \alpha_nu_n$$

of $\mathfrak{A}$ satisfies

$$(18) \qquad \phi(\lambda; \alpha_1, \ldots, \alpha_n) = 0 .$$

It is particularly important to notice that this means that the minimum function of $\mathfrak{A}$ is now an equation satisfied by every a of $\mathfrak{A}$ after the proper replacement, and that (18) is divisible by the minimum function of a.

EXERCISES

1. The minimum function of an algebra $\mathfrak{A}$ over $\mathfrak{F}$ divides its characteristic function. Show that this implies that every scalar root of (18) for a quantity a of $\mathfrak{A}$ is a root of the minimum function $\psi(\lambda)$ of a and that (18) is thus a product of factors in $\mathfrak{F}[\lambda]$ of $\psi(\lambda)$.

2. Prove that if the minimum function $\psi(\lambda)$ of a in $\mathfrak{A}$ is irreducible the polynomial (18) is a power of $\psi(\lambda)$.

4. Direct products and sums. We shall obtain the matrix interpretation of two important operations with algebras. Let $\mathfrak{A}$ *be an algebra with a unity element* and order n over $\mathfrak{F}$ so that

$$\mathfrak{A} = (u_1, \ldots, u_n)$$

over $\mathfrak{F}$, where we may assume with no loss of generality that

$$u_1 = 1$$

is the unity element of $\mathfrak{F}$. We identify the elements a of $\mathfrak{F}$ with the elements $a \cdot u_1 = a \cdot 1$ of $\mathfrak{A}$ with no loss of generality and have

$$\mathfrak{A} \geqq \mathfrak{F} .$$

Let also $\mathfrak{B}$ be an algebra

$$\mathfrak{B} = (v_1, \ldots, v_q) , \qquad (v_1 = 1)$$

so that $\mathfrak{B} \geqq \mathfrak{F}$ and assume that $\mathfrak{A}$ and $\mathfrak{B}$ have no quantities in common except the quantities of $\mathfrak{F}$. We form an algebra

$$\mathfrak{A} \times \mathfrak{B} = (w_1, \ldots, w_{nq}) \qquad (w_1 = 1), \tag{19}$$

of order nq over $\mathfrak{F}$ which is called the *direct product* of $\mathfrak{A}$ and $\mathfrak{B}$. This algebra has a set of multiplication constants defined by those of $\mathfrak{A}$ and $\mathfrak{B}$ and the properties

$$w_{(k-1)n+j} = u_j v_k = v_k u_j \qquad (j = 1, \ldots, n;\ k = 1, \ldots, q). \tag{20}$$

The algebra $\mathfrak{A} \times \mathfrak{B}$ consists of all sums of all products of quantities of $\mathfrak{A}$ by quantities of $\mathfrak{B}$. The quantities of $\mathfrak{A}$ are all commutative with those of $\mathfrak{B}$ and the order of $\mathfrak{A} \times \mathfrak{B}$ is the product of the orders of $\mathfrak{A}$ and $\mathfrak{B}$. Notice that $\mathfrak{A}$ and $\mathfrak{B}$ are subalgebras of $\mathfrak{A} \times \mathfrak{B}$ and that this is due particularly to our assumption that $\mathfrak{A}$, $\mathfrak{B}$, and $\mathfrak{A} \times \mathfrak{B}$ have the same unity element.

The direct sum of two algebras is defined in a somewhat analogous but now additive fashion. We again let

$$\mathfrak{A} = (u_1, \ldots, u_n) , \qquad \mathfrak{B} = (v_1, \ldots, v_m) , \tag{21}$$

with no assumption now about u_1, v_1, and let the linear set

$$(u_1, \ldots, u_n, v_1, \ldots, v_m) = \mathfrak{A} \oplus \mathfrak{B} \tag{22}$$

have order $n + m$ and be such that

$$u_i v_j = v_j u_i = 0 \qquad (i = 1, \ldots, n;\ j = 1, \ldots, m). \tag{23}$$

Then the multiplication defined by (23) and that in $\mathfrak{A}$ and $\mathfrak{B}$ makes $\mathfrak{A} \oplus \mathfrak{B}$ an algebra of order $n + m$ over $\mathfrak{F}$ called the *direct sum* of $\mathfrak{A}$ and $\mathfrak{B}$.

It is easy to show explicitly by matrices what the direct sum means. We let $\mathfrak{A}_0$ and $\mathfrak{B}_0$ be any algebra of matrices equivalent over $\mathfrak{F}$ to $\mathfrak{A}$, $\mathfrak{B}$, respectively, under correspondences $a \longleftrightarrow A$, $b \longleftrightarrow B$ for a in $\mathfrak{A}$, A in $\mathfrak{A}_0$, b in $\mathfrak{B}$, B in $\mathfrak{B}_0$. Then the algebra of all matrices

$$\begin{pmatrix} A & 0 \\ 0 & B \end{pmatrix}$$

is equivalent to $\mathfrak{A} \oplus \mathfrak{B}$. It has the quantities

$$\begin{pmatrix} A & 0 \\ 0 & 0 \end{pmatrix} \tag{24A}$$

as the quantities of its subalgebra $\mathfrak{A}$, and

$$\begin{pmatrix} 0 & 0 \\ 0 & B \end{pmatrix} \tag{24B}$$

as the quantities of $\mathfrak{B}$.

5. Direct products of total matric algebras. The direct product of two algebras may also be described in terms of matrices. We first define total matric algebras.

DEFINITION. *An algebra $\mathfrak{M}_s$ of order s^2 over $\mathfrak{F}$ is called an s-rowed total matric algebra if $\mathfrak{M}_s$ is equivalent to the algebra of all s-rowed square matrices with elements in $\mathfrak{F}$.*

We have seen that every algebra $\mathfrak{A}$ of order m over $\mathfrak{F}$ is a subalgebra of a total matric algebra $\mathfrak{M}_s$, that is, that given in the regular representation of $\mathfrak{A}$. The unity elements of $\mathfrak{A}$ and $\mathfrak{M}_s$ are the same and similarly their zero elements are the same. We find the corresponding total matric algebra $\mathfrak{M}_t$ for $\mathfrak{B}$ and the reader will notice that the direct sum of $\mathfrak{A}$ and $\mathfrak{B}$ was formed in (24A, B) by forming the direct sum of $\mathfrak{M}_s$ and $\mathfrak{M}_t$ as a subalgebra of the algebra $\mathfrak{M}_{s+t}$ of all $(s + t)$-rowed square matrices with elements in $\mathfrak{F}$. We shall similarly form $\mathfrak{A} \times \mathfrak{B}$ as a subalgebra of $\mathfrak{M}_s \times \mathfrak{M}_t = \mathfrak{M}_{st}$, the st-rowed total matric algebra.

To form $\mathfrak{M}_s \times \mathfrak{M}_t$ we let $m = st$ and consider the algebra $\mathfrak{M}_m$ of all m-rowed square matrices with elements in $\mathfrak{F}$, where $s > 1$, $t > 1$ are integers. Every m-rowed square matrix C has the form

$$(25) \qquad C = (b_{ij}) \qquad (i, j = 1, \ldots, s)$$

where the b_{ij} range over all t-rowed square matrices with elements in $\mathfrak{F}$. In particular let

$$(26) \qquad E_{ij} \qquad (i, j = 1, \ldots, s)$$

be the s-rowed square matrix with the t-rowed identity matrix in the ith row and jth column and t-rowed zero matrices elsewhere. Our rule for multiplying matrices implies that

$$E_{ij}E_{kl} = \delta_{jk}E_{il} \qquad (i, j, k, l = 1, \ldots, s)$$

where δ_{jk} is the *Kronecker delta*, that is, a symbol whose value is unity when $j = k$ and zero when $j \neq k$. The algebra

$$(27) \qquad \mathfrak{A}_0 = \mathfrak{M}_s = (E_{11}, E_{12}, \ldots, E_{s-1\,s}, E_{ss})$$

is evidently an s-rowed total matric algebra.

Let now

$$(28) \qquad G_{pq} = \text{diag}\,\{\epsilon_{pq}, \epsilon_{pq}, \ldots, \epsilon_{pq}\} \qquad (p, q = 1, \ldots, t).$$

Then G_{pq} is an s-rowed diagonal matrix whose diagonal elements are t-rowed equal matrices ϵ_{pq}, and ϵ_{pq} is a square matrix with unity in the pth row and qth column and zeros elsewhere. The algebra

$$(29) \qquad \mathfrak{B}_0 = \mathfrak{M}_t = (G_{11}, G_{12}, \ldots, G_{tt})$$

is evidently a t-rowed total matric algebra.

Algebras $\mathfrak{A}_0$ and $\mathfrak{B}_0$ are subalgebras of the algebra

$$(30) \qquad \mathfrak{C} = \mathfrak{M}_m\,.$$

The unity elements of $\mathfrak{A}_0$, $\mathfrak{B}_0$, $\mathfrak{C}$ are all the same and in fact are all the m-rowed identity matrix. We have obtained our desired representation and merely have to show that

$$(31) \qquad \mathfrak{C} = \mathfrak{A}_0 \times \mathfrak{B}_0\,.$$

To see this notice that if b_{ij} is any t-rowed square matrix the matrix

$$B_{ij} = \text{diag}\,\{b_{ij}, \ldots, b_{ij}\} \qquad (i, j = 1, \ldots, s) \tag{32}$$

is in $\mathfrak{B}_0$. We sometimes speak of B_{ij} as the *direct product of* b_{ij} *by the identity matrix* $E = E_{11} + E_{22} + \ldots + E_{ss}$ of $\mathfrak{A}_0$. It is evident that

$$B_{ij}E_{ij} = E_{ij}B_{ij} \tag{33}$$

is an s-rowed square matrix with b_{ij} in the ith row and jth column and zero matrices elsewhere. But then the general matrix of $\mathfrak{C} = \mathfrak{M}_m$ given by (25) has the form

$$C = \sum_{i,\,j}^{1,\ldots,s} B_{ij}E_{ij}\,,$$

and is a sum of products of quantities of $\mathfrak{B}_0$ by quantities of $\mathfrak{A}_0$. By (33) every quantity of $\mathfrak{A}_0$ is commutative with every quantity of $\mathfrak{B}_0$. The order of $\mathfrak{C}$ is s^2t^2, the product of the orders of $\mathfrak{A}_0$ and $\mathfrak{B}_0$. This proves that $\mathfrak{C} = \mathfrak{A}_0 \times \mathfrak{B}_0$.

We have obtained a representation of $\mathfrak{M}_s \times \mathfrak{M}_t$ in which $\mathfrak{M}_s$ and $\mathfrak{M}_t$ occupy different roles. It is evident that these roles may be interchanged. Thus we have obtained st-rowed square matrices as s-rowed square matrices whose elements are themselves t-rowed square matrices. We may also write them as t-rowed square matrices whose elements are s-rowed square matrices. This implies a correspondence which is a special case of

Theorem 6. *Let $\mathfrak{M}$ be the algebra of all* m*-rowed square matrices with elements in $\mathfrak{F}$. Then if $\mathfrak{M}_s$ is an* s*-rowed total matric subalgebra of $\mathfrak{M}$ with the same unity element as $\mathfrak{M}$ we have* m = st *and there exists an inner automorphism of $\mathfrak{M}$ which carries $\mathfrak{M}_s$ into the algebra $\mathfrak{A}_0$ of* (27).

For, let $\mathfrak{M}_s$ have what is generally called an *ordinary total matric algebra basis* e_{ij} with $i, j = 1, \ldots, s$ and

$$e_{ij}e_{ab} = \delta_{ja}e_{ib} \qquad (i, j, a, b = 1, \ldots, s).$$

The unity element of $\mathfrak{M}_s$ is

$$1 = e_{11} + \ldots + e_{ss} = I\,,$$

the identity matrix of $\mathfrak{M}$, and the e_{ii} are idempotent matrices such that $e_{ii}e_{jj} = 0$ for $i \neq j$. Apply Theorem 4.12 to see that the rank r_i of e_{ii} has the property $r_1 + \ldots + r_s = m$. Now $e_{ii} = e_{ij}e_{jj}e_{ji}$ so that we have

$r_i \leqq r_j$, and ultimately $r_i = r_j = t$. Thus the rank of every e_{ii} is t and $m = st$ as desired.

By Section 4.7 we may find an inner automorphism of $\mathfrak{M}$ which carries the e_{ii} into the E_{ii}. This means that there exists a non-singular matrix T with elements in $\mathfrak{F}$ such that

$$T^{-1}e_{ii}T = E_{ii}\,.$$

We apply T and now assume that $e_{ii} = E_{ii}$. Then the e_{ij} for $i \neq j$ are not yet determined but

$$e_{ij}e_{jk} = e_{ik}\,, \qquad E_{ii}e_{ij} = e_{ij}E_{jj} = e_{ij}\,.$$

This implies that e_{ij} is an s-rowed square matrix with zero matrices except in the ith row and jth column and the quantity there a matrix $\epsilon_{ij} \neq 0$. We may therefore write

$$(34) \qquad e_{ij} = [\text{diag}\,\{\epsilon_{ij}, \epsilon_{ij}, \ldots, \epsilon_{ij}\}]\cdot E_{ij}\,,$$

for all values of i and j, where ϵ_{ii} is the t-rowed identity matrix. But $e_{ij}e_{jk} = e_{ik}$ so that since the E_{ij} are commutative with the matrices (28) we have

$$(35) \qquad e_{ij}e_{jk} = [\text{diag}\,\{\epsilon_{ij}\epsilon_{jk}, \ldots, \epsilon_{ij}\epsilon_{jk}\}]E_{ij}E_{jk} = e_{ik}$$

and by (34) obtain

$$(36) \qquad \epsilon_{ij}\epsilon_{jk} = \epsilon_{ik} \qquad (i, j, k = 1, \ldots, s)$$

Since $\epsilon_{ij}\epsilon_{ji} = \epsilon_{ii} = I_t$, the matrices ϵ_{ij} are non-singular. We now write

$$T = \text{diag}\,\{I_t, \epsilon_{21}, \ldots, \epsilon_{s1}\}$$

which is non-singular and commutative with the E_{ii} and have

$$T^{-1}E_{ii}T = E_{ii}\,, \qquad T^{-1}e_{1j}T = T^{-1}\begin{pmatrix} 0 & 0 & \ldots & \epsilon_{1j} & 0 & \ldots & 0 \\ 0 & 0 & \ldots & 0 & 0 & \ldots & 0 \\ . & . & \ldots & . & . & \ldots & . \\ 0 & 0 & \ldots & 0 & 0 & \ldots & 0 \end{pmatrix} T = E_{1j}$$

by a trivial computation. We may therefore use this inner automorphism and carry the e_{1j} to the E_{1j}. This implies that for the new basis we have $\epsilon_{1j} = I_t$. Since always $\epsilon_{j1} = \epsilon_{1j}^{-1}$ we have $\epsilon_{i1} = I_t$, $\epsilon_{ij} = \epsilon_{i1}\epsilon_{1j} = I_t$, $E_{ij} = e_{ij}$ as desired. This gives our theorem.

The algebra of all quantities of $\mathfrak{C} = \mathfrak{M}_m$ commutative with every quantity of its subalgebra $\mathfrak{A} = \mathfrak{M}_s$ given by (27) is algebra $\mathfrak{B}$ of (29). For, as we have shown, $\mathfrak{C} = \mathfrak{A} \times \mathfrak{B}$, every C of $\mathfrak{C}$ has the form $C = \Sigma B_{ij}E_{ij}$ with B_{ij} in $\mathfrak{B}$. If $CE_{ab} = E_{ab}C$ for all a and b then $\sum_{i=1}^{s} B_{ip}E_{iq} = \sum_{j=1}^{s} E_{pj}B_{qj}$ by direct computation. The fact that $\mathfrak{C} = \mathfrak{A} \times \mathfrak{B}$ then implies that $B_{qj} = 0$ unless $j = q$, $B_{ip} = 0$ unless $p = i$, and $B_{ii} = B_{qq}$ for all i and q. Hence $B_{ii} = B_{11} = B$ is in $\mathfrak{B}$ and $C = \Sigma B_{ii}E_{ii} = B$ is in $\mathfrak{B}$. We may now prove

Theorem 7. *Let $\mathfrak{C}$ be a total matric algebra and $\mathfrak{A}$ be a total matric subalgebra of $\mathfrak{C}$ with the same unity quantity as $\mathfrak{C}$. Then the subset $\mathfrak{B}$ of all quantities of $\mathfrak{C}$ commutative with every quantity of $\mathfrak{A}$ is a total matric algebra and*

$$\mathfrak{C} = \mathfrak{A} \times \mathfrak{B}.$$

If also

$$\mathfrak{C} = \mathfrak{A}_0 \times \mathfrak{B}_0$$

where $\mathfrak{A}_0$ is equivalent to $\mathfrak{A}$ then there is an inner automorphism of $\mathfrak{C}$ carrying $\mathfrak{A}_0$ into $\mathfrak{A}$ and this carries $\mathfrak{B}_0$ into $\mathfrak{B}$.

For we may carry $\mathfrak{A}$ into the algebra $\mathfrak{M}_s$ of (27) by an inner automorphism of $\mathfrak{C}$. This of course means that we are thinking of the abstract algebra $\mathfrak{C}$ in terms of its representation as the equivalent algebra of all m-rowed square matrices. Algebra $\mathfrak{B}$ is then carried into an equivalent algebra which we designate by $\mathfrak{M}_t$ and we have proved above that $\mathfrak{M}_t$ is a total matric algebra. Hence $\mathfrak{B}$ is a total matric algebra. Since $\mathfrak{C} = \mathfrak{M}_s \times \mathfrak{M}_t$ the inverse of our inner automorphism carries $\mathfrak{C}$ into itself, $\mathfrak{M}_s$ into $\mathfrak{A}$, $\mathfrak{M}_t$ into $\mathfrak{B}$, and $\mathfrak{C} = \mathfrak{A} \times \mathfrak{B}$ (by the fact that we are using an automorphism of $\mathfrak{C}$). Then if $\mathfrak{A}_0 \times \mathfrak{B}_0 = \mathfrak{C}$ we carry $\mathfrak{A}_0$ into $\mathfrak{M}_s$ and $\mathfrak{B}_0$ into $\mathfrak{B}_1$. The quantities of $\mathfrak{B}_1$ are commutative with every quantity of $\mathfrak{M}_s$ so that $\mathfrak{B}_1 \leqq \mathfrak{M}_t$. But the orders of $\mathfrak{B}_1$ and $\mathfrak{M}_t$ are both t^2 and $\mathfrak{B}_1 = \mathfrak{M}_t$, $\mathfrak{B}_0$ is equivalent to $\mathfrak{B}$. The product of the inner automorphism of $\mathfrak{C}$ which carries $\mathfrak{A}_0$ into $\mathfrak{M}_s$ by that carrying $\mathfrak{M}_s$ into $\mathfrak{A}$ carries $\mathfrak{A}_0$ into $\mathfrak{A}$ and $\mathfrak{B}_0$ into $\mathfrak{B}$.

The result above states that the expression of a total matric algebra $\mathfrak{C}$ as a direct product of total matric subalgebras of fixed orders is unique apart from an inner automorphism of $\mathfrak{C}$. Our construction also implies without further argument

Theorem 8. *The direct product of two total matric algebras is a total matric algebra.*

We may now obtain

Theorem 9. *Let $\mathfrak{A}$ be an algebra of order* n *over $\mathfrak{F}$ with a unity element and $\mathfrak{M}$ be a total matric subalgebra of $\mathfrak{A}$ whose unity element is that of $\mathfrak{A}$. Then*

$$\mathfrak{A} = \mathfrak{M} \times \mathfrak{B}$$

where $\mathfrak{B}$ *has the same unity element as* $\mathfrak{A}$ *and is the algebra of all quantities of* $\mathfrak{A}$ *commutative with every quantity of* $\mathfrak{M}$.

For we let $\mathfrak{M}_n$ be the total matric algebra which contains the regular representation of $\mathfrak{A}$. By Theorem 1.9 algebra $\mathfrak{M}_n$ is equivalent to a total matric algebra $\mathfrak{C}$ with $\mathfrak{A}$ as a subalgebra and, since the unity element of $\mathfrak{A}$ corresponds to the identity matrix of its regular representation, the unity element of $\mathfrak{C}$ is that of $\mathfrak{A}$. By Theorem 10.7 the algebra $\mathfrak{C} = \mathfrak{M} \times \mathfrak{M}_t$ where $\mathfrak{M}_t$ is a total matric algebra. For every C of $\mathfrak{C}$ we have

$$\sum_{p=1}^{s} E_{pq}CE_{rp} = B_{qr} \qquad (q, r = 1, \ldots, s), \tag{37}$$

where we use (25) and

$$C = \sum B_{ij}E_{ij}, \qquad B_{ij} = \text{diag}\,\{b_{ij}, \ldots, b_{ij}\}.$$

Then the B_{qr} are in $\mathfrak{M}_t$. The E_{pq} are in $\mathfrak{A}$ and if C is in $\mathfrak{A}$ so are the corresponding B_{ij} of (37). We conclude that every quantity of $\mathfrak{A}$ is a sum of products of quantities of $\mathfrak{M}$ by quantities in the algebra $\mathfrak{B}$ of all quantities of $\mathfrak{A}$ commutative with every quantity of $\mathfrak{M}$. For the B_{ij} are in $\mathfrak{M}_t$ and in $\mathfrak{A}$ and hence in $\mathfrak{B}$. Now $\mathfrak{C} = \mathfrak{M} \times \mathfrak{M}_t$ and we may take a basis of $\mathfrak{B}$ as a partial basis of $\mathfrak{M}_t$, $\mathfrak{C}$ contains $\mathfrak{M} \times \mathfrak{B}$. Since $\mathfrak{B} \leqq \mathfrak{A}$, $\mathfrak{M} \leqq \mathfrak{A}$ we have $\mathfrak{M} \times \mathfrak{B} \leqq \mathfrak{A}$ whereas we have just shown that $\mathfrak{A} \leqq \mathfrak{M} \times \mathfrak{B}$. Thus $\mathfrak{A} = \mathfrak{M} \times \mathfrak{B}$ as desired.

We shall obtain a generalization and consequence (Theorem 10.12) of Theorem 10.9 which is of fundamental importance as a tool in the theory of algebras. It is astonishing how many important results are directly obtainable by the application of this really simple theorem.

6. The degree of a total matric algebra. A total matric algebra $\mathfrak{A}$ of order $m^2 = n$ over $\mathfrak{F}$ is equivalent to the algebra $\mathfrak{M}_m$ of all m-rowed square matrices with elements in $\mathfrak{F}$. We may thus replace $\mathfrak{A}$ by $\mathfrak{M}_m$ when we compute the degree of $\mathfrak{A}$.

The general element of $\mathfrak{M}_m$ is the matrix

$$X = (x_{ij}) \qquad (i, j = 1, \ldots, m),$$

where the x_{ij} are independent indeterminates over $\mathfrak{F}$. We may easily prove

Theorem 10. *The degree of an* m*-rowed total matric algebra is* m *and its minimum function is*

$$\phi(\lambda; x_{11}, \ldots, x_{ij}, \ldots, x_{mm}) = |\lambda I_m - X|\,. \tag{38}$$

For we have seen that $\phi(\lambda; x_{11}, \ldots, x_{mm})$ has the property that if we replace $x_{11}, \ldots, x_{mm}$ by elements $y_{11}, \ldots, y_{mm}$ in $\mathfrak{F}$ the corresponding matrix

$$Y = (y_{ij})$$

satisfies $\phi(\lambda; y_{11}, \ldots, y_{mm}) = 0$. Hence ϕ is divisible by the minimum function of Y. But the fundamental lemma used in the proof of Theorem 4.4 states that there exists an m-rowed matrix whose characteristic function is its minimum function and this latter polynomial has degree m. Hence the degree of algebra $\mathfrak{A}$ is at least m. Since $\phi(\lambda; x_{11}, \ldots, x_{mm})$ must divide the $|\lambda I_m - X|$ and the former has degree at least m, and the latter has degree m, it follows that they both have degree m and are identical.

7. Quadrate algebras. There are many scalar extensions of an algebra $\mathfrak{A}$ of order n over $\mathfrak{F}$ and in general their structure is not the same as that of $\mathfrak{A}$. These scalar extensions are algebras $\mathfrak{A}_{\mathfrak{K}}$ over a field $\mathfrak{K}$ over $\mathfrak{F}$. They are defined by the same linear set and multiplication table as $\mathfrak{A}$ but the coefficients are allowed to vary over the scalar extension field $\mathfrak{K}$ instead of its subfield $\mathfrak{F}$.

DEFINITION. *An algebra $\mathfrak{A}$ of order* n *over $\mathfrak{F}$ is called a quadrate algebra* if there exists a scalar extension field $\mathfrak{K}$ of finite degree over $\mathfrak{F}$ called a splitting field of $\mathfrak{A}$ such that*

$$\mathfrak{A}_{\mathfrak{K}} \text{ over } \mathfrak{K}$$

is a total matric algebra.

Our definition of direct product implies that we are now really considering the structure of the direct product

$$\mathfrak{A} \times \mathfrak{K},$$

which is an algebra over $\mathfrak{F}$. *But we are considering its properties as an algebra over* $\mathfrak{K}$, and are stating that

$$\mathfrak{A} \times \mathfrak{K} = \mathfrak{M} \times \mathfrak{K}$$

where $\mathfrak{M}$ is a total matric algebra over $\mathfrak{F}$. For the algebra of all m-rowed square matrices with elements in $\mathfrak{K}$ is the direct product of $\mathfrak{K}$ over $\mathfrak{F}$ and the algebra of all m-rowed square matrices with elements in the given subfield $\mathfrak{F}$ of $\mathfrak{K}$.

* In what follows we shall assume that $\mathfrak{F}$ is an infinite field.

Theorem 11. *The order of a quadrate algebra* $\mathfrak{A}$ *is* $n = m^2$ *and its degree is* m. *The minimum function* $\phi(\lambda; x_1, \ldots, x_n)$ *of* $\mathfrak{A}$ *may be carried into the characteristic function of a general* m*-rowed square matrix by a non-singular linear transformation on the indeterminates* $x_1, \ldots, x_n$ *with coefficients in any splitting field* $\mathfrak{K}$ *of* $\mathfrak{A}$.

For let $u_1, \ldots, u_n$ be a basis of $\mathfrak{A}$ over $\mathfrak{F}$ and e_{ij} $(i, j = 1, \ldots, m)$ be an ordinary matric basis of the total matric algebra $\mathfrak{A}_{\mathfrak{K}}$. Then the minimum function of $\mathfrak{A}_{\mathfrak{K}}$ is the characteristic function of the matrix $\sum_{i,j=1}^{m} \xi_{ij}e_{ij}$ where the ξ_{ij} are independent indeterminates. We have seen that the minimum function of $\mathfrak{A}_{\mathfrak{K}}$ is independent of the basis used to represent $\mathfrak{A}_{\mathfrak{K}}$ as a linear set of order m^2 over $\mathfrak{K}$ and we have $n = m^2$ and see that there exists a non-singular linear transformation with coefficients in $\mathfrak{K}$ which carries the characteristic function of the general matrix

$$(\xi_{ij})$$

into the minimum function of $x = x_1u_1 + \ldots + x_nu_n$. Here the $x_1, \ldots, x_n$ are independent indeterminates over $\mathfrak{K}$ and are m^2 linear combinations with coefficients in $\mathfrak{K}$ of $\xi_{11}, \ldots, \xi_{mm}$. However, the minimum function of x is independent of $\mathfrak{K}$ and depends only on the elements in the matrices of any algebra of matrices equivalent to $\mathfrak{A}_{\mathfrak{K}}$. We may take $u_1, \ldots, u_n$ to be the matrices of the regular representation of $\mathfrak{A}$ and see that the minimum function of x has leading coefficient unity, other coefficients in $\mathfrak{F}[x_1, \ldots, x_n]$ and is the minimum function of algebra $\mathfrak{A}$. This proves Theorem 11 and gives a clear conception of what happens to the minimum function under the scalar extension to a splitting field $\mathfrak{K}$.

As a corollary of Theorem 10.9 we have

Theorem 12. *Let an algebra* $\mathfrak{A}$, *of order* n *over* $\mathfrak{F}$ *and with a unity element, have a quadrate subalgebra* $\mathfrak{B}$ *over* $\mathfrak{F}$ *with the same unity element as* $\mathfrak{A}$. *Then*

$$\mathfrak{A} = \mathfrak{B} \times \mathfrak{C}$$

where $\mathfrak{C}$ *is the subalgebra of* $\mathfrak{A}$ *of all quantities of* $\mathfrak{A}$ *commutative with every quantity of* $\mathfrak{B}$ *and* $\mathfrak{C}$ *contains the unity element of* $\mathfrak{A}$.

For let $\mathfrak{K}$ be a splitting field of $\mathfrak{B}$, and form $\mathfrak{A}_{\mathfrak{K}}$. By Theorem 10.9 we have $\mathfrak{A}_{\mathfrak{K}} = \mathfrak{B}_{\mathfrak{K}} \times \mathfrak{C}_0$. The field $\mathfrak{K}$ is a linear set

$$(g_1, \ldots, g_r)$$

of order r over $\mathfrak{F}$ and every quantity of $\mathfrak{A}_{\mathfrak{K}}$ has the unique form

$$a = a_1g_1 + \ldots + a_rg_r$$

with a_i in $\mathfrak{A}$. If $\mathfrak{B}$ has order m^2 then so does $\mathfrak{B}_{\mathfrak{K}}$ and

$$n = m^2 q , \tag{39}$$

where q is the order of $\mathfrak{C}_0$. Every quantity c of $\mathfrak{C}_0$ has the form $c_1 g_1 + \ldots + c_r g_r$, c_i in $\mathfrak{A}$, and $cb = bc$ for every b of $\mathfrak{B}_{\mathfrak{K}}$ implies that $cb = bc$ for every b of $\mathfrak{B}$,

$$bc_1 g_1 + \ldots + bc_r g_r = c_1 b g_1 + \ldots + c_r b g_r .$$

This proves that every $c_i b = bc_i$ and that each c_i is in the subalgebra of $\mathfrak{A}$ called $\mathfrak{C}$ in the theorem. Hence $\mathfrak{C}_0 \leqq \mathfrak{C}_{\mathfrak{K}}$. Since $\mathfrak{C}$ is obviously in $\mathfrak{C}_0$ we have $\mathfrak{C}_0 \geqq \mathfrak{C}_{\mathfrak{K}}$ so that $\mathfrak{C}_0 = \mathfrak{C}_{\mathfrak{K}}$ and $\mathfrak{C}_0$ has a basis of $\mathfrak{C}$ over $\mathfrak{F}$ as a basis over $\mathfrak{K}$. If this basis is $v_1, \ldots, v_q$ and $u_1, \ldots, u_{m^2}$ is a basis of $\mathfrak{B}$ over $\mathfrak{F}$ we have the system $v_1 u_1, \ldots, v_q u_{m^2}$ as a basis of $\mathfrak{A}_{\mathfrak{K}}$ over $\mathfrak{K}$ where the $v_i u_j$ are in $\mathfrak{A}$. Their linear independence in $\mathfrak{K}$ implies their linear independence in $\mathfrak{F}$. They are thus a basis of $\mathfrak{A}$ over $\mathfrak{F}$ and evidently $\mathfrak{A} = \mathfrak{B} \times \mathfrak{C}$.

The quantities of a quadrate algebra $\mathfrak{A}$ may be thought of as n-rowed square matrices with elements in a splitting field $\mathfrak{K}$ over $\mathfrak{F}$ of $\mathfrak{A}$. If a is in $\mathfrak{A}$ and the characteristic function of a has distinct roots this function $f(x)$ is the only non-trivial invariant factor of a and a_0 in $\mathfrak{A}$ is similar to a in $\mathfrak{A}$ if and only if $f(a_0) = 0$. But if $g_1, \ldots, g_r$ are a basis of $\mathfrak{K}$ over $\mathfrak{F}$ we have

$$a_0 = bab^{-1} , \qquad b = b_1 g_1 + \ldots + b_r g_r$$

with b_i in $\mathfrak{A}$. Since a and a_0 are in $\mathfrak{A}$ we have $(a_0 b_1 - b_1 a) g_1 + \ldots + (a_0 b_r - b_r a) g_r = 0$, $a_0 b_i = b_i a$. The matrix

$$b(\xi_1, \ldots, \xi_r) = b_1 \xi_1 + \ldots + b_r \xi_r$$

has determinant $d(\xi_1, \ldots, \xi_r) \neq 0$ in the indeterminates ξ_i over $\mathfrak{K}$. We apply Theorem 7.30 to obtain $\xi_{10}, \ldots, \xi_{r0}$ in $\mathfrak{F}$ such that $d(\xi_{10}, \ldots, \xi_{r0}) \neq 0$. Then $b_0 = b(\xi_{10}, \ldots, \xi_{r0})$ is a regular element of the algebra $\mathfrak{A}$ since it is a non-singular matrix. We see that $b_0 a b_0^{-1} = a_0$ and have proved a result which we shall state as

Theorem 13. *Let the minimum function* $\mathrm{f}(\lambda)$ *of a quantity* a *of a quadrate algebra* $\mathfrak{A}$ *over an infinite field* $\mathfrak{F}$ *have degree equal to the degree* n *of* $\mathfrak{A}$ *and all distinct roots. Then there exists an inner automorphism of* $\mathfrak{A}$ *carrying* a_0 *of* $\mathfrak{A}$ *into* a *if and only if* $\mathrm{f}(\mathrm{a}_0) = 0$.

This result may be applied to separable subfields of degree n of $\mathfrak{A}$ and gives

Theorem 14. *Let $\mathfrak{A}$ be a quadrate algebra of degree* n *over an infinite field $\mathfrak{F}$. Then if $\mathfrak{K}$ and $\mathfrak{K}_0$ are equivalent separable subfields of degree* n *over $\mathfrak{F}$ of $\mathfrak{A}$ there exists an inner automorphism of $\mathfrak{A}$ carrying $\mathfrak{K}_0$ into $\mathfrak{K}$.*

Quadrate algebras are probably the most interesting types of algebras. We shall give what is considered the most important case of a quadrate algebra in Section 10.10. We now pass to a deeper discussion of the properties of matrices whose characteristic roots are all distinct.

EXERCISES

1. Let $\mathfrak{D}$ be a quadrate algebra of degree m over $\mathfrak{F}$ represented by its regular representation as a subalgebra of a total matric algebra $\mathfrak{M}$ of degree $n = m^2$ over $\mathfrak{F}$. Prove that $\mathfrak{M} = \mathfrak{D} \times \mathfrak{D}_0$ where $\mathfrak{D}_0$ is quadrate of degree m.

2. Use Theorem 10.11 to prove that there exist quantities a in any quadrate algebra $\mathfrak{A}$ of degree n over $\mathfrak{F}$ for which the polynomial (18) has distinct roots. Show that then (18) is the minimum function of a.

3. Let $\mathfrak{K}$ be a splitting field of a quadrate algebra $\mathfrak{A}$ of degree n over $\mathfrak{F}$ so that the quantities of $\mathfrak{A}$ may be taken to be n-rowed square matrices a with elements in $\mathfrak{K}$. Then the characteristic function of a is the polynomial (18). Show that if the minimum function of a is a separable irreducible polynomial the non-trivial invariant factors of a are all equal to its minimum function.

4. Use Ex. 3 and the method of proof of Theorem 10.13 to remove the restriction on the degree of the separable field $\mathfrak{K}$ of Theorem 10.14.

5. Prove that if a quadrate algebra $\mathfrak{A}$ of degree n over $\mathfrak{F}$ contains n supplementary idempotent elements e_{ii} it is a total matric algebra. Hint: Show that if $\mathfrak{K}$ is a splitting field of $\mathfrak{A}$ we may choose an ordinary matric basis e_{ij} of $\mathfrak{A}_{\mathfrak{K}}$ with e_{ii} in $\mathfrak{A}$. Then $E_{ij} = e_{ii}ae_{jj} = \alpha_{ij}e_{ij}$ are in $\mathfrak{A}$ for every a in $\mathfrak{A}$ and $a = \sum_{i,j} E_{ij}$. Show that then particular E_{ij} may be chosen to form a total matric basis of $\mathfrak{A}$ over $\mathfrak{F}$.

8. Matrices with separable characteristic equations. The invariant factors of a matrix A with a separable characteristic function have been investigated and we have seen that $|\lambda I - A|$ is the only non-trivial invariant factor. This result implied Theorem 10.13. But we may easily obtain some deeper further results.

The characteristic roots $\alpha_1, \ldots, \alpha_n$ of an n-rowed square matrix A with elements in $\mathfrak{F}$ define the root field $\mathfrak{K} = \mathfrak{F}(\alpha_1, \ldots, \alpha_n)$ of the equation

$$f(\lambda) = |\lambda I - A| = \lambda^n + a_1\lambda^{n-1} + \ldots + a_n = 0 \quad (a_i \text{ in } \mathfrak{F}). \tag{40}$$

We consider the case where $f(\lambda)$ is a separable polynomial so that $\alpha_1, \ldots, \alpha_n$ are all distinct and $\mathfrak{K}$ is separable over $\mathfrak{F}$. The minimum function of A

has all of the α_i as roots and must be the characteristic function of A. Thus A is similar in $\mathfrak{F}$ to

$$(41) \qquad A = \begin{pmatrix} 0 & 1 & 0 & \cdots & 0 \\ 0 & 0 & 1 & \cdots & 0 \\ . & . & . & \cdots & . \\ 0 & 0 & 0 & \cdots & 1 \\ -a_n & -a_{n-1} & -a_{n-2} & \cdots & -a_1 \end{pmatrix}$$

by the fundamental lemma of Section 4.4.

The matrix A has all simple elementary divisors in $\mathfrak{K}$ and is similar in $\mathfrak{K}$ to the diagonal matrix

$$(42) \qquad \alpha = \text{diag}\,\{\alpha_1, \ldots, \alpha_n\}\,.$$

We easily compute the matrix V which carries A into

$$(43) \qquad V^{-1}AV = \alpha\,.$$

For in fact V is the Vandermonde matrix

$$(44) \qquad V = (\alpha_j^{i-1}) \qquad (i, j = 1, \ldots, n)\,.$$

Here, of course, i is the row subscript, and we may prove by direct multiplication that

$$(45) \qquad V\alpha = (\alpha_j^i) = AV\,,$$

where we have used the fact that $b_i = -a_n - a_{n-1}\alpha_i - \ldots - a_1\alpha_i^{n-1} = \alpha_i^n$ since $f(\alpha_i) = 0$. Now the matrix product

$$(46) \qquad T = VV' = (\sigma_{i+j-2}) \qquad (i, j = 1, \ldots, n)$$

of V by its transpose V' has the property that the element in its ith row and jth column is σ_{i+j-2} where

$$(47) \qquad \sigma_k = \sum_{p=1}^{n} \alpha_p^k$$

is symmetric in the α_p and is in $\mathfrak{F}$. The determinant of T is

$$(48) \qquad |V|^2 = \prod_{\substack{i,j=1,\ldots,n \\ i<j}} (\alpha_i - \alpha_j)^2 = D(f)$$

where $D(f)$ is the discriminant of $f(\lambda)$ and is not zero. Hence both V and T are non-singular.

If B is any matrix with elements b_{ij} in $\mathfrak{F}$ we see that

$$V'BV = (\alpha_i^{t-1})(b_{tk})(\alpha_j^{k-1}) = (\gamma_{ij}) \qquad (i, j = 1, \ldots, n)\,,$$

where $\gamma_{ij} = \sum_{t,k}^{1,\ldots,} \alpha_i^{t-1} b_{tk} \alpha_j^{k-1}$. We let x and y be independent indeterminates over $\mathfrak{F}$ and see that if we define the polynomial

$$\gamma(x, y) = \sum_{t,k}^{1,\ldots,n} x^{t-1} b_{tk} y^{k-1}$$

with coefficients in $\mathfrak{F}$ then the quantity in the ith row and jth column o $V'BV$ is $\gamma(\alpha_i, \alpha_j)$.

We now consider an n-rowed square matrix B with elements in $\mathfrak{F}$ and form

$$V^{-1}BV = V'(VV')^{-1}BV = V'CV\,,$$

where

$$C = T^{-1}B = (c_{tk}) \qquad (t, k = 1, \ldots, n) \tag{49}$$

has elements in $\mathfrak{F}$. Define

$$b(x, y) = \sum_{t,k}^{1,\ldots,n} x^{t-1} c_{tk} y^{k-1}\,. \tag{50}$$

Then if $\beta = V^{-1}BV$ we have

$$\beta = (b(\alpha_i, \alpha_j)) \qquad (i, j = 1, \ldots, n)\,. \tag{51}$$

Conversely, if β has the element $b(\alpha_i, \alpha_j)$ in its ith row and jth column we may define the c_{tk} in $\mathfrak{F}$ by (50), put $C = (c_{tk})$, $B = TC$, and have $V\beta V^{-1} = B$, a matrix with elements in $\mathfrak{F}$. We have therefore obtained a result which is an important tool and we now state it as

Theorem 15. *A matrix β with elements in $\mathfrak{F}(\alpha_1, \ldots, \alpha_n)$ has the form* $V^{-1}BV$ *where* B *has elements in $\mathfrak{F}$ and* V *is given by* (44) *if and only if there exists a polynomial* (50) *with* c_{tk} *in $\mathfrak{F}$ such that* (51) *holds.*

Theorem 15 may be used to prove

Theorem 16. *Let the characteristic equation of an* n*-rowed square matrix* A *with elements in $\mathfrak{F}$ have all distinct roots. Then an* n*-rowed square matrix* B *with elements in $\mathfrak{F}$ is commutative with* A *if and only if* B *is a polynomial of* $\mathfrak{F}[A]$.

For $BA = AB$ is equivalent to $V^{-1}BVV^{-1}AV = V^{-1}AVV^{-1}BV = \beta\alpha = \alpha\beta$. But

$$\alpha\beta = (a_i b(a_i, a_j)) = \beta\alpha = (b(a_i, a_j)a_j)$$

if and only if

$$(a_i - a_j)b(a_i, a_j) = 0 \qquad (i, j = 1, \ldots, n) . \tag{52}$$

The quantities $a_i - a_j \neq 0$ for $i \neq j$ are in the field $\mathfrak{F}(a_1, \ldots, a_n)$ and (52) is equivalent to $b(a_i, a_j) = 0$ for all $i \neq j$. Then $b(a_i, a_i) = b(a_i)$ is a polynomial in a_i with coefficients in $\mathfrak{F}$ and the corresponding polynomial in the matrix α is

$$\beta = b(\alpha) = \text{diag}\,\{b(a_1), \ldots, b(a_n)\} . \tag{53}$$

It is easy to see that $V^{-1}b(A)V = b(\alpha)$ and thus that

$$b(\alpha) = V^{-1}BV = V^{-1}b(A)V , \qquad B = b(A) .$$

This proves Theorem 10.16 and also shows that the precise polynomial $b(A)$ may be obtained by forming $V'(T^{-1}B)V$ and computing the element $b(a_1, a_1)$ in its first row and column.

EXERCISES

1. Show that if A is any n-rowed square matrix with distinct characteristic roots $a_1, \ldots, a_n$ the quantities

$$e_{ii} = [(A - a_1)(A - a_2) \ldots (A - a_{i-1})(A - a_{i+1}) \ldots (A - a_n)] \cdot [(a_i - a_1) \ldots (a_i - a_{i-1})(a_i - a_{i+1}) \ldots (a_i - a_n)]^{-1}$$

are n supplementary idempotent matrices. Hint: Use (43), (42).

2. Use Exercise 1 above, Exercises 2, 5 of Section 10.7 to show that any quadrate algebra has a *separable* splitting field over an infinite field $\mathfrak{F}$.

9. The cyclic representation of a total matric algebra. We may use the results just obtained to obtain a new representation of a total matric algebra. Consider an irreducible monic polynomial $f(\lambda)$ of $\mathfrak{F}[\lambda]$ with the stem field of $f(\lambda)$ a cyclic field of degree n over $\mathfrak{F}$ and generating automorphism S. Then $S^n = I$ and every integer k has the form

$$k = qn + r , \qquad S^k = S^r ,$$

where $0 \leqq r < n$. The roots of $f(\lambda)$ may now be represented by

$$a_i = a_1^{S^{i-1}} \qquad (i = 1, \ldots, n) \tag{54}$$

where we define $a_{k+1} = a_1^{S^k} = a_1^{S^r} = a_{r+1}$.

The polynomial $f(y)$ has $y - a_1^S$ as a factor and we may write

(55) $$f(y) = (y - a_2)g(a_1, y)$$

where $a_2 = a_1^S$ is in $\mathfrak{F}(a_1)$, $g(a_1, y) = (y - a_1)(y - a_3) \dots (y - a_n)$. The polynomial

$$g(x, y)$$

has degree at most $n - 1$ in each indeterminate and coefficients in $\mathfrak{F}$. Our factorization implies that

(56) $$\gamma(a_1) = g(a_1, a_2) \neq 0, \quad g(a_1, a_r) = 0$$

for $r = 1, 3, 4, \dots, n$. The polynomial $\gamma(a_1)$ has a polynomial inverse $\delta(a_1)$ in $\mathfrak{F}[a_1]$ and $h(a_1, y) = \delta(a_1)g(a_1, y)$ may be expressed as a polynomial in a_1 and y of degree at most $n - 1$ in each symbol. Thus there exists a polynomial $h(x, y)$ of degree at most $n - 1$ in each indeterminate and coefficients in $\mathfrak{F}$ such that

(57) $$h(a_1, a_2) = 1, \qquad h(a_1, a_r) = 0 \qquad (r = 1, 3, 4, \dots, n).$$

Applying the automorphism S^i to (57) we obtain the

LEMMA. *There exists a polynomial*

(58) $$\mathrm{h}(\mathrm{x}, \mathrm{y})$$

of degree at most $\mathrm{n} - 1$ *in each indeterminate* x *and* y *and with coefficients in* $\mathfrak{F}$ *such that*

(59) $$\mathrm{h}(a_{\mathrm{i}}, a_{\mathrm{i}+1}) = 1, \qquad \mathrm{h}(a_{\mathrm{i}}, a_{\mathrm{j}}) = 0 \qquad (\mathrm{i} = 1, \dots, \mathrm{n})$$

for any $\mathrm{j} \neq \mathrm{i} + 1$.

The matrix

(60) $$\eta = \begin{pmatrix} 0 & 1 & 0 & \dots & 0 \\ 0 & 0 & 1 & \dots & 0 \\ . & . & . & \dots & . \\ 0 & 0 & 0 & \dots & 1 \\ 1 & 0 & 0 & \dots & 0 \end{pmatrix}$$

has the property $\eta^n = 1$ by our fundamental lemma of Section 4.4. An evident computation gives

(61) $$\eta a = a^S\eta, \quad a = \mathrm{diag}\,\{a_1, \dots, a_n\}, \quad a^S = \mathrm{diag}\,\{a_2, \dots, a_n, a_1\},$$

as we saw in equation (4.53). If $\theta(a)$ is any polynomial in a, then

$$\theta(a) = \text{diag}\,\{\theta(a_1), \ldots, \theta(a_n)\}\,, \tag{62}$$

so that if $a_2 = \theta(a_1)$ then $a^S = \theta(a)$. The matrix A of (41) similar to our a generates a cyclic field $\mathfrak{F}[A] = \mathfrak{F}(A)$ of degree n over $\mathfrak{F}$ equivalent to $\mathfrak{F}(a_1)$ under the correspondence $\psi(a_1) \longleftrightarrow \psi(A)$ for any polynomial $\psi(\lambda)$ of degree at most $n-1$ of $\mathfrak{F}[\lambda]$. Moreover

$$\psi(A) \longleftrightarrow \psi(A^S) \tag{63}$$

is a generating automorphism of $\mathfrak{F}[A]$ over $\mathfrak{F}$ where $A^S = \theta(A) = Va^SV^{-1}$. But our lemma and Theorem 10.15 imply that there exists a matrix Y with elements in $\mathfrak{F}$ such that $\eta = V^{-1}YV$. Then

$$Y = V\eta V^{-1}\,, \qquad YA = A^SY\,, \qquad Y^n = 1\,. \tag{64}$$

Every diagonal matrix is a polynomial in a with coefficients in $\mathfrak{K} = \mathfrak{F}(a_1)$ by Exercise 1 of Section 10.8 and hence every square matrix is a linear combination with coefficients in $\mathfrak{K}$ of

$$a^i\eta^j \qquad (i, j = 0, 1, \ldots, n-1) \tag{65}$$

by Theorem 4.15. Thus the n^2 matrices (65) are linearly independent in $\mathfrak{K}$ and so the similar matrices

$$A^iY^j = Va^i\eta^jV^{-1}$$

must also be linearly independent in $\mathfrak{K}$. But then the A^iY^j are linearly independent in $\mathfrak{F}$ and, since they are n^2 matrices of the linear set $\mathfrak{M}_n$ of order n^2 over $\mathfrak{F}$, they form a basis of $\mathfrak{M}_n$. We have proved

Theorem 17. *Let $\mathfrak{M}_n$ be the algebra of all* n*-rowed square matrices with elements in a field $\mathfrak{F}$ and let* $f(\lambda) = 0$ *be an irreducible separable equation of degree* n *with coefficients in $\mathfrak{F}$ and the cyclic group with respect to $\mathfrak{F}$. Then there exist matrices* A *and* Y *in $\mathfrak{M}_n$ such that*

$$f(A) = 0\,, \qquad \mathfrak{Z} = \mathfrak{F}[A]$$

is a cyclic field of degree n *over $\mathfrak{F}$ with generating automorphism* S,

$$Y^n = 1\,, \qquad Y^iZ = Z^{S^i}Y^i \qquad (i = 1, 2, \ldots, n)\,, \tag{66}$$

for every Z *of $\mathfrak{Z}$. Every matrix of $\mathfrak{M}_n$ is uniquely expressible in the form*

$$Z_0 + Z_1Y + \ldots + Z_{n-1}Y^{n-1} \qquad (Z_i \textit{ in } \mathfrak{Z}). \tag{67}$$

This new representation of $\mathfrak{M}_n$ is called its *cyclic representation.*

EXERCISES

1. Review our important bases of $\mathfrak{M}_n$ over $\mathfrak{F}$ by stating our two other forms of bases.

2. Every separable irreducible quadratic equation is cyclic. Find the cyclic representations of $\mathfrak{M}_2$ in the cases where $\mathfrak{F}$ does not or does have characteristic two by computing Y explicitly.

10. Cyclic algebras. The algebra $\mathfrak{M}$ of all n-rowed square matrices is a linear associative algebra of order n^2 over $\mathfrak{F}$. Let $\gamma \neq 0$ be any quantity of $\mathfrak{F}$ and

$$\mathfrak{K} = \mathfrak{F}(\gamma^{1/n}) .$$

The field $\mathfrak{K}$ has degree at most n over $\mathfrak{F}$ and $\mathfrak{M}_{\mathfrak{K}}$ over $\mathfrak{K}$ has a representation (66), (67) with Y replaced by $Y_0 = \gamma^{1/n}Y$, $Y_0^n = \gamma$.

Define a linear set of order n^2 over $\mathfrak{F}$ whose elements are given by (67) and

$$(68) \qquad Y^n = \gamma , \qquad Y^iZ = Z^{S^i}Y^i \qquad (i = 1, 2, \dots , n) ,$$

for any Z of a cyclic field $\mathfrak{Z}$ of degree n over $\mathfrak{F}$ and generating automorphism S. The linear set is an algebra $\mathfrak{D}$ of order n^2 over $\mathfrak{F}$ and by our computation above

$$\mathfrak{D}_{\mathfrak{K}} = \mathfrak{M}_{\mathfrak{K}} .$$

It follows that $\mathfrak{D}$ is a quadrate algebra of order n^2 over $\mathfrak{F}$ with $\mathfrak{K} = \mathfrak{F}(\gamma^{1/n})$ as a splitting field. As we have seen (for arbitrary quadrate algebras) the algebra $\mathfrak{D}$ has degree n.

We shall call $\mathfrak{D}$ a *cyclic algebra.* Such algebras are probably the most important types of algebras. In fact it has been shown in modern literature* that if $\mathfrak{F}$ is an algebraic (number) field of finite degree over the field $\mathfrak{R}$ of all rational numbers then the only quadrate algebras over $\mathfrak{F}$ are cyclic algebras. The proof is of course beyond the scope of this text.

A cyclic algebra $\mathfrak{D}$ is uniquely defined by the cyclic field $\mathfrak{Z}$, its generating automorphism S, and the defining element γ of $\mathfrak{F}$. We shall therefore write

$$(69) \qquad \mathfrak{D} = (\mathfrak{Z}, S, \gamma) .$$

* Cf. Deuring's *Algebren* for proof.

One of the most obvious transformations on $\mathfrak{D}$ is obtained by replacing S by any other generating automorphism S^r of $\mathfrak{Z}$ over $\mathfrak{F}$. This is accomplished by replacing Y by Y^r and we have

$$\mathfrak{D} = (\mathfrak{Z}, S, \gamma) = (\mathfrak{Z}, S^r, \gamma^r) \tag{70}$$

for any integer r prime to n. This latter restriction on r is necessary since S^r generates $[S]$ if and only if r is prime to the order n of the cyclic group $[S]$.

The only quantities of $\mathfrak{D}$ commutative with every Z of $\mathfrak{Z}$ are quantities of $\mathfrak{Z}$. For

$$Z(Z_0 + Z_1Y + \ldots + Z_{n-1}Y^{n-1}) = (Z_0 + Z_1Y + \ldots + Z_{n-1}Y^{n-1})Z$$

if and only if $ZZ_i = Z_iZ^{S^i}$ $(i = 0, \ldots, n-1)$. For $i > 0$ we have $Z - Z^{S^i} \neq 0$, $(Z - Z^{S^i})Z_i = 0$, so that $Z_i = 0$ for $i > 0$, and our result is proved.

The unique form (67) of the quantities of $\mathfrak{D}$ remains a unique expression for these quantities when we replace Y by $Y_0 = aY$ where a is any non-zero quantity of $\mathfrak{Z}$. For (67) states that $1, Y, Y^2, \ldots, Y^{n-1}$ are left linearly independent* in $\mathfrak{Z}$ and this is of course true of $1, aY, (aa^S)Y^2, \ldots, aa^S \ldots a^{S^{n-1}}Y^{n-1}$ which differ from $1, Y, \ldots, Y^{n-1}$ by left coefficients in $\mathfrak{Z}$. The replacement of Y by Y_0 replaces $\gamma = Y^n$ by

$$(aY)^n = aa^S \ldots a^{S^{n-1}}Y^n = \gamma N_{\mathfrak{Z}|\mathfrak{F}}(a)\,.$$

We have proved the first part of

Theorem 18. *The cyclic algebras* $(\mathfrak{Z}, \mathrm{S}, \gamma)$ *and* $(\mathfrak{Z}, \mathrm{S}, \delta)$ *for the same* $\mathfrak{Z}$ *and* S *are equivalent if and only if there exists an* a *in* $\mathfrak{Z}$ *such that*

$$\delta = \gamma \mathrm{N}_{\mathfrak{Z}|\mathfrak{F}}(\mathrm{a})\,. \tag{71}$$

To prove the last part let $(\mathfrak{Z}, S, \delta)$ be equivalent to $(\mathfrak{Z}, S, \gamma)$. The correspondence defining this equivalence states that there is a subfield $\mathfrak{Z}_0$ of $(\mathfrak{Z}, S, \gamma)$ equivalent to $\mathfrak{Z}$ and a quantity Y_0 such that $Y_0{}^iZ_0 = Z_0{}^{S^i}Y_0$ for every Z_0 of $\mathfrak{Z}_0$, $Y_0^n = \delta$. By Theorem 10.14 and our tacit assumption that $\mathfrak{F}$ is an infinite field we may carry $\mathfrak{Z}_0$ into $\mathfrak{Z}$ by an inner automorphism of $\mathfrak{D} = (\mathfrak{Z}, S, \gamma)$. This carries Y_0 into Y_1 such that $Y_1Z = Z^SY_1$ for every Z of $\mathfrak{Z}$. Now $YZ = Z^SY$, $Y^{-1} = \gamma^{-1}Y^{n-1}$ is in $\mathfrak{D}$, $ZY^{-1} = Y^{-1}Z^S$, $(Y^{-1}Y_1)Z = Z(Y^{-1}Y_1)$. But we have seen that then $Y^{-1}Y_1 = b$ is in $\mathfrak{Z}$ and $Y_1 = Yb = aY$ where $a = b^S$ is in $\mathfrak{Z}$. Hence $Y_1^n = N_{\mathfrak{Z}|\mathfrak{F}}(a)\gamma = \delta$.

* By this we mean that no sum $a_0 + a_1Y + \ldots + a_{n-1}Y^{n-1}$ with a_i in $\mathfrak{Z}$ is zero unless the a_i are all zero.

Theorem 18 implies the immediate consequence

Theorem 19. *A cyclic algebra* $(\mathfrak{Z}, S, \delta)$ *is a total matric algebra if and only if*

$$\delta = N_{\mathfrak{Z}|\mathfrak{F}}(a) \tag{72}$$

for a *in* $\mathfrak{Z}$.

For in Theorem 10.17 we proved that the algebra $\mathfrak{M}_n$ of all n-rowed square matrices is a cyclic algebra $(\mathfrak{Z}, S, 1)$. Now $(\mathfrak{Z}, S, \delta)$ is a total matric algebra if and only if it is equivalent to $\mathfrak{M}_n$ and (72) is thus an immediate consequence of (71) with $\gamma = 1$.

There are many other interesting properties of cyclic algebras but we leave them for discussion in a more advanced text.

11. The matrices commutative with a subfield of $\mathfrak{M}_n$. We have shown that if A has distinct characteristic roots then $AB = BA$ if and only if B is a polynomial in A. This is also true if the characteristic function is irreducible. We shall in fact prove

Theorem 20. *Let* $\mathfrak{Z}$ *be a field, of degree* m *over* $\mathfrak{F}$, *of* n*-rowed square matrices with elements in* $\mathfrak{F}$ *such that the unity element of* $\mathfrak{Z}$ *is the* n*-rowed identity matrix. Then*

$$n = qm ,$$

and the algebra of all n*-rowed square matrices in* $\mathfrak{M}_n$ *over* $\mathfrak{F}$ *commutative with every quantity of* $\mathfrak{Z}$ *is a total matric algebra* $\mathfrak{M}_q$ *of degree* q *(order* q^2*) over* $\mathfrak{Z}$.

The statement immediately preceding our theorem is the case where $\mathfrak{Z} = \mathfrak{F}(A)$ has degree n, $q = 1$, $\mathfrak{M}_q$ over $\mathfrak{Z}$ is $\mathfrak{Z}$ itself. To prove our theorem we notice that it is trivial when $n = 1$ and we make an induction on n. If $\mathfrak{Z}$ has any proper subfield over $\mathfrak{F}$ it has some proper subfield $\mathfrak{Y} > \mathfrak{F}$ which itself has no proper subfield over $\mathfrak{F}$ except $\mathfrak{F}$ itself. Then $\mathfrak{Y} \geqq \mathfrak{F}[Y]$ where Y is any quantity in $\mathfrak{Y}$ and our hypothesis implies that if Y is not in $\mathfrak{F}$ then $\mathfrak{Y} = \mathfrak{F}[Y]$ is a simple extension of $\mathfrak{F}$ whose degree over $\mathfrak{F}$ is the degree of the minimum function of Y. The matrix Y generates the field $\mathfrak{Y}$ and this minimum function $\phi(\lambda)$ has degree ρ and is an irreducible monic polynomial of $\mathfrak{F}[\lambda]$. By Theorem 4.4 the matrix Y is similar in $\mathfrak{F}$ to

$$Y = \text{diag}\,\{y, \ldots, y\} \tag{73}$$

where y is given by equation (4.6) for $\phi(\lambda)$. Hence the order $n = \rho\sigma$, and since $\mathfrak{Y}$ is a subfield of $\mathfrak{Z}$, $m = \rho\tau$.

Every n-rowed square matrix B has the form

$$B = (b_{ij}) \qquad (i, j = 1, \ldots, \sigma) , \tag{74}$$

with square matrices b_{ij} of ρ columns, and $BY = YB$ for Y given by (73) if and only if

$$b_{ij}y = yb_{ij} \qquad (i, j = 1, \ldots, \sigma) . \tag{75}$$

If y is separable over $\mathfrak{F}$ this implies that the b_{ij} are polynomials in $\mathfrak{F}[y]$ by Theorem 10.16 and this is also true in the inseparable case where $y^\rho = \gamma$ in $\mathfrak{F}$ by Theorem 4.16. We now write $b_{ij} = b_{ij}(y)$ and have

$$B = \sum_{i, j=1}^{\sigma} b_{ij}(Y)E_{ij} \tag{76}$$

for E_{ij} an ordinary matric basis of the total matric subalgebra $\mathfrak{M}_\sigma$ of $\mathfrak{M}_n$. The degree of $\mathfrak{M}_\sigma$ over the field $\mathfrak{Y} = \mathfrak{F}[Y]$ is less than n and $\mathfrak{M}_\sigma$ over $\mathfrak{Y}$ contains $\mathfrak{Z}$ of degree τ over $\mathfrak{Y}$. By the hypothesis of our induction $\sigma = \tau q$, $n = \rho\sigma = \rho\tau q = mq$. Also the algebra of all matrices of $\mathfrak{M}_\sigma$ over $\mathfrak{Y}$ commutative with every Z of $\mathfrak{Z}$ is $\mathfrak{M}_q$ over $\mathfrak{Z}$. Every matrix of $\mathfrak{M}_n$ over $\mathfrak{F}$ commutative with every Z of $\mathfrak{Z}$ is commutative with every quantity of $\mathfrak{Y}$, is in $\mathfrak{M}_\sigma$ over $\mathfrak{Y}$, and is thus in $\mathfrak{M}_q$ over $\mathfrak{Z}$ as desired. This completes our induction and proves our theorem.

EXERCISES

1. State the analogue of Theorem 10.20 for quadrate algebras.

2. Use Theorem 10.12 and Ex. 1 of Section 10.7 to prove the analogue above. Hint: If $\mathfrak{Z}$ is a subfield of $\mathfrak{D}$ then the algebra of all quantities of $\mathfrak{D} \times \mathfrak{D}_0$ commutative with $\mathfrak{Z}$ is $\mathfrak{D}_0 \times \mathfrak{Q}$, $\mathfrak{Q}$ the corresponding subalgebra of $\mathfrak{D}$.

12. The polynomial algebra $\mathfrak{F}[A]$. We have already considered total matric algebras and quadrate algebras and have seen that their structure depends partly on the structure of the algebras generated by their individual quantities. It is therefore natural to conclude our introduction to the theory of algebras by the study of the algebra

$$\mathfrak{F}[A] = (I, A, \ldots, A^{m-1}) \tag{77}$$

over $\mathfrak{F}$ which consists of all polynomials

$$a_0 + a_1A + \ldots + a_{m-1}A^{m-1} \qquad (a_i \text{ in } \mathfrak{F}) .$$

The multiplication table of $\mathfrak{F}[A]$ is completely defined by the minimum function

$$\phi(\lambda) = \lambda^m + b_1\lambda^{m-1} + \ldots + b_m \qquad (b_i \text{ in } \mathfrak{F}) ,$$

of A and $\mathfrak{F}[A]$ is a commutative algebra with a unity element. We are assuming that A is in some algebra, and there is no loss of generality if we assume that A is an arbitrary n-rowed square matrix with elements in $\mathfrak{F}$. Thus A is in $\mathfrak{M}_n$ over $\mathfrak{F}$. We now use Theorem 4.6 to prove

Theorem 21. *Let* $\phi(\lambda) = \mathrm{p}(\lambda) \cdot \mathrm{q}(\lambda)$ *where* $\mathrm{p}(\lambda)$ *and* $\mathrm{q}(\lambda)$ *are relatively prime. Then* $\mathfrak{F}[A]$ *is the direct sum of two of its subalgebras* $\mathfrak{F}[\mathrm{B}_0]$ *and* $\mathfrak{F}[\mathrm{C}_0]$, *and there exists an inner automorphism of* $\mathfrak{M}_n$ *carrying* A *into*

$$(78)\quad \mathrm{A} = \mathrm{B}_0 + \mathrm{C}_0 = \begin{pmatrix} \mathrm{B} & 0 \\ 0 & \mathrm{C} \end{pmatrix}, \quad \mathrm{B}_0 = \begin{pmatrix} \mathrm{B} & 0 \\ 0 & 0 \end{pmatrix}, \quad \mathrm{C}_0 = \begin{pmatrix} 0 & 0 \\ 0 & \mathrm{C} \end{pmatrix}.$$

A *matrix* G *is commutative with* A *if and only if*

$$(79)\qquad \mathrm{G} = \begin{pmatrix} \mathrm{G}_1 & 0 \\ 0 & \mathrm{G}_2 \end{pmatrix}, \qquad \mathrm{G}_1\mathrm{B} = \mathrm{B}\mathrm{G}_1\,, \qquad \mathrm{G}_2\mathrm{C} = \mathrm{C}\mathrm{G}_2\,.$$

For A may be taken to have the form (78) where $p(\lambda)$ is the minimum function of B and $q(\lambda)$ is the minimum function of C. Since $p(\lambda)$ and $q(\lambda)$ are relatively prime we have

$$p(\lambda)p_0(\lambda) + q(\lambda)q_0(\lambda) = 1\,.$$

Thus $p(B) = 0$, $q(C) = 0$ implies that

$$I_n = p(A)p_0(A) + q(A)q_0(A) = \begin{pmatrix} q(B) \cdot q_0(B) & 0 \\ 0 & p(C) \cdot p_0(C) \end{pmatrix}.$$

It follows that if

$$h(\lambda) = q(\lambda)q_0(\lambda)\,, \qquad k(\lambda) = p(\lambda)p_0(\lambda)\,,$$

then

$$h(B) = I_r\,, \quad h(C) = 0\,, \quad k(C) = I_{n-r}\,, \quad k(B) = 0\,,$$

where r is the order of B. Hence

$$e_1 = h(A) = \begin{pmatrix} I_r & 0 \\ 0 & 0 \end{pmatrix}, \qquad e_2 = k(A) = \begin{pmatrix} 0 & 0 \\ 0 & I_{n-r} \end{pmatrix}, \qquad e_1 + e_2 = I_n$$

are in $\mathfrak{F}[A]$ and $e_1e_2 = e_2e_1 = 0$,

$$B_0 = e_1A\,, \quad C_0 = e_2A\,, \quad A = B_0 + C_0$$

as desired. Every polynomial in A has the form $q(A) = q(B_0) + q(C_0)$ and $q(B_0) = e_1q(A)$, $q(C_0) = e_2q(A)$, $e_1e_2 = e_2e_1 = 0$. This form is evidently unique and $q_1(B_0) \cdot q_2(C_0) = q_2(C_0) \cdot q_1(B_0) = 0$ for any quantities q_1

and q_2 of $\mathfrak{F}[B_0]$, $\mathfrak{F}[C_0]$, respectively. This proves that $\mathfrak{F}[A] = \mathfrak{F}[B_0] \oplus \mathfrak{F}[C_0]$ is the desired direct sum. Let $GA = AG$ so that G is commutative with every quantity of $\mathfrak{F}[A]$. Then e_1 and e_2 are in $\mathfrak{F}[A]$,

$$G = \begin{pmatrix} G_1 & G_3 \\ G_4 & G_2 \end{pmatrix}, \qquad e_1G = \begin{pmatrix} G_1 & G_3 \\ 0 & 0 \end{pmatrix} = Ge_1 = \begin{pmatrix} G_1 & 0 \\ G_4 & 0 \end{pmatrix}$$

and $G_3 = G_4 = 0$,

$$G = \begin{pmatrix} G_1 & 0 \\ 0 & G_2 \end{pmatrix}.$$

The second part of (79) follows from a trivial computation.

Theorem 10.21 reduces the problems of the structure of $\mathfrak{F}[A]$ and of the matrices commutative with A to the case where the minimum function of A is a power $[f(\lambda)]^r$ of an irreducible monic $f(\lambda)$ of $\mathfrak{F}[\lambda]$. The case where $f(\lambda)$ is inseparable does not occur in the classic literature and has not yet been studied in general. We shall therefore assume that $f(\lambda)$ is always separable. The difficulty will be seen when we prove the following

Lemma. *Let* $f(\lambda)$ *be an irreducible separable polynomial of* $\mathfrak{F}[\lambda]$ *and* r *be any integer. Then there exists a polynomial* $g(\lambda)$ *such that* $f[g(\lambda)]$ *is divisible by* $[f(\lambda)]^r$.

The lemma is true for $r = 1$ since then we take $g(\lambda) = \lambda$. Let it be true for $r - 1$ and thus assume that $f[g_{r-1}(\lambda)] = q(\lambda) \cdot [f(\lambda)]^{r-1}$. Put $g(\lambda) = g_{r-1}(\lambda) + [f(\lambda)]^{r-1}\mu(\lambda)$ and see that the Taylor expansion implies that

$$f[g(\lambda)] = f[g_{r-1}(\lambda)] + \mu(\lambda)[f(\lambda)]^{r-1}f'[g_{r-1}(\lambda)] + [f(\lambda)]^r s(\lambda)$$

by Exercise 1 of Section 7.9. Then

$$f[g(\lambda)] = \{q(\lambda) + \mu(\lambda)f'[g_{r-1}(\lambda)]\}[f(\lambda)]^{r-1} + [f(\lambda)]^r s(\lambda).$$

The polynomial $f'(\lambda)$ is prime to $f(\lambda)$ by Theorem 7.20, and hence there exist polynomials $a(\lambda)$ and $b(\lambda)$ such that $fa + f'b = 1$. Put

$$\mu(\lambda) = [f(\lambda) - q(\lambda)]b(\gamma), \qquad \gamma = g_{r-1}(\lambda),$$

and obtain $\mu(\lambda)f'(\gamma) + q(\lambda) = [f(\lambda) - q(\lambda)][1 - f(\gamma)a(\gamma)] + q(\lambda) = f(\lambda) \cdot \{1 - f(\gamma)a(\gamma) + a(\gamma)[q(\lambda)]^2[f(\lambda)]^{r-2}\}$, where we have replaced λ by γ in the identity $fa + f'b = 1$. Thus each term of the right member of $f[g(\lambda)]$ is divisible by $[f(\lambda)]^r$ as desired.

The lemma above will be used to prove

Theorem 22. *Let* A *be an* n*-rowed square matrix with elements in* $\mathfrak{F}$ *whose minimum function is* $[f(\lambda)]^r$ *where* $f(\lambda)$ *is a separable irreducible quantity of*

$\mathfrak{F}[\lambda]$. *Then* $\mathfrak{F}[\mathrm{A}]$ *has a subfield* $\mathfrak{Z} = \mathfrak{F}(\mathrm{X})$ *which is a stem field of* $\mathrm{f}(\lambda) = 0$ *and*

$$\mathfrak{F}[\mathrm{A}] = \mathfrak{Z}[\mathrm{N}],$$

where N *is a nilpotent matrix of index* r.

For we find a $g(\lambda)$ by the lemma such that $f[g(\lambda)]$ is divisible by $[f(\lambda)]^r$. Then if

$$X = g(A)$$

we have $f[g(A)] = [f(A)]^r q(A) = 0$,

$$f(X) = 0.$$

Since $f(\lambda)$ is irreducible, the subalgebra $\mathfrak{F}[X]$ of $\mathfrak{F}[A]$ is a field $\mathfrak{Z} = \mathfrak{F}(X)$. The order of $\mathfrak{F}[A]$ is the degree of the minimum function of A, and this degree is mr where m is the degree of $f(\lambda)$. Put

$$N = f(A).$$

Then $N^r = 0$ and no lower power of N is zero since $[f(A)]^t = 0$ implies that $[f(\lambda)]^t$ is divisible by the minimum function of A. The algebra $\mathfrak{Z}[N]$ has order r over $\mathfrak{Z}$ and a basis $1, N, N^2, \ldots, N^{r-1}$ over $\mathfrak{Z}$ where the unity element 1 is the identity matrix. It is obvious that this is equivalent to the fact that

$$1, X, \ldots, X^{m-1}, N, NX, \ldots, NX^{m-1}, \ldots, N^{r-1}, N^{r-1}X, \ldots, N^{r-1}X^{m-1}$$

are a basis of $\mathfrak{Z}[N]$ over $\mathfrak{F}$ and that this latter algebra has order mr over $\mathfrak{F}$. Since $\mathfrak{F}[A] \geqq \mathfrak{Z}[N]$ and both have order mr, they are equal.

The algebra of all n-rowed square matrices commutative with A is contained in the algebra of all n-rowed square matrices commutative with X. By Theorem 10.20

$$n = mq,$$

and this latter algebra is a total matric algebra $\mathfrak{M}_q$ over $\mathfrak{F}$. The matrix N is in $\mathfrak{M}_q$ and $BA = AB$ implies that $BX = XB$, $BN = NB$ where B is thus in $\mathfrak{M}_q$. Conversely if B in $\mathfrak{M}_q$ is commutative with N then B is commutative with every quantity of $\mathfrak{Z}(N)$ and $BA = AB$.

Theorem 23. *A matrix* B *is commutative with* A *of Theorem* 10.22 *if and only if* B *is in the total matric algebra* $\mathfrak{M}_q$ *defined for* $\mathfrak{Z}$ *in Theorem* 10.20 *and is commutative with* N.

We are considering the structure of $\mathfrak{F}[A]$ in the case where the irreducible factors of the characteristic function of A are separable. Theorems 10.21 and 10.22 then reduce our problem to the case where A is nilpotent and thus solve it. For now $\mathfrak{F}[A] = \mathfrak{F}[N]$ is what is called a *nilpotent algebra with a unity element adjoined* and $\mathfrak{F}[A]$ is the sum of the nilpotent algebra $(N, N^2, \ldots, N^{r-1})$ of order $r - 1$ over $\mathfrak{F}$ and the field consisting of all scalar matrices (and equivalent to $\mathfrak{F}$). Similarly, Theorems 10.21 and 10.23 reduce the problem of finding all matrices commutative with A to the case where $A = N$ is nilpotent of index r. We now pass to a similar matrix

$$N = \operatorname{diag}\{N_1, \ldots, N_t\}, \tag{80}$$

where N_j is nilpotent of order and index r_j, and

$$r = r_1 \geqq r_2 \geqq \ldots \geqq r_t > 0. \tag{81}$$

By Section 4.6 every nilpotent matrix of order and index $s + 1$ is similar to a matrix of the form

$$Q_s = \begin{pmatrix} 0_{s1} & I_s \\ 0_{11} & 0_{1s} \end{pmatrix} \tag{82}$$

where 0_{ab} is the zero matrix of a rows and b columns and I_s is an s-rowed identity matrix. By a simple inductive computation which the reader should verify

$$Q_s^{k+1} = \begin{pmatrix} 0_{s-k,\,k} & I_{s-k} \\ 0_{k,\,k} & 0_{k,\,s-k} \end{pmatrix}. \tag{83}$$

By Theorem 4.16 the only matrices commutative with Q_s are polynomials of the algebra $\mathfrak{F}[Q_s]$. We now prove the

Lemma. *Let* $s > t$ *and*

$$BQ_s = Q_tB, \qquad CQ_t = Q_sC.$$

Then

$$B = (0_{t,\,s-t}, B_0), \qquad C = \begin{pmatrix} C_0 \\ 0_{s-t,\,t} \end{pmatrix},$$

where B_0 *and* C_0 *are polynomials of* $\mathfrak{F}[Q_t]$.

For a simple induction shows that

$$BQ_s^k = Q_t^kB, \qquad CQ_t^k = Q_s^kC \tag{84}$$

for any integer k. We write

$$B = (B_1, B_0) ,$$

where B_0 is a t-rowed square matrix and B_1 has t rows and $s - t$ columns. Then $Q_t^{t+1} = 0$ and by (84), (83)

$$BQ_s^{t+1} = 0 = (B_1, B_0)\begin{pmatrix} 0_{s-t,\,t} & I_{s-t} \\ 0_{t,\,t} & 0_{t,\,s-t} \end{pmatrix} = (0, B_1) .$$

Hence $B_1 = 0$. But evidently Q_s may be partitioned to have the forms

$$\begin{pmatrix} Y & Z \\ 0 & Q_t \end{pmatrix}, \qquad \begin{pmatrix} Q_t & Z_0 \\ 0 & Y_0 \end{pmatrix}$$

for matrices Y, Z, Y_0, Z_0 and

$$BQ_s = (0, B_0)\begin{pmatrix} Y & Z \\ 0 & Q_t \end{pmatrix} = (0, B_0Q_t) = Q_t(0, B_0) = (0, Q_tB_0) .$$

Thus $Q_tB_0 = B_0Q_t$, B_0 is in $\mathfrak{F}[Q_t]$. Analogously,

$$C = \begin{pmatrix} C_0 \\ C_1 \end{pmatrix}, \qquad Q_s^{t+1}C = \begin{pmatrix} C_1 \\ 0 \end{pmatrix}, \qquad C_1 = 0, \qquad CQ_t = \begin{pmatrix} C_0Q_t \\ 0 \end{pmatrix}$$
$$= \begin{pmatrix} Q_t & Z_0 \\ 0 & Y_0 \end{pmatrix}\begin{pmatrix} C_0 \\ 0 \end{pmatrix} = \begin{pmatrix} Q_tC_0 \\ 0 \end{pmatrix}$$

and C_0 is in $\mathfrak{F}[Q_t]$.

The results above imply

Theorem 24. *Let*

$$B = (B_{ij}) \qquad (i, j = 1, \ldots, t) ,$$

where B_{ij} *is a matrix of* r_i *rows and* r_j *columns of elements of* $\mathfrak{F}$. *Then* B *is commutative with the matrix* N *of* (80) *if and only if* B_{ij} *is in* $\mathfrak{F}[N_i]$ *when* $r_i = r_j$, *and*

$$B_{ij} = \begin{pmatrix} B_{ij0} \\ 0 \end{pmatrix}, \qquad B_{ji} = (0, B_{ji0}) \qquad (r_i > r_j)$$

with B_{ij0} *and* B_{ji0} *in* $\mathfrak{F}[N_j]$.

For $BN = NB$ is equivalent to $B_{ij}N_j = N_iB_{ij}$ and thus $B_{ji}N_i = N_jB_{ji}$. When $r_i > r_j$ we take $r_i = s$, $r_j = t$ and apply our lemma.

CHAPTER XI

INTRODUCTION TO THE TRANSCENDENTAL THEORY OF FIELDS

1. Archimedean ordered fields. The reader undoubtedly has an intuitive notion of the meaning of the term "real number." Thus a real number may be thought of as a measure in terms of a given unit of the distance of a point on a straight line from the zero point. Such a notion is usually adequate for an elementary treatise on the Calculus but is of course not precise.

Certain fields called p-adic number fields have recently assumed great importance in modern Algebra and the Theory of Numbers. We need a postulational treatment of the definitions and properties of these fields and a fundamental concept in these definitions is that of a real number. It is therefore elegant so to phrase these definitions as to include a definition of the real number system in terms of rational numbers. We shall do this. Our definitions necessarily involve infinite sequences and what is essentially the notion of a limit. Thus we have left purely algebraic notions and are now studying what may be called the transcendental theory of fields.

The notion of order in the real number system is evident in the above intuitive definition. This concept was generalized in Section 5.9 where ordered fields were defined. We shall now impose a further restriction which brings us more closely to our intuitive notion of a real number.

DEFINITION. *An ordered field* $\mathfrak{T}$ *is said to be archimedean ordered if for every* a *of* $\mathfrak{T}$ *there exists an integer* n $>$ a.

The definition above implies that for every $\epsilon > 0$ in $\mathfrak{T}$ we may find an integer $n > \epsilon^{-1}$. Then $\epsilon > n^{-1} > 0$ for the rational number n^{-1}. We may also find an integer m such that $2^m > n$ and thus $\epsilon > 2^{-m}$ if we desire. This property will be used in later computations.

The field $\mathfrak{R}$ of all rational numbers is an archimedean ordered field. We shall define a generalization of the notion of absolute value as a function

$$\phi \text{ on } \mathfrak{F} \text{ to } \mathfrak{T},$$

where $\mathfrak{T}$ is archimedean ordered, $\mathfrak{F}$ is a field. Then ϕ will define what we shall call a *derived field*

$$\mathfrak{F}_\phi$$

of $\mathfrak{F}$. The particular case where $\mathfrak{T}$ is $\mathfrak{R}$ and ϕ is ordinary absolute value then gives $\mathfrak{R}' = \mathfrak{R}_\phi$, the field of all real numbers. But we shall show that in our definition of ϕ every $\mathfrak{T}$ may be replaced by $\mathfrak{R}'$ and hence have ϕ on $\mathfrak{F}$ to the field of all real numbers. Thus we have a treatment of our functions ϕ on $\mathfrak{F}$ to $\mathfrak{R}'$ which includes the definition of $\mathfrak{R}'$ as a special case. But our principal interest is of course in the general $\mathfrak{F}$ and its derived $\mathfrak{F}_\phi$.

2. Elements of the theory of ideals. The elementary properties of ideals are of great importance in abstract algebra as well as in the theory of algebraic numbers. We shall define the concept of an algebraic ideal here and shall obtain some of the simpler properties. We first make the

DEFINITION. *A set $\mathfrak{M}$ of elements of a ring $\mathfrak{A}$ is called a modul of $\mathfrak{A}$ if*

$$\mathrm{a} - \mathrm{b}$$

is in $\mathfrak{M}$ for every a *and* b *of $\mathfrak{M}$.*

A modul $\mathfrak{M}$ has the property that $0 = a - a$, $0 - b = -b$, $a - (-b) = a + b$, $a - b$ are in $\mathfrak{M}$ for every a and b of $\mathfrak{M}$. Thus *a modul is simply an additive subgroup of the additive abelian group $\mathfrak{A}$.*

Every subgroup of an abelian group $\mathfrak{A}$ is a normal divisor of $\mathfrak{A}$. Thus a modul $\mathfrak{M}$ of $\mathfrak{A}$ may be used to define a difference group

$$\mathfrak{A} - \mathfrak{M}$$

whose elements are classes $[a]$* of congruent (equivalent) elements. The class $[a]$ consists of all quantities

$$a + m$$

with m in $\mathfrak{M}$, and $[a] = [b]$ if and only if $a - b$ is in $\mathfrak{M}$. This is as in our earlier definitions for groups and the set $\mathfrak{A}$–$\mathfrak{M}$ of all classes $[a]$ now forms a group with respect to the commutative operation defined by

$$[a] + [b] = [a + b] \,.$$

We made the additive abelian group $\mathfrak{A}$ into a ring in Chapter I by defining a second operation of multiplication. Similarly a modul $\mathfrak{M}$ will be made into a subring of $\mathfrak{A}$ if $\mathfrak{M}$ is closed with respect to the operation of multiplication of $\mathfrak{A}$. For this was shown in Theorem 1.8. We shall do more than this however and shall make the

DEFINITION. *A modul $\mathfrak{M}$ of $\mathfrak{A}$ is called a left ideal (or left invariant subring) if*

$$\mathrm{xm}$$

is in $\mathfrak{M}$ for every x *of $\mathfrak{A}$ and* m *of $\mathfrak{M}$.*

* This notation for a class is the same used in earlier chapters for cyclic groups $[S]$. But there S was an automorphism or a group element and no confusion will result.

Right ideals (*right invariant subrings*) are defined analogously, and we call $\mathfrak{M}$ an *ideal* (*invariant subring*) if it is both a right and a left ideal. When $\mathfrak{A}$ is commutative every right ideal is a left ideal so that every right or left ideal is an ideal. We shall not use the alternative terminology of invariant subring here.

If $\mathfrak{M}$ is an ideal the difference group $\mathfrak{A} - \mathfrak{M}$ will become a ring when we define the product

$$[a] \cdot [b] = [ab]$$

of the classes of $\mathfrak{A}$–$\mathfrak{M}$. For it is sufficient that this definition be independent of the particular representatives a, b, that is $[a + m_1] \cdot [b + m_2] = [ab + m_1b + am_2 + m_1m_2] = [ab]$. This is true since $\mathfrak{M}$ is an ideal, $m_1b + am_2 + m_1m_2$ is in $\mathfrak{M}$ for every m_1 and m_2 of $\mathfrak{M}$. This ring is called the *difference ring* $\mathfrak{A} - \mathfrak{M}$.

We shall now assume that $\mathfrak{A}$ is a ring with a unity element 1. Then $\mathfrak{A}$ contains the ideal

$$(1) = \mathfrak{A}$$

which is called the *unit ideal.* We say that 1 *generates* $\mathfrak{A}$ and generalize this concept as follows. Let $\mathfrak{Q}$ be a subset of $\mathfrak{A}$ and $\mathfrak{M}$ be the set of all finite sums

$$x_1a_1y_1 + \ldots + x_ra_ry_r \qquad (x_i,\ y_i \text{ in } \mathfrak{A},\ a_i \text{ in } \mathfrak{Q}) .$$

Then $\mathfrak{M}$ is an ideal said to be generated by $\mathfrak{Q}$. This result is an evident consequence of our definition of an ideal.

When $\mathfrak{Q}$ consists of a single element a of $\mathfrak{A}$ the generated ideal is called a *principal ideal*

$$(a) ,$$

(read "principal a") of $\mathfrak{A}$. Thus (a) consists of all sums of terms xay with x and y in our ring $\mathfrak{A}$ with a unity element. In particular the ideal (0) is called the *zero ideal* and consists of zero alone, and the unit ideal is *principal 1.*

When $\mathfrak{Q}$ consists of a finite number of elements $a_1, \ldots, a_r$ of $\mathfrak{A}$ we say that $\mathfrak{M}$ has a *finite basis* and write $\mathfrak{M} = (a_1, \ldots, a_r)$. This is the case that occurs in the theory of algebraic integers.

The product of two subsets $\mathfrak{L}$ and $\mathfrak{M}$ of a ring $\mathfrak{A}$ with a unity element is the set of all products lm with l in $\mathfrak{L}$ and m in $\mathfrak{M}$. If $\mathfrak{L}$ and $\mathfrak{M}$ are now ideals, we define their product

$$\mathfrak{L}\mathfrak{M}$$

to be the ideal generated by the product of the two sets. Thus $\mathfrak{L}\mathfrak{M}$ consists of all finite sums $\Sigma l_i m_i$ with l_i in $\mathfrak{L}$, m_i in $\mathfrak{M}$. The operation so defined is associative, and

$$\mathfrak{A}\mathfrak{M} = \mathfrak{M}\mathfrak{A} = \mathfrak{M},$$

so that $\mathfrak{A}$ is the identity ideal for this operation. When $\mathfrak{A}$ is commutative the ideal product is commutative. We easily verify that for commutative rings

$$(0)\mathfrak{M} = \mathfrak{M}(0) = 0, \quad (a)(b) = (ab), \quad (a)^n = (a^n).$$

3. Ideals in a commutative ring. Let $\mathfrak{A}$ be a commutative ring with a unity element. If a is any quantity of $\mathfrak{A}$ we say that a is *congruent to zero modulo an ideal* $\mathfrak{M}$ if $a - 0 = a$ is in $\mathfrak{M}$. We shall also say that a is divisible by $\mathfrak{M}$. If all quantities of an ideal $\mathfrak{L}$ are divisible by $\mathfrak{M}$ we say that $\mathfrak{L}$ *is divisible by* $\mathfrak{M}$. Thus the statement "$\mathfrak{L}$ is divisible by $\mathfrak{M}$" means "$\mathfrak{L}$ is contained in $\mathfrak{M}$." We re-emphasize this notion by stating that $\mathfrak{M}$ divides $\mathfrak{L}$ means that $\mathfrak{M}$ contains $\mathfrak{L}$. When also $\mathfrak{L} \neq \mathfrak{M}$ we call $\mathfrak{M}$ *a proper divisor of* $\mathfrak{L}$, and $\mathfrak{L}$ *a proper multiple* of $\mathfrak{M}$.

The unit ideal contains every ideal and thus divides every ideal. We have, however, defined our property of divisibility independently of that of product and *we do not state that if* $\mathfrak{M}$ *divides* $\mathfrak{L}$ *then* $\mathfrak{L} = \mathfrak{M}\mathfrak{N}$ *with* $\mathfrak{N}$ *an ideal.* It is true, however, that if $\mathfrak{L} = \mathfrak{M}\mathfrak{N}$, then $\mathfrak{M}$ divides $\mathfrak{L}$. For every quantity of $\mathfrak{L}$ is a sum $\Sigma m_i n_i$ and is in the ideal $\mathfrak{M}$.

The greatest common divisor of two ideals $\mathfrak{L}$ and $\mathfrak{M}$ is the ideal $(\mathfrak{L}, \mathfrak{M})$ generated by the set of all elements of $\mathfrak{L}$ and of $\mathfrak{M}$. It is a greatest common divisor, first in the sense that it contains both $\mathfrak{L}$ and $\mathfrak{M}$, and hence divides both. Next if $\mathfrak{C}$ divides $\mathfrak{L}$ and $\mathfrak{M}$ then $\mathfrak{C}$ contains $\mathfrak{L}$ and $\mathfrak{M}$, $\mathfrak{C}$ contains $(\mathfrak{L}, \mathfrak{M})$, $\mathfrak{C}$ divides $(\mathfrak{L}, \mathfrak{M})$. These are the usual formal properties of a g.c.d.

The least common multiple of $\mathfrak{L}$ and $\mathfrak{M}$ is their intersection $\mathfrak{D}$. For $\mathfrak{D}$ is evidently an ideal and $\mathfrak{D} \leqq \mathfrak{L}$, $\mathfrak{D} \leqq \mathfrak{M}$ so that $\mathfrak{D}$ is divisible by both $\mathfrak{L}$ and $\mathfrak{M}$. Any ideal divisible by $\mathfrak{L}$ and $\mathfrak{M}$ is contained in both $\mathfrak{L}$ and $\mathfrak{M}$ and is in $\mathfrak{D}$, is divisible by $\mathfrak{D}$. Thus the product $\mathfrak{L}\mathfrak{M}$ is a common multiple of $\mathfrak{L}$ and $\mathfrak{M}$ and is divisible by $\mathfrak{D}$.

DEFINITION. *An ideal* $\mathfrak{P}$ *not the unit or zero ideal is called a divisorless ideal if* $\mathfrak{P}$ *has no proper divisor except the unit ideal.*

LEMMA 1. *Let* $\mathfrak{P}$ *be a divisorless ideal and* a *in* $\mathfrak{A}$ *be not in* $\mathfrak{P}$. *Then there exists a quantity* h *in* $\mathfrak{A}$ *and a quantity* g *of* $\mathfrak{P}$ *such that*

$$1 = \mathrm{ha} + \mathrm{g}.$$

For we are actually proving that $\mathfrak{A}$ is the greatest common divisor $\mathfrak{Q} = (\mathfrak{P}, (a))$ of $\mathfrak{P}$ and (a). Thus $\mathfrak{Q}$ is the set of all quantities of the form $ha + g$ with h in $\mathfrak{A}$ and g in $\mathfrak{P}$. Since $\mathfrak{Q}$ contains $\mathfrak{P}$ it divides $\mathfrak{P}$. Also $\mathfrak{Q}$ is a proper divisor of $\mathfrak{P}$ since $\mathfrak{Q}$ contains a not in $\mathfrak{P}$. Thus $\mathfrak{Q}$ is the unit ideal $\mathfrak{A} = (1)$, $1 = ha + g$ is in $\mathfrak{A} = \mathfrak{Q} = (\mathfrak{P}, (a))$.

LEMMA 2. *A divisorless ideal $\mathfrak{P}$ divides a product* ab *of quantities of $\mathfrak{A}$ if and only if $\mathfrak{P}$ divides* a *or* b.

For if $\mathfrak{P}$ divides a then a is in $\mathfrak{P}$, ab is in $\mathfrak{P}$, $\mathfrak{P}$ divides ab. Conversely if $\mathfrak{P}$ divides ab and not a we apply Lemma 1 and obtain $1 = ha + g$, $b = hab + gb$ is a sum of elements of $\mathfrak{P}$ and is in $\mathfrak{P}$. Hence $\mathfrak{P}$ divides b.

The results above are used to prove

Theorem 1. *The difference ring $\mathfrak{A} - \mathfrak{M}$ of a commutative ring $\mathfrak{A}$ with a unity element and an ideal $\mathfrak{M}$ of $\mathfrak{A}$ is a field if and only if $\mathfrak{M}$ is a divisorless ideal.*

For the ring $\mathfrak{A} - \mathfrak{M}$ is a commutative ring with a unity element $[1]$, and is a field if and only if the equation $\alpha x = \beta$ is solvable for every $\alpha \neq 0$ and β of $\mathfrak{A} - \mathfrak{M}$. If $\mathfrak{M}$ is a divisorless ideal and $\alpha \neq [0]$ then $\alpha = [a]$ with a not in $\mathfrak{M}$. By Lemma 1 we have $b = (bh)a + bg$ so that $\beta = [b] = [bh] \cdot [a] = \alpha x$, where $x = [bh]$ is in $\mathfrak{A} - \mathfrak{M}$ as desired. Conversely, if $\mathfrak{A} - \mathfrak{M}$ is a field and $\mathfrak{Q}$ is a proper divisor of $\mathfrak{M}$ there exists an a in $\mathfrak{Q}$ and not in $\mathfrak{M}$. Hence $[a] \neq 0$, and for every b of $\mathfrak{A}$ there exists an x in $\mathfrak{A}$ such that $[ax] = [b]$. But ax is in $\mathfrak{Q}$ for every x, so that $\mathfrak{Q}$ contains every b of $\mathfrak{A}$, and $\mathfrak{Q} = \mathfrak{A}$ is the unit ideal. This proves that the only proper divisor of $\mathfrak{M}$ is the unit ideal, $\mathfrak{M}$ is a divisorless ideal.

DEFINITION. *A commutative ring with a unity element is called a principal ideal ring if every ideal of $\mathfrak{A}$ is a principal ideal.*

Every two quantities of a principal ideal ring have a greatest common divisor. For if a and b are in $\mathfrak{A}$ the set of all quantities $\lambda a + \mu b$ with λ and μ in $\mathfrak{A}$ is an ideal and hence a principal ideal (c), $\lambda a + \mu b = \nu c$. Thus a and b are in (c), $a = \nu_1 c$, $b = \nu_2 c$, and c divides both a and b. Also c is in the ideal generated by a and b and $c = \lambda a + \mu b$, every common divisor of a and b divides c.

The set $\mathfrak{R}$ of all rational numbers contains the ring $\mathfrak{E}$ of all rational integers. This latter ring is a principal ideal ring. For let $\mathfrak{M}$ be an ideal of $\mathfrak{E}$ and q be the least positive integer of $\mathfrak{M}$. Then if b is in $\mathfrak{M}$ and $b = aq + r$, $0 \leqq r < q$, we have aq in $\mathfrak{M}$, $r = b - aq$ in $\mathfrak{M}$. Hence $r = 0$, $b = aq$, $\mathfrak{M} = (q)$.

4. Fields with a valuation. Consider a field $\mathfrak{F}$ and a function ϕ on $\mathfrak{F}$ to an archimedean ordered field $\mathfrak{T}$. Then

$$\text{(1)} \qquad \phi(a) \text{ is in } \mathfrak{T}$$

for every a of $\mathfrak{F}$. The function ϕ is said to define a *valuation of* $\mathfrak{F}$ if

$$\phi(0) = 0\,, \quad \phi(a) > 0 \text{ for every } a \neq 0 \text{ of } \mathfrak{F}\,, \tag{2}$$

$$\phi(ab) = \phi(a)\phi(b) \qquad (a, b \text{ in } \mathfrak{F})\,, \tag{3}$$

$$\phi(a + b) \leqq \phi(a) + \phi(b) \qquad (a, b \text{ in } \mathfrak{F})\,. \tag{4}$$

From (2) we have $\phi(1) > 0$, $\phi(-1) > 0$. Then apply (3) and obtain $[\phi(1)]^2 = [\phi(-1)]^2 = \phi(1)$. This implies that

$$\phi(1) = \phi(-1) = 1 \text{ and } \phi(-a) = \phi(a)\,. \tag{5}$$

We also see that (4) implies that

$$\phi(n) \leqq n \tag{6}$$

for every integer n.

Replace b by $b - a$ in (4) and obtain $\phi(b) \leqq \phi(a) + \phi(b - a)$. Similarly, $\phi(a) \leqq \phi(b) + \phi(a - b)$. Since $\phi(b - a) = \phi(a - b)$ and $|\phi(a) - \phi(b)| = \phi(a) - \phi(b)$ or $\phi(b) - \phi(a)$ we have

$$|\phi(a) - \phi(b)| \leqq \phi(a - b)\,. \tag{7}$$

The single bars of course represent absolute value in $\mathfrak{T}$.

The valuation function ϕ is said to provide a *non-trivial* valuation of $\mathfrak{F}$ if

$$\phi(a_0) \neq 0, 1 \tag{8}$$

for some a_0 of $\mathfrak{F}$. The trivial valuation in which we assign to every non-zero element of $\mathfrak{F}$ the valuation 1 is evidently a valuation of any field and cannot possibly characterize any special properties of $\mathfrak{F}$. We shall therefore restrict our further attention to non-trivial valuations. We would actually make (8) one of the postulates for a valuation ϕ except that it is natural to desire that ϕ shall also provide a valuation of every subfield of $\mathfrak{F}$. This our ϕ does and the reader should verify this. But (8) may not hold since we have in particular the

Theorem 2. *The only valuation of a finite field is the trivial valuation.*

For if $\mathfrak{F}$ is a finite field there exists an integer $q > 0$ such that $x^q = 1$ for every $x \neq 0$ of $\mathfrak{F}$. Then $[\phi(x)]^q = 1$ so that since $\mathfrak{T}$ is ordered $\phi(x) = 1$, ϕ is the trivial valuation.

We now make the

DEFINITION. *Two fields $\mathfrak{F}$ and $\mathfrak{H}$ with valuation functions ϕ, ψ respectively are said to be analytically equivalent if they are equivalent under a correspondence*

$$\mathrm{f} \longleftrightarrow \mathrm{h} \qquad (\mathrm{f} \text{ in } \mathfrak{F}, \mathrm{h} \text{ in } \mathfrak{H})$$

such that for corresponding f *and* h *we have*

$$\phi(\mathrm{f}) = \psi(\mathrm{h}) .$$

We shall also use the

DEFINITION. *Two valuation functions ϕ and ψ of a field $\mathfrak{F}$ are said to provide equivalent valuations of $\mathfrak{F}$ if*

$$\phi(\mathrm{a}) > \phi(\mathrm{b})$$

for a *and* b *in $\mathfrak{F}$ if and only if*

$$\psi(\mathrm{a}) > \psi(\mathrm{b}) .$$

This terminology will be justified later (in Theorem 12.1).

5. Regular sequences of $\mathfrak{F}$. An infinite sequence

$$a = \{a_n\} \tag{9}$$

of quantities a_n of a field $\mathfrak{F}$ with a valuation function ϕ on $\mathfrak{F}$ to $\mathfrak{T}$ is called a *regular sequence* if for every $\epsilon > 0$ in $\mathfrak{T}$ there exists an integer n_ϵ such that

$$\phi(a_p - a_q) < \epsilon \qquad (p > n_\epsilon,\ q > n_\epsilon) . \tag{10}$$

Regular sequences are sometimes called *Cauchy sequences* or *fundamental sequences.*

A regular sequence has bounded elements. For (10) and (4) together with $a_p = (a_p - a_q) + a_q$ give

$$\phi(a_p) \leqq \phi(a_q) + \phi(a_p - a_q) < \phi(a_q) + \epsilon . \qquad (p, q > n_\epsilon) .$$

We take $q = n_\epsilon + 1$ and $\mu = \phi(a_q) + \epsilon$ and obtain

$$\phi(a_p) < \mu \qquad (p > n_\epsilon) . \tag{11}$$

This states that $\mu + \sum_{i=1}^{n_\epsilon} \phi(a_i)$ is an upper bound of the valuations of the elements of $a = \{a_n\}$.

If

(12) $$\alpha = \{a_n\}, \quad \beta = \{b_n\}$$

are regular we may choose integers n_1, n_2 such that $\phi(a_p) < \mu_1$ for $p > n_1$ and $\phi(b_p) < \mu_2$ for $p > n_2$. But then

(13) $$\phi(a_p) < \mu_1, \quad \phi(b_p) < \mu_2 \qquad (p > n_0),$$

where n_0 is the larger of n_1 and n_2. In a similar fashion we may select any $\epsilon_1 > 0$, $\epsilon_2 > 0$ and then always choose an integer n such that

(14) $$\phi(a_p - a_q) < \epsilon_1, \quad \phi(b_p - b_q) < \epsilon_2 \quad (p > n, q > n).$$

We now use these properties.

The *product*

(15) $$\alpha\beta = \gamma = \{c_n\}, \quad c_n = a_n b_n$$

of two regular sequences is a regular sequence. For we use (13) and if $\epsilon > 0$ is given take

$$\epsilon_1 = \frac{\epsilon}{2\mu_2}, \quad \epsilon_2 = \frac{\epsilon}{2\mu_1},$$

where $n > n_0$ in (14). But then by (3) we have

$$\phi(a_p b_p - a_q b_p) = \phi(b_p) \cdot \phi(a_p - a_q) < \phi(b_p)\frac{\epsilon}{2\mu_2} < \frac{\epsilon}{2} \quad (p > n, q > n),$$

and similarly

$$\phi(a_q b_p - a_q b_q) < \frac{\epsilon}{2} \qquad (p > n, q > n).$$

However, $a_p b_p - a_q b_q = (a_p b_p - a_q b_p) + (a_q b_p - a_q b_q)$ and thus

$$\phi(a_p b_p - a_q b_q) < \frac{\epsilon}{2} + \frac{\epsilon}{2} = \epsilon \qquad (p > n, q > n),$$

so that $\alpha\beta$ is regular.

The *sum*

(16) $$\alpha + \beta = \{a_n + b_n\}$$

of two regular sequences is regular. For we take $\epsilon_1 = \frac{1}{2}\epsilon$, $\epsilon_2 = \frac{1}{2}\epsilon$ in (14) and have

$$\phi(a_p + b_p - a_q - b_q) \leqq \phi(a_p - a_q) + \phi(b_p - b_q) < \epsilon \qquad (p > n,\ q > n)\ .$$

We also notice that any sequence

$$\delta = \{d_n\}\ , \qquad d = d_1 = d_2 = \ldots \tag{17}$$

in $\mathfrak{F}$ is a regular sequence all of whose elements are equal to d. Thus we define

$$\delta\alpha = \{da_n\} = d\alpha\ ,$$

which is regular for any d of $\mathfrak{F}$. In particular.

$$-\alpha = \{-a_n\}\ , \qquad 0 = \{0\}\ , \tag{18}$$

are regular and so is

$$\alpha - \beta\ ,$$

for any regular α and β. If α is regular and so is $\alpha - \beta$ then β is regular. For

$$\alpha - \beta = \{c_n\}, \qquad b_n = a_n - c_n\ ,$$

where $\alpha = \{a_n\}$, $\gamma = \{c_n\}$ are regular, $\beta = \alpha - \gamma$ is regular.

Addition and multiplication are commutative and associative in $\mathfrak{F}$ and our definitions (15), (16) imply that they are commutative and associative in the set $\mathfrak{A}$ of all regular sequences. Evidently $\mathfrak{A}$ is an additive abelian group with (18) holding and $\mathfrak{A}$ is closed with respect to multiplication. The distributive law of our field $\mathfrak{F}$ implies that

$$\alpha(\beta + \gamma) = \{a_n(b_n + c_n)\} = \{a_nb_n + a_nc_n\} = \alpha\beta + \alpha\gamma\ . \tag{19}$$

The element

$$1 = \{1\}$$

is a unity element of $\mathfrak{A}$ and we have proved

Theorem 3. *The set $\mathfrak{A}$ of all regular sequences of a field $\mathfrak{F}$ with a valuation ϕ is a commutative ring with a unity element.*

The ring $\mathfrak{A}$ is not in general a field. We shall however define an ideal $\mathfrak{N}$ of $\mathfrak{A}$ and shall obtain a field $\mathfrak{A} - \mathfrak{N}$.

6. The derived field of $\mathfrak{F}$. A sequence $\tau = \{t_n\}$ is called a *null sequence* if for every $\epsilon > 0$ in $\mathfrak{T}$ there exists an n_ϵ such that

$$\phi(t_p) < \epsilon \qquad (p > n_\epsilon) . \tag{20}$$

Then n_ϵ may be so chosen that $\phi(t_p) < \epsilon/2$ for $p > n_\epsilon$, and

$$\phi(t_p - t_q) \leqq \phi(t_p) + \phi(t_q) < \epsilon \qquad (p > n_\epsilon, q > n_\epsilon) \cdot \tag{21}$$

Thus a null sequence is regular. Moreover, if α is regular and $\alpha - \beta$ is a null sequence then $\tau = \alpha - \beta$ is regular, $\beta = \alpha - \tau$ is regular. Hence it is sufficient to state that α is regular and prove that $\alpha - \beta$ is a null sequence when we wish to prove that β is in our ring $\mathfrak{A}$ of regular sequences and differs from α by a null sequence.

We shall show that the set $\mathfrak{N}$ of all null sequences is a divisorless ideal of $\mathfrak{A}$. Then by Theorem 11.1 the difference ring $\mathfrak{A} - \mathfrak{N}$ is a field.

Let $\tau = \{t_n\}$, $\sigma = \{s_n\}$ be in $\mathfrak{N}$, and choose an integer n_ϵ so that

$$\phi(t_p) < \tfrac{1}{2}\epsilon , \qquad \phi(s_q) < \tfrac{1}{2}\epsilon \qquad (p > n_\epsilon, q > n_\epsilon) .$$

Then

$$\phi(t_p - s_p) < \epsilon \qquad (p > n_\epsilon) \tag{22}$$

and $\tau - \sigma$ is in $\mathfrak{N}$. If $\alpha = \{a_p\}$ is in $\mathfrak{A}$ we have proved the existence of a quantity μ in the archimedean ordered field $\mathfrak{T}$ and an integer n_0 such that $\phi(a_p) < \mu$ for $p > n_0$. Choose $n_\epsilon > n_0$ and such that $\phi(t_p) < \epsilon\mu^{-1}$, and obtain $\phi(t_p a_p) < \epsilon$ for $p > n_\epsilon$. Thus $\alpha\tau$ is in $\mathfrak{N}$ for every α of $\mathfrak{A}$ and τ of $\mathfrak{N}$. We have proved that $\mathfrak{N}$ is an ideal of $\mathfrak{A}$.

If $\mathfrak{Q}$ is a proper divisor of $\mathfrak{N}$ then $\mathfrak{Q}$ contains a regular sequence $\alpha = \{a_n\}$ which is not a null sequence. Thus for every $\epsilon > 0$ (7) implies that

$$\delta = \phi(a_p) - \epsilon < \phi(a_q) \qquad (p > n_0, q > n_0) .$$

Since α is not a null sequence we may select an ϵ and a $p_0 > n_0$ such that $\phi(a_{p_0}) \geqq 2\epsilon$. Take $p = p_0$ and have

$$\phi(a_q) > \delta > 0 \qquad (q > n_0) . \tag{23}$$

The quantity $a = a_{n_0+1}$ is in $\mathfrak{F}$ and $\phi(a) > \delta > 0$ so that $a \neq 0$. The sequence τ defined by

$$t_p = a_p - a , \quad t_q = 0 \quad (p = 1, \ldots , n_0; q > n_0)$$

is a null sequence and $\gamma = \alpha - \tau$ is in $\mathfrak{O}$. However $\gamma = \{c_n\}$ has the property that $\phi(c_n) > \delta > 0$ for all values of n and there exists a sequence $\beta = \{b_n\}$, $b_n = c_n^{-1}$ such that $\gamma\beta$ is the sequence all of whose elements are unity. But β is regular since if we choose n_0 so that $\phi(c_p - c_q) < \epsilon\delta^2$ for $p > n_0$, $q > n_0$, we have

$$\phi(c_p^{-1} - c_q^{-1}) = \phi(c_p^{-1}c_q^{-1})\phi(c_p - c_q) < \epsilon \quad (p > n_0\,,\, q > n_0)\,,$$

since $\phi(c_p) > \delta$. Hence $\{1\}$ is in $\mathfrak{O}$ and so is every $\alpha\{1\}$ with α in $\mathfrak{A}$. Thus $\mathfrak{O}$ is the unit ideal, $\mathfrak{N}$ is a divisorless ideal, $\mathfrak{A} - \mathfrak{N}$ is a field by Theorem 11.1.

The elements of $\mathfrak{A} - \mathfrak{N}$ are classes of equivalent regular sequences. Each class $A = [\alpha]$ consists of all regular sequences

$$\alpha + \tau \qquad (\tau \text{ in } \mathfrak{N})\,,$$

equivalent (or congruent modulo $\mathfrak{N}$) to α. In particular $\mathfrak{A} - \mathfrak{N}$ contains the classes $[\alpha]$ where α is equivalent to a regular sequence of all equal elements a. The set $\mathfrak{F}_0$ of all such classes is equivalent to our field $\mathfrak{F}$ under a correspondence

$$a_0 = [\{a\}] \longleftrightarrow a \qquad (a_0 \text{ in } \mathfrak{F}_0,\ a \text{ in } \mathfrak{F})\,.$$

We use the method of Theorem 1.9 to replace $\mathfrak{F}_0$ by $\mathfrak{F}$ and have replaced all classes $[\alpha]$ with α equivalent to $\{a\}$ by a. The resulting field

$$\mathfrak{F}_\phi$$

is called the *derived field* of $\mathfrak{F}$. It is a field over $\mathfrak{F}$ and all of its quantities not in $\mathfrak{F}$ are classes $[\alpha]$, where α is not equivalent to a sequence whose elements are all equal.

7. Certain conventions about derived fields. We have defined $\mathfrak{F}_\phi$ to be a field some of whose quantities are quantities of $\mathfrak{F}$ and the remaining ones classes of regular sequences with elements in $\mathfrak{F}$. This convention was adopted so as to make $\mathfrak{F}_\phi$ contain $\mathfrak{F}$ as a subfield. We now make a further agreement. Represent *every* element A of $\mathfrak{F}_\phi$ by the notation

$$A = [\alpha]$$

where α is a regular sequence. When α is equivalent to a sequence $\{a_n\}$ with all the $a_n = a$ in $\mathfrak{F}$ then A is the quantity a of $\mathfrak{F}$. Thus we identify all classes $A = [\alpha]$, where α is equivalent to a sequence with all equal elements a, with a. In particular if any sequence of A is a null sequence the element

$$A = 0.$$

Let $\mathfrak{K}$ be a field with a valuation ϕ and

$$\mathfrak{K} > \mathfrak{F}.$$

The subfield $\mathfrak{F}$ of $\mathfrak{K}$ also has ϕ as a valuation function. Then

$$\mathfrak{K}_\phi, \qquad \mathfrak{F}_\phi$$

are defined. In general $\mathfrak{K}_\phi$ does not contain $\mathfrak{F}_\phi$. For the quantities of $\mathfrak{K}_\phi$ are the elements of $\mathfrak{K}$ and classes $B_0 = [\beta]$ where the sequences of B_0 have the form

$$\beta + \sigma$$

with σ any null sequence of $\mathfrak{K}$. The case where β is a regular sequence α of $\mathfrak{F}$ implies that in $\mathfrak{F}_\phi$

$$A = [\alpha]$$

consists of all $\alpha + \tau$ with τ a null sequence of $\mathfrak{F}$. Now there are generally null sequences of $\mathfrak{K}$ which are not null sequences of $\mathfrak{F}$. Thus the correspondence

$$A_0 = [\alpha] \text{ of } \mathfrak{K}_\phi \longleftrightarrow A = [\alpha] \text{ of } \mathfrak{F}$$

between $\mathfrak{F}_\phi$ and the set of all A_0 with α a regular sequence of $\mathfrak{F}$ provides an equivalence between $\mathfrak{F}_\phi$ and a subfield of $\mathfrak{K}_\phi$. However, the quantities of this subfield are not quantities of $\mathfrak{F}_\phi$ in general.

We shall agree to replace this subfield of $\mathfrak{K}_\phi$ equivalent to $\mathfrak{F}_\phi$ by $\mathfrak{F}_\phi$ whenever we are talking about a field $\mathfrak{K} > \mathfrak{F}$ with a valuation. Thus we define $\mathfrak{K}_\phi$ so that

$$\mathfrak{K}_\phi \geqq \mathfrak{F}_\phi.$$

We now see that the quantities of $\mathfrak{K}_\phi$ consist of the quantities of $\mathfrak{K}$, the quantities of $\mathfrak{F}_\phi$, and classes

$$[\beta]$$

where β is not equivalent either to a regular sequence of $\mathfrak{F}$ or to a regular sequence of $\mathfrak{K}$ with equal elements.

We shall later obtain a valuation Φ of $\mathfrak{F}_\phi$ which preserves that of $\mathfrak{F}$. This is similarly true of $\mathfrak{K}_\phi$ if $\mathfrak{K} > \mathfrak{F}$. We then easily see that the above replacement replaces our original $\mathfrak{K}_\phi$ by a new field which is analytically equivalent to $\mathfrak{K}_\phi$.

DEFINITION. *A field $\mathfrak{F}$ with a valuation ϕ is called complete if $\mathfrak{F}_\phi = \mathfrak{F}$, that is, every regular sequence of $\mathfrak{F}$ has a limit in $\mathfrak{F}$.*

One of our goals is to prove that the above-mentioned valuation Φ has the property

$$(\mathfrak{F}_\phi)_\Phi = \mathfrak{F}_\phi ,$$

that is, that *the derived field of any $\mathfrak{F}$ is complete with respect to Φ.*

8. Limits in a complete field. Let $\mathfrak{F}$ be a complete field with respect to a valuation ϕ on $\mathfrak{F}$ to Ω. Then a sequence $\alpha = \{a_n\}$ of quantities a_n of $\mathfrak{F}$ is said to have a *limit* a in $\mathfrak{F}$ if for every $\epsilon > 0$ in Ω there exists an n_ϵ such that

$$\phi(a_p - a) < \epsilon \qquad (p > n_\epsilon) . \tag{24}$$

This will be designated by writing

$$\lim_{n \to \infty} a_n = a . \tag{25}$$

We of course have

Theorem 4. *The* $\lim\limits_{n \to \infty} a_n = a$ *if and only if* $\{a_n\}$ *is a regular sequence.*

For the definition above states that the sequence $\tau = \{a_n - a\}$ is a null sequence. Since $a = \{a\}$ is regular so is $\{a_n\} = a + \tau$. Conversely if $\alpha = \{a_n\}$ is regular the class $[\alpha]$ is a quantity a of the complete field $\mathfrak{F}$, that is α is equivalent to a sequence $a = \{a\}$. Hence $\alpha - a$ is a null sequence and (24) holds.

We shall designate the property (25) when we say that the sequence $\alpha = \{a_n\}$ converges to a. This will of course mean that α differs from the sequence whose elements are all a by a null sequence, and that $[\alpha] = a$ in the complete field $\mathfrak{F}$. We see trivially that α uniquely determines a.

9. The case $\mathfrak{F} = \mathfrak{T}$. Let $\mathfrak{F}$ be archimedean ordered. We take $\mathfrak{F} = \mathfrak{T}$,

$$\phi(a) = |a| = a \text{ or } -a \text{ in } \mathfrak{F} .$$

Designate the derived field $\mathfrak{F}_\phi$ of $\mathfrak{F}$ with respect to this simple valuation by $\mathfrak{F}'$. We shall deduce an archimedean ordering of $\mathfrak{F}'$ preserving that of $\mathfrak{F}$ and hence a valuation of $\mathfrak{F}'$ preserving that of $\mathfrak{F}$.

DEFINITION. *A regular sequence* $\alpha = \{a_n\}$ *is called positive if there exists an* $\epsilon > 0$ *in* $\mathfrak{F}$ *and an integer* n_ϵ *such that*

$$a_p > \epsilon \qquad (p > n_\epsilon) . \tag{26}$$

This definition will now be used to provide an archimedean ordering of $\mathfrak{F}'$. In fact we prove

Theorem 5. *Let* $\mathfrak{F}'_P$ *be the set of all classes* A *every sequence of which is positive, and* $\mathfrak{F}'_N$ *be all remaining* $A \neq 0$ *of* $\mathfrak{F}'$. *Then this classification of* $\mathfrak{F}'$ *into positive and negative classes and zero provides an archimedean ordering of* $\mathfrak{F}'$ *in which the ordering of* $\mathfrak{F}$ *is preserved.*

For, let $\tau = \{t_n\}$ be a null sequence. Then for every $\epsilon > 0$ there exists an n_0 such that

$$|t_p| < \epsilon \qquad (p > n_0) . \tag{27}$$

If τ is positive the above definition shows that $t_p > \epsilon$ for $p > n_\epsilon$. But we choose ϵ in (27) to be that of (26) and take n larger than both n_ϵ and n_0 so that $|t_p| < \epsilon$, $t_p > \epsilon$ $(p > n)$, which is impossible. Thus no null sequence is positive and the zero element of $\mathfrak{F}'$ is not in $\mathfrak{F}'_P$.

Next let $A = [\alpha]$ where α is positive. We shall actually prove that $\alpha + \tau$ is always positive for every null sequence τ. Select n_0 so that $|t_p| < \frac{1}{2}\epsilon$ for ϵ as in (26) and $p > n_0$ and choose $n > n_0, n_\epsilon$. Then $a_p + t_p > \frac{1}{2}\epsilon$ for $p > n$ since $a_p > \epsilon$, $a_p + t_p \geqq a_p - |t_p| > \frac{1}{2}\epsilon$ when $|t_p| < \frac{1}{2}\epsilon$. Thus $\alpha + \tau$ is positive as desired. Hence if any α of A is positive A is in $\mathfrak{F}'_P$.

We next show that every A of $\mathfrak{F}'_N$ has the property that $-A$ is in $\mathfrak{F}'_P$. For now $A = [\alpha]$ where α is not a null sequence and is not positive. Then (23) holds and there exists an integer n_0 and a $\delta > 0$ such that $|a_q| > \delta$ for $q > n_0$. But now δ is in $\mathfrak{F}$ and

$$a_q > \delta > 0 \text{ or } -a_q > \delta > 0 \qquad (q > n_0) .$$

If $a_q > \delta > 0$ for $q > n_0$ the sequence α is positive, a contradiction. If $-a_q > \delta > 0$ for $q > n_0$ then $-\alpha$ is positive as desired. There remains the case where for every $n_1 > n_0$ there exists an integer $p > n_1$ and an integer $q > n_1$ such that

$$a_p > \delta , \qquad -a_q > \delta ,$$

and thus

$$a_p - a_q = |a_p - a_q| > \delta .$$

But α is regular and there exists an $n_1 > n_0$ such that

$$|a_p - a_q| < \delta ,$$

for every $p > n_1$ and $q > n_1$. This is a contradiction and we have proved our second ordering postulate of Section 5.9.

There remains only III of Section 5.9 since our definition states that $\mathfrak{F}'_N$ consists of all elements in $\mathfrak{F}'$ not zero and not in $\mathfrak{F}'_P$ and this is the postulate I. Now $A = [\alpha]$, $B = [\beta]$ implies that $A + B = [\alpha + \beta]$, $AB = [\alpha\beta]$. If $\alpha > 0$, $\beta > 0$ we find n_1, n_2, δ_1, δ_2 such that $a_p > \delta_1 > 0$, $b_q > \delta_2 > 0$ for $p > n_1$, $q > n_2$. We select the larger n of n_1 and n_2, and have $a_p > \delta_1$, $b_p > \delta_2$, so that $a_p + b_p > \delta_1 + \delta_2 > 0$, $a_pb_p > \delta_1\delta_2 > 0$ for $p > n$.

Finally, if a is in $\mathfrak{F}$ the sequence $a, a, \ldots$ defines an element A of $\mathfrak{F}'$ which is either in $\mathfrak{F}'_P$, is zero, or is in $\mathfrak{F}'_N$, according as $a > 0$, $a = 0$, or $a < 0$. Thus our ordering of $\mathfrak{F}'$ preserves that of $\mathfrak{F}$.

We shall use the notation $A > B$ as in Section 5.9. Then for every A of $\mathfrak{F}'$ there exists an a in $\mathfrak{F}$ such that

$$a > A .$$

For $A = [\alpha]$ where $\alpha = \{a_n\}$ is regular and by (11) there exists a quantity b in $\mathfrak{F}$ and an integer n_0 such that $|a_p| < b$ for $p > n_0$. We choose $c > 0$ in $\mathfrak{F}$ and put $a = b + c$ so that $a - |a_p| > c > 0$. Then the difference

$$\{a\} - \{a_n\}$$

is a positive sequence, since $|a - a_p| \geqq |a - |a_p|| > c > 0$ for every $p > n_0$ by (7). Hence $a > A$. But $\mathfrak{F}$ is archimedean ordered and there exists an integer $n > a$. Thus $n > A$ and $\mathfrak{F}'$ is archimedean ordered.

The archimedean ordered field $\mathfrak{F}$ contains a subfield ordered equivalent to the field $\mathfrak{R}$ of all rational numbers and there is no loss of generality if we assume that $\mathfrak{F} \geqq \mathfrak{R}$. We do this and prove the important

Theorem 6. *Every quantity* A *of the derived field* $\mathfrak{F}'$ *of an archimedean ordered field* $\mathfrak{F}$ *may be represented by a class*

$$\mathrm{A} = [\alpha] , \qquad \alpha = \{\mathrm{a}_n\} ,$$

where the a_n *are rational numbers.*

It is sufficient to prove this result when A is positive since if $A = 0$ we take the $a_n = 0$ and if A is negative we make the proof for $-A = [-\alpha] > 0$, $-\alpha = \{-a_n\}$, with rational a_n. Let β_0 be a representative of the class $A > 0$ so that the elements of β_0 are $b_p > \epsilon > 0$ for $p > n_0$. The sequences $\tau = \{t_p\}$ which have $t_p = 0$ for $p > n_0$ are null sequences for arbitrary $t_1, t_2, \ldots, t_{n_0}$. We may therefore replace β_0 by $\beta = \beta_0 + \tau$ where β now has the property that every $b_n > \epsilon > 0$. There exists an integer greater than $2^n b_n$ and we choose h_n to be the smallest such integer, that is, an integer such that $h_n - 1 \leqq 2^n b_n < h_n$. By addition of $1 - h_n$ we have

$$0 \leqq 2^n b_n - (h_n - 1) < 1 .$$

Multiply by 2^{-n} and obtain

$$0 \leqq b_n - a_n < 2^{-n} \tag{28}$$

where

$$a_n = 2^{-n}(h_n - 1) \tag{29}$$

is rational. For every $\epsilon > 0$ there exists an integer n_ϵ such that $0 < 2^{-p} < \epsilon$ for all $p > n_\epsilon$. We apply (28) to obtain

$$0 \leqq b_p - a_p < \epsilon \qquad (p > n_\epsilon) ,$$

so that the sequence with terms $b_p - a_p$ is a null sequence. Since $\beta = \{b_n\}$ is regular so is $\alpha = \{a_n\}$ and α is equivalent to β. This proves that $A = [\alpha]$ where α is a regular sequence of rational numbers.

10. The real number system. The field $\mathfrak{R}$ of all rational numbers was shown in Section 5.9 to have a unique archimedean ordering and thus a particular valuation $\phi(a) = |a|$ for every rational a. We defined $\mathfrak{R}_\phi = \mathfrak{R}'$ for this ϕ. We now make the

DEFINITION. *The real number system is the derived field $\mathfrak{R}'$ of the field $\mathfrak{R}$ of all rational numbers. Its quantities are called real numbers.*

Archimedean ordered fields were defined abstractly and we assumed the existence of only one such field, the field $\mathfrak{R}$ of all rational numbers. We have now obtained the field $\mathfrak{R}'$ of all real numbers as the derived field $\mathfrak{R}'$ of $\mathfrak{R}$ with respect to a valuation with the ordered field $\mathfrak{T}$ the rational number field itself. But we now show that $\mathfrak{R}'$ is essentially a field containing every archimedean ordered field and this will allow us henceforth to define valuations ϕ on $\mathfrak{F}$ to $\mathfrak{R}'$.

Theorem 7. *The derived field $\mathfrak{F}'$ of any archimedean ordered field $\mathfrak{F}$ is complete and is ordered equivalent to the field $\mathfrak{R}'$ of all real numbers.*

For $\mathfrak{F}$ is non-modular and we may assume with no loss of generality that $\mathfrak{F}$ contains the subfield $\mathfrak{R}$ of $\mathfrak{R}'$. The only ordering of $\mathfrak{R}$ is the natural order and the derived field $\mathfrak{R}'$ of $\mathfrak{R}$ with respect to the valuation of Section 11.9 for $\mathfrak{F}$ is ordered equivalent to the field of all real numbers. As in Section 11.7 we may assume that $\mathfrak{F}' \geqq \mathfrak{R}'$. But by Theorem 11.6 every regular sequence of quantities of $\mathfrak{F}$ is equivalent to a sequence of rational numbers. Our convention about $\mathfrak{F}'$ in Section 11.7 then states that the choice $\mathfrak{F}' \geqq \mathfrak{R}'$ implies that every quantity of $\mathfrak{F}'$ is in $\mathfrak{R}'$, $\mathfrak{F}' = \mathfrak{R}'$. We have of course obtained this result by replacing the original $\mathfrak{F}'$ of classes of regular sequences of quantities of $\mathfrak{F}$ by an ordered equivalent field. Put $\mathfrak{F} = \mathfrak{R}'$. Evidently $\mathfrak{F}' = (\mathfrak{R}')' = \mathfrak{R}'$ is complete.

We have now arrived at the goal of the first part of our program. Real numbers have been determined by rational numbers and we may now replace the $\mathfrak{T}$ of values of valuation functions ϕ by an equivalent field whose elements are real numbers. Since now $\mathfrak{T} \leqq \mathfrak{R}'$ we may define a valuation ϕ to be a function on $\mathfrak{F}$ to $\mathfrak{R}'$ and in fact one whose functional values are in the set of non-negative real numbers. We shall use the elementary properties of real numbers. In particular we assume that the $\lim_{n \to \infty} \rho^n = 0$ if $0 < \rho < 1$, ρ is a real number. This is of course justified in elementary real analysis.

11. A valuation of $\mathfrak{F}_\phi$. We consider an arbitrary field $\mathfrak{F}$ with a valuation ϕ on $\mathfrak{F}$ to the field $\mathfrak{R}'$ of all real numbers, and wish to define a function

$$\Phi \text{ on } \mathfrak{F}_\phi \text{ to } \mathfrak{R}',$$

where $\mathfrak{F}_\phi$ is the derived field of $\mathfrak{F}$ defined by ϕ, such that

$$\Phi(a) = \phi(a) \tag{30}$$

for every a of $\mathfrak{F}$.

Let A be in $\mathfrak{F}_\phi$ so that we may represent A by $[\alpha]$ where $\alpha = \{a_n\}$ is a regular sequence of the class A. Then for every real positive ϵ there exists an n_ϵ such that

$$\phi(a_p - a_q) < \epsilon \qquad (p > n_\epsilon,\ q > n_\epsilon).$$

By (7) we have

$$|\phi(a_p) - \phi(a_q)| < \epsilon \qquad (p > n_\epsilon,\ q > n_\epsilon), \tag{31}$$

so that the sequence of real numbers

$$\Gamma = \{\phi(a_p)\}$$

is a regular sequence of $\mathfrak{R}'$ and defines a real number

$$\Phi(A) = [\Gamma] = \lim_{n \to \infty} \phi(a_n). \tag{32}$$

We shall prove

Theorem 8. *The function Φ defined in* (32) *is a function on $\mathfrak{F}_\phi$ to $\mathfrak{R}'$ and provides a valuation of $\mathfrak{F}_\phi$ in which* (30) *holds.*

For let $\beta = \alpha + \tau$ where τ is a null sequence $\{t_n\}$. Then by our definitions it is true that for every real positive ϵ there exists an integer n_0 such that $\phi(t_p) < \epsilon$ for $p > n_0$. Thus $\beta = \{b_n\}$, $t_n = b_n - a_n$,

$$|\phi(b_p) - \phi(a_p)| \leqq \phi(b_p - a_p) < \epsilon \qquad (p > n_0) .$$

The sequence

$$T + \Gamma = \{\phi(b_p)\}$$

is a sequence of non-negative numbers equivalent to Γ and $\Phi(A) = [\Gamma + T]$ is independent of the particular representative α of A. This proves that $\Phi(A)$ is on $\mathfrak{F}_\phi$ to $\mathfrak{R}'$. When $\alpha = \{a_n\}$, $a_n = a$, we have $\Gamma = \{\phi(a)\}$, $[\Gamma] = \phi(a) = \Phi(A)$ as desired. It remains to prove (2), (3), (4), (8).

Property (8) holds for Φ since ϕ is a non-trivial valuation, $\phi(a_0) \neq 1$ for some $a_0 \neq 0$ in $\mathfrak{F}$, $\Phi(a_0) = \phi(a_0) \neq 1$ for $a_0 \neq 0$ of $\mathfrak{F}_\phi \geqq \mathfrak{F}$. Also $\Phi(0) = \phi(0) = 0$. If $A \neq 0$ and $\alpha = \{a_n\}$ is a sequence of A it is not a null sequence. By (23) there exists an n_0 and a real number $\delta > 0$ such that

$$\phi(a_p) > \delta > 0 \qquad (p > n_0) .$$

The real sequence $\Gamma = \{\phi(a_p)\}$ is thus positive and $\Phi(A) = [\Gamma] > 0$. This proves (2).

Let $A = [\alpha]$, $B = [\beta]$ so that $AB = [\alpha\beta]$ and $\Phi(AB) = [\{\phi(a_n b_n)\}] = [\{\phi(a_n) \cdot \phi(b_n)\}] = [\{\phi(a_n)\} \cdot \{\phi(b_n)\}] = \Phi(A) \cdot \Phi(B)$. Similarly, the property $\phi(a_n + b_n) \leqq \phi(a_n) + \phi(b_n)$ implies that $\Phi(A + B) \leqq \Phi(A) + \Phi(B)$, so that we have (3), (4) and our theorem.

We may now prove

Theorem 9. *The derived field $\mathfrak{F}_\phi$ of a field $\mathfrak{F}$ with a valuation ϕ is complete with respect to the valuation Φ of* (32).

For let $\{A^{(p)}\}$ be a regular sequence of quantities $A^{(p)} = [\alpha^{(p)}]$ of $\mathfrak{F}_\phi$. Each

$$\alpha^{(p)} = \{a_n^{(p)}\}$$

is a regular sequence of quantities $a_n^{(p)}$ of $\mathfrak{F}$. This implies that for every $\epsilon > 0$ there exists an integer n_0 such that

$$\phi(a_n^{(p)} - a_m^{(p)}) < \epsilon \qquad (n > n_0,\ m > n_0) .$$

Take $\epsilon = 2^{-(p+1)}$, $m = n_0 + 1$ and define

$$a^{(p)} = a_m^{(p)} .$$

If we replace ϕ by Φ we have

$$\Phi(a_n^{(p)} - a^{(p)}) < 2^{-(p+1)} \qquad (n > n_0) .$$

We next use the notation $\mathfrak{a}^{(p)} = \{a_\mu^{(p)}\}$ and notice that

$$A^{(p)} - a_n^{(p)} = [\{a_\mu^{(p)} - a_n^{(p)}\}] .$$

But for every $\epsilon > 0$ there exists an $n_\epsilon > n_0$ such that

$$\phi(a_\mu^{(p)} - a_n^{(p)}) < \tfrac{1}{2}\epsilon \qquad (\mu > n_\epsilon,\ n > n_\epsilon) ,$$

and for every n, $\Phi(A^{(p)} - a_n^{(p)})$ is the $\lim\limits_{\mu\to\infty} \phi(a_\mu^{(p)} - a_n^{(p)})$. We fix $n > n_\epsilon$ and see that each $\phi(a_\mu^{(p)} - a_n^{(p)}) < \frac{1}{2}\epsilon$ so that certainly the limit

$$\Phi(A^{(p)} - a_n^{(p)}) < \epsilon \qquad (n > n_\epsilon) .$$

We again take $\epsilon = 2^{-(p+1)}$ and have $A^{(p)} - a^{(p)} = A^{(p)} - a_n^{(p)} + a_n^{(p)} - a^{(p)}$, so that

$$\Phi(A^{(p)} - a^{(p)}) \leqq \Phi(A^{(p)} - a_n^{(p)}) + \Phi(a_n^{(p)} - a^{(p)}) < 2^{-p}$$

for all values of p.

For every $\epsilon > 0$ there exists an integer p_ϵ such that $2^{-p} < \epsilon$ for $p > p_\epsilon$. Hence $\Phi(A^{(p)} - a^{(p)}) < \epsilon$ for $p > p_\epsilon$ and the sequence

$$\mathfrak{a}_0 = \{a^{(p)}\}$$

is a regular sequence equivalent to $\{A^{(p)}\}$. The class $[\mathfrak{a}_0]$ is the general quantity of the derived field of $\mathfrak{F}_\phi$, where $\mathfrak{a}_0$ is a sequence of quantities of $\mathfrak{F}$ and $[\mathfrak{a}_0]$ is in $\mathfrak{F}_\phi$ by our convention made earlier. Thus $\mathfrak{F}_\phi = (\mathfrak{F}_\phi)_\Phi$ is complete.

If $\mathfrak{F}_\phi$ is the derived field of $\mathfrak{F}$ and $\mathfrak{K}$ is any complete field containing $\mathfrak{F}$ we have agreed to take $\mathfrak{K}_\phi \geqq \mathfrak{F}_\phi$. But $\mathfrak{K} = \mathfrak{K}_\phi$ so that we always have $\mathfrak{K} \geqq \mathfrak{F}_\phi$. This result may be stated as

Theorem 10. *The derived field $\mathfrak{F}_\phi$ of any field $\mathfrak{F}$ is the smallest complete extension of $\mathfrak{F}$ with respect to ϕ in the sense of analytic equivalence.*

12. Algebraic extensions. We shall prove

Theorem 11. *Let $\mathfrak{K}$ be algebraic of finite degree* n *over $\mathfrak{F}$ and let $\mathfrak{K}$ have a valuation ϕ such that $\mathfrak{F}_\phi = \mathfrak{F}$. Then $\mathfrak{K}_\phi = \mathfrak{K}$ is also complete.*

For $\mathfrak{K}$ is obtainable from $\mathfrak{F}$ by a finite number of successive simple extensions $\mathfrak{F}_{i+1} = \mathfrak{F}_i(\xi_i)$ of $\mathfrak{F}$. It is evident that it is sufficient to prove the theorem for the case where $\mathfrak{K} = \mathfrak{F}(\xi)$ is a simple extension of $\mathfrak{F}$, ξ is a root of an irreducible equation with coefficients in $\mathfrak{F}$ and degree n, and $1, \xi, \ldots, \xi^{n-1}$ form a basis of $\mathfrak{K}$ over $\mathfrak{F}$. Every quantity η of $\mathfrak{K}$ has the form

$$\eta = a_0 + a_1\xi + \ldots + a_{n-1}\xi^{n-1} \qquad (a_i \text{ in } \mathfrak{F}) , \tag{33}$$

and if t is the largest integer for which $a_t \neq 0$ we call t the degree of η.

Let $\mathfrak{K}_\phi$ be the derived field of $\mathfrak{K}$ so that if $A = [a]$ is in $\mathfrak{K}_\phi$ we have $a = \{\eta_p\}$, η_p in $\mathfrak{K}$. Define a_{ip} as in (33) for the η_p. If the sequences

$$\{a_{ip}\} \qquad (i = 0, \ldots, n-1) ,$$

are all regular sequences of $\mathfrak{F}$ there are quantities a_i of $\mathfrak{F}$ such that $a_i = [\{a_{ip}\}]$. For $\mathfrak{F}$ is complete. But define η by (33) for these a_i and see that

$$\phi(\eta - \eta_p) \leqq \phi(a_0 - a_{0p}) + \phi(a_1 - a_{1p})\phi(\xi) + \ldots + \phi(a_{n-1} - a_{n-1\,p})\phi(\xi^{n-1}) .$$

Since $\xi \neq 0$ we have $\phi(\xi) = \gamma > 0$. Thus for every $\epsilon > 0$ we may choose integers n_i such that

$$\phi(a_i - a_{ip}) < \frac{\epsilon}{n\gamma^i} \qquad (p > n_i) .$$

We take $n_\epsilon > n_i$ $(i = 0, \ldots, n-1)$ and see that

$$\phi(\eta - \eta_p) < \epsilon \qquad (p > n_\epsilon) ,$$

so that $\{\eta - \eta_p\}$ is a null sequence. But then $A = [a] = [\beta]$, where $\beta = \{\eta\}$ has all its quantities equal to η in $\mathfrak{K}$ and, by the convention we have adopted, $\eta = A$ is in $\mathfrak{K}$. Moreover, the degree of A is the largest subscript i for which $\{a_{ip}\}$ is not a null sequence. Thus A is in $\mathfrak{K}$ when the coefficient sequences are all regular.

We now let $\{\eta_p\}$ be regular and prove that $\mathfrak{K} = \mathfrak{K}_\phi$ by proving that all the corresponding sequences $\{a_{ip}\}$ of coefficients are regular. Suppose this has been proved true for the η_p all of degree at most $t - 1$ as it has been for $t = 0$, and assume that $\{\eta_p\}$ is regular, the degrees of the η_p are at most t, and that one sequence $\{a_{ip}\}$ of coefficients is not regular. If $\{a_{tp}\}$ is regular so is $\{a_{tp}\xi^t\}$ and so is $\{\eta_p - a_{tp}\xi^t\}$ whose elements have degrees at most $t - 1$, a contradiction. Hence if we define

$$b_p \equiv a_{tp}$$

the sequence $\{b_p\}$ is not regular. Then there exists a real number $\delta > 0$, an increasing sequence of integers

$$n_1 < n_2 < n_3 < \ldots ,$$

and a set of integers $q_1, q_2, \ldots$, such that

$$\phi(b_{n_i} - b_{n_i+q_i}) > \delta > 0 \qquad (i = 1, 2, \ldots) . \tag{34}$$

This is an evident statement of the hypothesis that $\{b_p\}$ is not regular. We define the quantities

$$\psi_i = \psi_i(\xi) = \frac{\eta_{n_i} - \eta_{n_i+q_i}}{b_{n_i} - b_{n_i+q_i}} \qquad (i = 1, \ldots) \tag{35}$$

of $\mathfrak{K}$. They are polynomials in ξ with coefficients in $\mathfrak{F}$ and leading coefficient unity.

The sequence $\{\eta_p\}$ is regular and for every positive real ϵ there exists an integer p_0 such that

$$\phi(\eta_p - \eta_q) < \delta\epsilon \qquad (p > p_0,\ q > p_0) . \tag{36}$$

The n_i are an infinite increasing sequence of integers and thus there exists an integer h for which $n_h > p_0$. We choose $p = n_i > n_h$, $q = n_i + q_i > p_0$ in (36), and see that by (35) and (34)

$$\phi(\psi_i) = \frac{\phi(\eta_p - \eta_q)}{\phi(b_p - b_q)} < \frac{\phi(\eta_p - \eta_q)}{\delta} < \epsilon \qquad (i > h) . \tag{37}$$

Hence the sequence ψ_i is a regular sequence and is in fact a null sequence. But by (35) the polynomials ψ_i have leading coefficient unity and our proof above shows that there exists a quantity ψ of degree t and leading coefficient unity, such that $\{\psi\}$ is equivalent to $\{\psi_i\}$. However, $\psi \neq 0$ and $\{\psi\}$ is not a null sequence. This contradiction completes our induction and proves our theorem.

We now let $\mathfrak{K}$ be any field with a valuation ϕ and assume that $\mathfrak{K} = \mathfrak{F}(\xi)$ is algebraic of degree n over $\mathfrak{F}$. Then we have defined $\mathfrak{K}_\phi$ so that $\mathfrak{K}_\phi$ contains $\mathfrak{F}_\phi$ and $\mathfrak{K}$ and thus $\mathfrak{F}_\phi(\xi)$. By the theorem above the field $\mathfrak{F}_\phi(\xi)$ is complete. But by Theorem 11.10 $\mathfrak{F}_\phi(\xi) \geqq \mathfrak{K}_\phi$. Hence we have

Theorem 12. *Let $\mathfrak{K}$ be a field with a valuation ϕ and let $\mathfrak{K} = \mathfrak{F}(\xi)$ be algebraic of degree* n *over $\mathfrak{F}$. Then*

$$\mathfrak{K}_\phi = \mathfrak{F}_\phi(\xi) .$$

A field $\mathfrak{K} = \mathfrak{F}(\xi)$ which is a simple algebraic extension of degree n of $\mathfrak{F}$ is defined by a root ξ of $f(x) = 0$ where $f(x)$ is a monic irreducible polynomial of degree n of $\mathfrak{F}[x]$. We saw in Chapter VII that if $\mathfrak{F}'$ is a field containing $\mathfrak{F}$ the composites $(\mathfrak{K}, \mathfrak{F}')$ are algebraic over $\mathfrak{F}'$ but are not all equivalent. Thus the derived field $\mathfrak{K}'$ of $\mathfrak{K}$ is not uniquely determined by $f(x)$ and $\mathfrak{F}' = \mathfrak{F}_\phi$. However when $\mathfrak{K}$ is normal over $\mathfrak{F}$ the field $\mathfrak{F}'(\xi)$ is unique in the sense of equivalence and $\mathfrak{K}' = \mathfrak{F}'(\xi)$ is uniquely determined by the valuation of $\mathfrak{F}'$ and the polynomial $f(x)$.

The above reasoning gives a construction of $\mathfrak{K}'$ without obtaining a valuation of $\mathfrak{K}$. For we assume that $\mathfrak{F}$ has a valuation and obtain the irreducible factors $f_i(x)$ of $f(x)$ in $\mathfrak{F}'[x]$. Each $f_i(x)$ defines an essentially unique (in the sense of equivalence) field $\mathfrak{F}'[\xi_i]$ and every $\mathfrak{K}'$ is a field $\mathfrak{F}'[\xi_i]$. Moreover these are all equivalent when $\mathfrak{K}$ is normal over $\mathfrak{F}$. We now pass to a consideration of the case $n = 2$.

13. Quadratic extensions of a complete field. Consider an equation

$$x^2 + \gamma x + \delta = 0\,, \tag{38}$$

with coefficients γ, δ in a complete field $\mathfrak{F}$ with a valuation ϕ. Then we may prove

Theorem 13. *The equation* (38) *is reducible in* $\mathfrak{F}$ *if*

$$\phi(\gamma^2) - 4\phi(\delta) > 0\,. \tag{39}$$

For if $\gamma = 0$ then $\phi(\gamma) = 0$ and $-\phi(\delta) > 0$ which is impossible. Hence $\gamma \neq 0$ has an inverse in $\mathfrak{F}$ and

$$1 > 4\phi(\delta\gamma^{-2})\,. \tag{40}$$

We define an infinite sequence u_n by

$$u_{-1} = 0\,, \qquad u_0 = 2\,, \qquad u_n = -\gamma u_{n-1} - \delta u_{n-2} \tag{41}$$

for all positive integral values of n. Then by (40) we have

$$\phi(\gamma)\phi(u_{n-2}) \geqq 4\phi(\gamma)\phi(\delta\gamma^{-2})\phi(u_{n-2}) = 4[\phi(\gamma)]^{-1}\phi(\delta u_{n-2})\,. \tag{42}$$

We shall show that

$$2\phi(u_n) \geqq \phi(\gamma)\phi(u_{n-1})\,. \tag{43}$$

This is true for $n = 1$ since $u_0 = 2$, $u_1 = -2\gamma$, $2\phi(-2\gamma) = 2\phi(-2)\phi(\gamma) = 2\phi(2)\phi(\gamma) > \phi(\gamma)\phi(2)$. Assume the truth of (43) for $n - 1$ and thus that

$2\phi(u_{n-1}) \geqq \phi(\gamma)\phi(u_{n-2})$. Use (42) to obtain $2\phi(u_{n-1}) \geqq 4[\phi(\gamma)]^{-1}\phi(\delta u_{n-2})$, $\phi(\gamma u_{n-1}) \geqq 2\phi(\delta u_{n-2})$. By (41) and (7) we have

$$2\phi(u_n) \geqq \phi(\gamma u_{n-1}) + \phi(\gamma u_{n-1}) - 2\phi(\delta u_{n-2}) \geqq \phi(\gamma u_{n-1})$$

as desired.

The valuations $\phi(u_n) > 0$ since $\phi(u_1) = \phi(2\gamma) \neq 0$ and we may apply (43). Thus every $u_n \neq 0$ and the quantities

$$a_n = \frac{u_n}{u_{n-1}} \neq 0 \tag{44}$$

are in $\mathfrak{F}$. The inequality (43) is equivalent to

$$\frac{\phi(\gamma)}{2\phi(a_n)} \leqq 1 , \quad \phi(a_n^{-1}) \leqq 2\phi(\gamma^{-1}) , \tag{45}$$

and by (41)

$$a_n = -\gamma - \delta a_{n-1}^{-1} . \tag{46}$$

This gives

$$a_n^2 + \gamma a_n + \delta\left(\frac{a_n}{a_{n-1}}\right) = 0 , \tag{47}$$

and also

$$\phi(a_n) \leqq \phi(\gamma) + 2\phi(\delta\gamma^{-1}) < \frac{M}{2} , \tag{48}$$

where M is a positive real number independent of n.

The equation (46) gives

$$a_p - a_q = -\delta(a_{p-1}^{-1} - a_{q-1}^{-1}) , \qquad \phi(a_p - a_q) = \phi(\delta)\,\frac{\phi(a_{p-1} - a_{q-1})}{\phi(a_{p-1})\phi(a_{q-1})} .$$

Define

$$\rho = 4\phi(\delta\gamma^{-2}) < 1 ,$$

and use $\phi(\delta) = \frac{1}{4}[\phi(\gamma)]^2\rho$ to get

$$\phi(a_p - a_q) = \rho\,\frac{\phi(\gamma)}{2\phi(a_{p-1})}\cdot\frac{\phi(\gamma)}{2\phi(a_{q-1})}\,\phi(a_{p-1} - a_{q-1}) \leqq \rho\phi(a_{p-1} - a_{q-1})$$

by (45). Then evidently

$$\phi(a_p - a_q) \leqq \rho^t \phi(a_{p-t} - a_{q-t}) \leqq \rho^t M \qquad (p > t,\ q > t)\,,$$

since $\phi(a_{p-t} - a_{q-t}) \leqq \phi(a_{p-t}) + \phi(_{q-t}) < M$. For every $\epsilon > 0$ there exists an integer t_0 such that $\rho^t < M^{-1}\epsilon$ for $t > t_0$. Then

$$\phi(a_p - a_q) < \epsilon \qquad (p > t_0,\ q > t_0)\,,$$

and $\alpha = \{a_n\}$ is regular. Since $\mathfrak{F}$ is complete the class $a = [\alpha]$ is in $\mathfrak{F}$. Also $\{a_n/a_{n-1}\}$ is regular and $[\{a_n/a_{n-1}\}]$ is the unity element of $\mathfrak{F}$. By (47) we have $a^2 + \gamma a + \delta = 0$, that is, (38) is reducible in $\mathfrak{F}$.

If $\mathfrak{F}$ is complete with respect to ϕ and the equation $x^2 + \gamma x + \delta = 0$ is irreducible in $\mathfrak{F}$ the field

$$\mathfrak{K} = \mathfrak{F}(\xi)\,, \qquad \xi^2 + \gamma\xi + \delta = 0$$

is a quadratic field over $\mathfrak{F}$. Every quantity of $\mathfrak{K}$ is uniquely expressible in the form

$$c = a + b\xi \qquad (a,\ b \text{ in } \mathfrak{F}).$$

The automorphism group of $\mathfrak{K}$ over $\mathfrak{F}$ is a cyclic group of order two with generating automorphism

$$c \longleftrightarrow \bar{c} = a + b(-\xi - \gamma)\,.$$

In particular

$$\xi + \bar{\xi} = -\gamma\,, \qquad \xi\bar{\xi} = \delta\,,$$

so that

$$c + \bar{c} = 2a - b\gamma\,, \qquad c\bar{c} = a^2 + b^2\xi\bar{\xi} + ab(\xi + \bar{\xi}) = a^2 + b^2\delta - \gamma ab$$

are in $\mathfrak{F}$ for any c of $\mathfrak{K}$. Since $\mathfrak{K}$ is a field and $c \longleftrightarrow \bar{c}$ we have $c\bar{c} = 0$ if and only if $c = 0$. Define

$$\Phi(c) = [\phi(c\bar{c})]^{1/2}\,, \tag{49}$$

so that $\Phi(c) = \phi(c)$ if $c = \bar{c}$ is in $\mathfrak{F}$, $\Phi(c) \neq 0$ for $c \neq 0$. Then we shall prove

Theorem 14. *The function Φ of (49) defines a valuation of $\mathfrak{K}$ such that*

$$\Phi(a) = \phi(a) \tag{50}$$

for every a *of* $\mathfrak{F}$. *The field* $\mathfrak{K}$ *is complete with respect to* Φ *and* (49) *is the only valuation of* $\mathfrak{K}$ *for which* (50) *holds.*

For $\Phi(cd) = [\phi(cd\overline{cd})]^{1/2} = [\phi(c\bar{c})]^{1/2} \cdot [\phi(d\bar{d})]^{1/2} = \Phi(c) \cdot \Phi(d)$ since $\overline{cd} = \bar{c}\bar{d}$, $\phi(ab) = \phi(a) \cdot \phi(b)$ in $\mathfrak{F}$. The equation (50) and the remaining valuation postulates have already been shown above except that we must prove

$$\Phi(c + d) \leqq \Phi(c) + \Phi(d)\,. \tag{51}$$

This inequality holds if $d = 0$ and, by division by $d \neq 0$ is equivalent to

$$\Phi(g + 1) \leqq \Phi(g) + 1 \tag{52}$$

for every g of $\mathfrak{K}$. When g is in $\mathfrak{F}$ we have (52) since Φ becomes ϕ. Let g be not in $\mathfrak{F}$ so that g generates the quadratic field $\mathfrak{K}$ of degree two over $\mathfrak{F}$ and

$$g^2 + \lambda g + \mu = 0$$

is irreducible in $\mathfrak{F}$. By Theorem 11.13

$$\phi(\lambda^2) \leqq 4\phi(\mu)\,.$$

But $g\bar{g} = \mu$, $g + \bar{g} = -\lambda$ and $(g + 1)(\bar{g} + 1) = \mu + 1 - \lambda$,

$$\begin{aligned}\Phi(g + 1) &= [\phi(\mu + 1 - \lambda)]^{1/2} \leqq [1 + \phi(\mu) + \phi(\lambda)]^{1/2} \\ &\leqq \{1 + \phi(\mu) + 2\,[\phi(\mu)]^{1/2}\}^{1/2} \leqq 1 + [\phi(\mu)]^{1/2} \\ &\leqq \Phi\,(g) + 1\end{aligned}$$

as desired. Thus Φ is a valuation of $\mathfrak{K}$ and $\mathfrak{K} = \mathfrak{K}_\Phi$ by Theorem 11.11.

Next let Ψ be any other valuation of $\mathfrak{K}$ such that $\Psi(a) = \phi(a)$ for a in $\mathfrak{F}$. Then

$$\Psi(c) \cdot \Psi(\bar{c}) = \Psi(c\bar{c}) = \phi(c\bar{c})\,,$$

and if $\Psi(c) \neq [\phi(c\bar{c})]^{1/2}$ we must have either $\Psi(c) > [\phi(c\bar{c})]^{1/2}$ or $\Psi(\bar{c}) > [\phi(c\bar{c})]^{1/2}$ for c in $\mathfrak{K}$. Since both c and $\bar{c}$ are in $\mathfrak{K}$ we may assume with no loss of generality that

$$\Psi(c) > [\phi(c\bar{c})]^{1/2} > \Psi(\bar{c})\,,$$

for some c in $\mathfrak{K}$ and not in $\mathfrak{F}$. Then

$$d = \bar{c}c^{-1}\,, \qquad \rho = \Psi(d) < 1\,,$$

where d is in $\mathfrak{K}$. The quantity

$$a_n = \frac{c^{n+1} + \bar{c}^{n-1}}{c^n + \bar{c}^n} = c\,\frac{1 + d^{n+1}}{1 + d^n}$$

is defined for all positive integral values of n since if $c^n + \bar{c}^n = 0$ then $c^n(1 + d^n) = 0$, $c \neq 0$, $1 + d^n = 0$, $d^n = -1$, $\Psi(d^n) = \rho^n = 1$ which is impossible when $\rho < 1$, $n > 0$. Also a_n is in $\mathfrak{F}$ since $c^n + \bar{c}^n$ is unaltered by all of the automorphisms of $\mathfrak{K}$ over $\mathfrak{F}$. The sequence

$$\{d^n\}$$

is a null sequence of $\mathfrak{K}$ since the $\lim_{n \to \infty} \rho^n = 0$ for real $\rho < 1$. Thus

$$\lim_{n \to \infty} a_n = c \lim_{n \to \infty} \frac{1 + d^{n+1}}{1 + d^n} = c$$

exists. Since $\mathfrak{F}$ is complete c is in $\mathfrak{F}$, a contradiction. Then $\Psi(c) = \Phi(c) = [\phi(c\bar{c})]^{1/2}$.

In the next chapter we shall consider a type of valuation in which it will always turn out that our derived field is analytically equivalent either to the field of all real or of all complex numbers. Thus a case of great importance is that where $\mathfrak{F} = \mathfrak{R}'$ is the field of all real numbers and $\mathfrak{K}$ is given by

$$\mathfrak{C} = \mathfrak{R}'(i)\,, \qquad i^2 + 1 = 0\,,$$

$\mathfrak{C}$ is the field of all complex numbers. Our result then shows that *the only valuation of* $\mathfrak{C}$ *which preserves the valuation* $\phi(a) = |a|$ *for real* a *is* $\Phi(c) = |c\bar{c}|^{1/2}$ *for complex* c, that is, $\Phi(c) = |c|$ in the ordinary sense of the absolute value of a complex number c.

CHAPTER XII

VALUATION FUNCTIONS

1. Introduction. The theory of Chapter XI forms a foundation for a more intensive study of fields with a valuation. Our primary interest is not in this subject for itself but in the resulting tool without which one is unable to study either the modern theory of algebraic numbers or its important applications to the theory of division algebras.

The most obvious problem that arises is that of determining the conditions that a given field $\mathfrak{F}$ shall have a non-trivial valuation. The next problem is that of finding all non-trivial valuations of $\mathfrak{F}$. Both problems are studied by a division of the investigation into two major cases. In the first case $\mathfrak{F}$ will be shown always to be analytically equivalent to a field of complex numbers, and this case will be completed. In the second case, however, no such general result can be obtained, and we shall limit our investigations to a very important case complete in the literature, that is, to the study of algebraic number fields. These are fields $\mathfrak{F}$ of finite degree over the field $\mathfrak{R}$ of all rational numbers. A complete discussion even in this case is beyond the scope of our text, but we shall at least give the fundamentals of the subject.

2. Lemmas from analysis. Much of our study will depend on certain results from that branch of mathematics called *analysis*. The theorems assumed are derivable as consequences of the definitions of real and complex numbers which were given as a section of our transcendental theory of fields. But it is not part of our program to develop elementary analysis here. We shall instead assume all of the well known elementary theorems which are required.

One of the notions of real analysis is that of the *real power*

$$a^{\beta},$$

where a is any positive real number and β is any real number. The case $\beta = 0$ gives $a^0 = 1$ and we shall assume that

$$\lim_{\beta \to 0} a^{\beta} = 1 .$$

In particular we assume

LEMMA 1. *Let* n *be an integer and* $a > 0$. *Then*

$$\lim_{n \to \infty} a^{1/n} = 1 .$$

Another important notion is given by

LEMMA 2. *Let* $b \neq 1$, $b > 0$, $a > 0$ *be real. Then there exists a real number*

$$\alpha = \log_b a\,,$$

called the **logarithm of** a **to the base** b, *such that*

$$b^\alpha = a\,.$$

Moreover, when a *ranges over all positive real numbers* α *ranges over all real numbers.*

As a consequence of Lemma 1 we may indicate a proof of a result already used in our proof of Theorem 11.14.

LEMMA 3. *The* $\lim_{n \to \infty} \rho^n = 0$ *if* $1 > \rho > 0$.

For let $\epsilon > 0$ and $\delta = 1 - \rho > 0$. By Lemma 1 there exists an integer n such that $\mu_p = |\epsilon^{1/p} - 1| < \delta$ for $p > n$. Thus $\epsilon^{1/p} = 1 + \mu_p$ or $1 - \mu_p$ and $\epsilon^{1/p} \geqq 1 - \mu_p > 1 - \delta = \rho$, $\epsilon > \rho^p$ for $p > n$. But then $\{\rho^n\}$ is a null sequence as desired.

As a corollary of Lemma 3 and the theory of indeterminate forms of elementary calculus we may prove

LEMMA 4. *Let* $1 > \rho > 0$. *Then*

$$\lim_{n \to \infty} \rho^n(n + 1) = 0\,.$$

For $\sigma = \rho^{-1}$ and $\lim_{n \to \infty} \sigma^{n-1} = \infty$. Hence by the differentiation method of the evaluation of indeterminate forms,

$$\lim_{n \to \infty} \frac{n+1}{\sigma^n} = \lim_{n \to \infty} \frac{1}{n\sigma^{n-1}} = 0\,.$$

LEMMA 5. *Let* m *and* n *be integers. Then*

$$\lim_{m \to \infty} [m(n + 1) + 1]^{1/m} = 1\,.$$

For $\log_e [m(n + 1) + 1]^{1/m} = 1/m \log_e [m(n + 1) + 1]$, where e is the base of natural logarithms. By differentiating numerator and denominator we obtain

$$\lim_{m \to \infty} \frac{n+1}{m(n+1)+1} = 0 = \lim_{m \to \infty} \log_e [m(n + 1) + 1]^{1/m}\,.$$

Hence the limit is unity in our lemma.

We next state without proof a result which is obtainable as a first analytic theorem of the real number system.

LEMMA 6. *Every set* $\mathfrak{M}$ *of real numbers with a lower bound has a greatest lower bound.*

The above lemma states that if there exists a real B_0 such that $a > B_0$ for every a of $\mathfrak{M}$, then there exists a real B such that for every $\epsilon > 0$ there exists an a_ϵ in $\mathfrak{M}$ with

$$0 \leqq a_\epsilon - B < \epsilon .$$

Of course B may be in $\mathfrak{M}$ and so we may take all the $a_\epsilon = B$. However, we may prove

LEMMA 7. *Let the greatest lower bound* B *of a set* $\mathfrak{M}$ *of real numbers be not in* $\mathfrak{M}$. *Then there exists a set of quantities*

$$a_1 > a_2 > a_3 > \dots$$

such that

$$\lim_{n \to \infty} a_n = B .$$

For B is not in $\mathfrak{M}$ and $a - B > 0$ for every a of $\mathfrak{M}$. We choose an a_0 such that $0 < a_0 - B < 1 = 2^0$. Let ϵ be the smaller of 2^{-n} and $a_{n-1} - B$ and find by an induction a sequence $a_n = a_\epsilon$ of quantities of $\mathfrak{M}$ such that $0 < a_\epsilon - B < \epsilon$. But then

$$0 < a_n - B < 2^{-n} , \qquad a_n - B < a_{n-1} - B ,$$

so that $a_{n-1} > a_n$ as desired. Also $\{a_n - B\}$ is a null sequence, and $\lim_{n \to \infty} a_n = B$.

We shall assume without proof the so-called Weierstrass-Bolzano Theorem:

LEMMA 8. *Let* $\mathfrak{M}$ *be an infinite set of distinct complex numbers* c *such that*

$$|c| < B$$

for a real number B *and every* c *of* $\mathfrak{M}$. *Then there exists an infinite set of distinct* c_i *in* $\mathfrak{M}$ *and a complex number* d *such that*

$$\lim_{n \to \infty} c_n = d .$$

The above results from analysis are well known to anyone who has even a brief knowledge of the subject. It is not part of our program to prove these theorems, but we shall use them in our derivation of fields with a certain type of valuation. We shall also assume the theorem called the *Fundamental Theorem of Algebra:*

LEMMA 9. *The field* $\mathfrak{C}$ *of all complex numbers is algebraically closed.*

This theorem has an algebraic counterpart stated as the

Theorem. *Every field* $\mathfrak{F}$ *is contained in an algebraically closed field.*

We shall not use this latter result but mention it because of its importance. However, all the lemmas above will be used in the following sections. We again recall that we have left the subject which might be called *pure algebra* and are now proving analytic theorems to be used in algebra and the theory of numbers.

In Section 11.4 we defined equivalent valuations. We stated there that ϕ and ψ are equivalent in case $\phi(a) > \phi(b)$ if and only if $\psi(a) > \psi(b)$. This is justified by

Theorem 1. *Let* ϕ *and* ψ *be equivalent. Then a sequence is regular according to* ϕ *if and only if it is regular according to* ψ*, is a null sequence according to* ϕ *if and only if it is null according to* ψ*. Moreover* $\mathfrak{F}_\phi = \mathfrak{F}_\psi$.

For $\phi(a) \neq \phi(1)$, $\phi(0)$ implies that $\psi(a) \neq 1, 0$. Hence ϕ is a trivial valuation of $\mathfrak{F}$ if and only if $\psi = \phi$ is trivial. Let ϕ be non-trivial and $\phi(a_0) \neq 1, 0$. If $\phi(a_0) > 1$ then $\phi(a_0^{-1}) < 1$ so that we may always assume the existence of an a_0 in $\mathfrak{F}$ such that

$$\rho = \phi(a_0) < 1\,.$$

Since ψ is equivalent to ϕ we also have

$$\sigma = \psi(a_0) < 1\,.$$

Consider a sequence $\{a_n\}$ regular with respect to ϕ. By Lemma 3 if $\epsilon > 0$ there exists an integer m such that

$$\sigma^m < \epsilon\,.$$

Since $\{a_n\}$ is regular there exists an integer n_0 such that

$$\phi(a_p - a_q) < \rho^m = \phi(a_0^m) \qquad (p > n_0,\, q > n_0)\,.$$

But then $\psi(a_p - a_q) < \psi(a_0^m) = \sigma^m < \epsilon$, $(p > n_0,\, q > n_0)$, and $\{a_n\}$ is regular with respect to ψ. Similarly if $\{a_n\}$ is a null sequence we make $\phi(a_p) < \rho^m$ and have $\psi(a_p) < \sigma^m < \epsilon$. The set $\mathfrak{F}_\phi$, consisting essentially of all classes $A = [a] = [a + \tau]$ where a is regular and τ is a null sequence, is then $\mathfrak{F}_\psi$.

3. The types of valuations. A valuation ϕ is called *archimedean* if $\phi(n) > 1$ for some integer n. Then ϕ is *non-archimedean* if $\phi(n) \leqq 1$ for every integer n. We prove

Theorem 2. *A valuation ϕ of a field $\mathfrak{F}$ is non-archimedean if and only if*

$$\phi(a + b) \leqq \text{maximum of } \phi(a), \phi(b) \tag{1}$$

for every a *and* b *of* $\mathfrak{F}$.

For if (1) holds we have $\phi(2) \leqq \phi(1)$, $\phi(3) \leqq \phi(1)$, $\phi(n) \leqq 1$. Conversely, let $\phi(n) \leqq 1$ for every integer n. If a, b are in $\mathfrak{F}$ we may assume that $\phi(a) \geqq \phi(b) > 0$ with no loss of generality since $b = 0$ implies that (1) holds. Then

$$[\phi(a + b)]^n = \phi[(a + b)^n] \leqq (n + 1)[\phi(a)]^n .$$

For the coefficients in the binomial expansion of $(a + b)^n$ are integers n_i and $\phi(n_i) \leqq 1$. Also the terms are $n_r a^r b^{n-r}$ such that $\phi(n_r a^r b^{n-r}) \leqq \phi(a^n)$, and there are $n + 1$ terms. Write $\rho = \phi(a)[\phi(a + b)]^{-1} > 0$ and obtain

$$1 \leqq \rho^n(n + 1) .$$

If $\rho < 1$ there exists an integer n such that $\rho^n(n + 1) < 1$ by Lemma 4. Thus $\rho \geqq 1$, $\phi(a + b) \leqq \phi(a)$, which is the maximum of $\phi(a)$, $\phi(b)$.

Non-archimedean valuations provide rather novel analytic theorems. One which we shall mention is a criterion for the convergence of infinite series. We consider an infinite sum

$$S = u_1 + u_2 + \ldots \qquad (u_i \text{ in } \mathfrak{F}_\phi) ,$$

define $S_n = u_1 + \ldots + u_n$, and say that *the series converges to*

$$S = \lim_{n \to \infty} S_n ,$$

if this limit exists in $\mathfrak{F}_\phi$. It is well known that when the series converges

$$\lim_{n \to \infty} \phi(u_n) = 0 , \tag{2}$$

that is $\{u_n\}$ is a null sequence. This is an obvious result since when $\{\phi(u_n)\}$ is not a null sequence there exists a $\delta > 0$ and an n such that $\phi(u_p) > \delta > 0$ for $p > n$ by (11.23). Then $\phi(S_p - S_{p-1}) = \phi(u_p) > \delta$ for all $p > n$, $\{S_n\}$ cannot be regular, and our condition has been proved necessary. However when ϕ is non-archimedean we may prove the condition sufficient and this is of course a rather unusual result.

For proof we see that when (2) holds we may write $\phi(S_p - S_q) = \phi(S_{q+r} - S_q) = \phi(u_{q+1} + \ldots + u_{q+r})$ when we take $p > q$ without loss of generality. Take $\phi(u_n) < \epsilon$ for $n > n_0$ and have

$$\phi(S_p - S_q) \leqq \max \phi(u_{q+i}) < \epsilon \qquad (p > q > n_0) .$$

Thus $\{S_n\}$ is regular and converges to a limit in the complete field $\mathfrak{F}_\phi$.

4. Archimedean valuations of $\mathfrak{R}$. Let $\mathfrak{R}$ be the field of all rational numbers and ϕ be a valuation of $\mathfrak{R}$ which is archimedean. Then there exists a prime p such that

$$\phi(p) > 1 .$$

For otherwise every $\phi(p) \leqq 1$, every $\phi(n) \leqq 1$. We let a be any positive integer and define an integral exponent

$$n = N(a) \tag{3}$$

with the property

$$p^n \leqq a < p^{n+1} . \tag{4}$$

Divide a by p^n and obtain

$$a = p^n a_n + b \qquad (0 \leqq b < p^n, 0 < a_n < p) ,$$

since $a < p^{n+1}$. Proceed similarly with b and ultimately obtain integers a_i such that

$$a = a_n p^n + a_{n-1}p^{n-1} + \ldots + a_1 p + a_0 \qquad (0 \leqq a_i < p,\ a_n \neq 0) . \tag{5}$$

We have $p^{mn} \leqq a^m < p^{m(n+1)}$ by (4) and thus

$$mn \leqq N(a^m) < m(n + 1)$$

for all integers m.

There exists a real number ρ such that

$$\phi(p) = p^\rho \qquad (0 < \rho \leqq 1) .$$

For, $p > 1$ and $1 < \phi(p) \leqq p$. We shall prove that for this given ρ we have

$$\phi(a) = |a|^\rho$$

for every rational number a. Thus $\phi(p) > 1$ determines every $\phi(a)$ uniquely in terms of a and $\phi(p)$. We now assume that a is a positive integer.

For every integer n we have $\phi(n) \leqq n$. Then for the integers a_i (5) implies that $\phi(a_i) \leqq a_i < p$ and $\phi(p^i) = p^{i\rho} \leqq p^{n\rho}$, so that since $p^n \leqq a$ we have $p^{n\rho} \leqq a^\rho$ and

$$\phi(a) < p(n + 1)p^{n\rho} \leqq p(n + 1)a^\rho .$$

Similarly,

$$\phi(a^m) \leqq p[N(a^m) + 1]a^{m\rho} < p[m(n + 1) + 1]a^{m\rho} ,$$

so that

$$\phi(a) < p^{1/m}[m(n + 1) + 1]^{1/m}a^\rho .$$

Applying Lemmas 1 and 5, we have $\phi(a) \leqq b_m a^\rho$ where $\lim_{m \to \infty} b_m = 1$. But then

$$\phi(a) \leqq a^\rho .$$

Write $a = p^{n+1} - d$. Then property (11.7) implies that

$$\phi(a) \geqq p^{\rho(n+1)} - \phi(d) .$$

Since $p^n \leqq a < p^{n+1}$ we have $p^n \leqq p^{n+1} - d < p^{n+1}$ and $- p^{n+1} < d - p^{n+1} \leqq -p^n$. Adding p^{n+1}, we have $0 < d \leqq p^n(p - 1)$. Since d is an integer $\phi(d) \leqq d^\rho$ and $d \leqq p^n(p - 1)$ gives

$$\phi(d) \leqq p^{\rho n}(p - 1)^\rho .$$

Substituting the quantity $p^{\rho n}(p - 1)^\rho$ for $\phi(d)$ in the difference above, we do not increase it and so obtain

$$\phi(a) \geqq p^{\rho(n+1)} - p^{\rho n}(p - 1)^\rho = p^{\rho n}\lambda ,$$

where

$$\lambda = p^\rho - (p - 1)^\rho$$

is a positive real number independent of n. The same process applied to a^m gives

$$[\phi(a)]^m = \phi(a^m) \geqq p^{\rho N}\lambda , \qquad N = N(a^m) .$$

Now the definition of N states that $a^m < p^{N+1}$, $a^{m\rho}p^{-\rho} < p^{\rho N}$, so that

$$[\phi(a)]^m \geqq a^{m\rho}\mu , \qquad \mu = \lambda p^{-\rho} .$$

The real positive number μ is independent of m and $\lim_{m \to \infty} \mu^{1/m} = 1$. Since $\phi(a) \geqq a^\rho \mu^{1/m}$ we have $\phi(a) \geqq a^\rho$, and thus

$$\phi(a) = a^\rho$$

for every positive integer a. If a is negative $\phi(a) = \phi(b) = b^\rho = |a|^\rho$, where $b = |a|$. When $a = bc^{-1}$ is rational $\phi(a) = \phi(b)[\phi(c)]^{-1} = |bc^{-1}|^\rho = |a|^\rho$ and we have proved

Theorem 3. *Let ϕ be an archimedean valuation of the field $\mathfrak{R}$ of all rational numbers. Then there exists a real number ρ such that $0 < \rho \leqq 1$ and*

$$\phi(\mathrm{a}) = \phi_\rho(\mathrm{a}) = |\mathrm{a}|^\rho \tag{6}$$

for every rational a.

As an immediate consequence of Theorem 12.3 we have

Theorem 4. *Let $\mathfrak{R}_\phi$ be the derived field of the field of all rational numbers. Then $\mathfrak{R}_\phi$ is the field $\mathfrak{R}'$ of all real numbers and the valuation Φ of Section* 11.11 *is*

$$\Phi(\mathrm{a}) = |\mathrm{a}|^\rho$$

for every real a.

For the valuations ϕ_ρ are all equivalent since $|a|^\rho \geqq |b|^\rho$ for rational a, b if and only if $|a| > |b|$. By Theorem 12.1 each $\mathfrak{R}_\phi$ is the field $\mathfrak{R}'$ defined by $\rho = 1$, that is the field of all real numbers. In (11.32) we defined $\Phi(A) = \lim_{n \to \infty} \phi(a_n)$ where A is a real number defined by the regular sequence $\{a_n\}$. But $|A| = \lim_{n \to \infty} |a_n|$ so that $\Phi(A) = \lim_{n \to \infty} |a_n|^\rho = |A|^\rho$.

5. Archimedean valued fields. We shall determine all fields with an archimedean valuation in the sense of analytic equivalence.

A field $\mathfrak{F}$ with an archimedean valuation is non-modular. For $\phi(n) > 1$ for some integer n and if $\mathfrak{F}$ has characteristic p it is true that $n^{p-1} = 1$. This well-known result of the theory of numbers implies that $[\phi(n)]^{p-1} = 1$, which is impossible since $[\phi(n)]^{p-1} > [\phi(n)]^{p-2} > \ldots > \phi(n) > 1$. Hence $\mathfrak{F}$ is non-modular and contains the field $\mathfrak{R}$ of all rational numbers. We write

$$\mathfrak{F} \geqq \mathfrak{R}\,.$$

Pass to the derived field $\mathfrak{F}_\phi$ of $\mathfrak{F}$. It contains $\mathfrak{R}' = \mathfrak{R}_\phi$. By Theorem 12.3 there exists a real number ρ such that $0 < \rho \leqq 1$ and

$$\phi(a) = |a|^\rho \tag{7}$$

for every rational a. Then the valuation function of the derived field $\mathfrak{F}_\phi$ may also be represented by ϕ and is of course taken to be the limit function of (11.32). This is true in $\mathfrak{R}_\phi$ also and we have (7) for every real a. We next have the

LEMMA. *Let $\mathfrak{K}$ be a complete field with respect to a valuation ϕ, and $\mathfrak{K} \geqq \mathfrak{C} = \mathfrak{R}'(\mathrm{i})$ where $\mathfrak{R}'$ is the field of all real numbers, $\mathrm{i}^2 = -1$ and (7) holds. Then*

$$\phi(\mathrm{c}) = |\mathrm{c}|^\rho \tag{8}$$

for every complex number c.

The above is of course an immediate application of Theorem 11.14. We apply this result to

$$\mathfrak{K} = \mathfrak{F}_\phi(i) \geqq \mathfrak{C}$$

when $i^2 + 1$ is reducible in $\mathfrak{F}_\phi$ so that $\mathfrak{K} = \mathfrak{F}_\phi$, and we have (8). Otherwise $\mathfrak{F}_\phi(i) = \mathfrak{K}$ is a quadratic field over $\mathfrak{F}$ and by Theorem 11.14 we may define the valuation

$$\phi(c) = [\phi(c\bar{c})]^{1/2} \tag{9}$$

for every c of $\mathfrak{K}$, where $c = a + bi$, $\bar{c} = a - bi$, $i^2 = -1$. Then (9) again holds for every complex number c. Hence in every case we have (9) for the subfield $\mathfrak{C}$ of $\mathfrak{K}$.

If $\mathfrak{C} \neq \mathfrak{K}$ there exists a ξ in $\mathfrak{K}$ such that

$$\phi(\xi - c) > 0 \tag{10}$$

for every complex number c. The set $\mathfrak{M}$ of positive real numbers (10) has a greatest lower bound m by Lemma 6, and either

$$m = \phi(\eta)\,, \qquad \eta = \xi - c_0 \qquad (c_0 \text{ in } \mathfrak{C})\,, \tag{11}$$

m is in $\mathfrak{M}$, or we may apply Lemma 7 to obtain an infinite sequence of complex numbers $c_1, c_2, \ldots$ such that

$$M = \phi(\xi - c_1) > \phi(\xi - c_2) > \ldots \tag{12}$$

and also

$$\lim_{n \to \infty} \phi(\xi - c_n) = m\,. \tag{13}$$

Now

$$|c_n|^p = \phi(c_n) = \phi(-c_n + \xi - \xi) \leqq \phi(\xi - c_n) + \phi(\xi) \leqq M + \phi(\xi),$$

so that the absolute values of the c_n are bounded. Apply Lemma 8 to prove the existence of an infinite subsequence

$$c_{n_i} \qquad (i = 1, \ldots)$$

and a complex number c such that

$$\lim_{i \to \infty} c_{n_i} = c.$$

The valuations $\phi(\xi - c_{n_i})$ are again a monotonic set as in (12) and we evidently have $\lim_{i \to \infty} \phi(\xi - c_{n_i}) = m$ since i may always be chosen so that $n_i > n_0$, $|\phi(\xi - c_{n_j}) - m| < \epsilon$ for $n_j > n_i$, that is, $j > i$. We replace the sequence of c_n by the subsequence (to avoid notational complexity) and now have (13) as well as

$$\lim_{n \to \infty} c_n = c. \tag{14}$$

But for every $\epsilon > 0$ there exists an integer n_ϵ such that $\phi(c - c_n) = |c - c_n|^p < \epsilon$ for $n > n_\epsilon$. Hence

$$\epsilon > \phi(c - c_n) = \phi[(\xi - c_n) - (\xi - c)] \geqq |\phi(\xi - c_n) - \phi(\xi - c)|, \quad n > n_\epsilon,$$

by (11.7). Thus the sequence $\{\phi(\xi - c_n) - \phi(\xi - c)\}$ is a null sequence, and in the complete field $\mathfrak{K}$

$$\lim_{n \to \infty} \phi(\xi - c_n) = \phi(\xi - c).$$

By (13) we have $\phi(\xi - c) = m$, a contradiction. This proves (11).

The quantity $m = \phi(\eta)$ has the property that $\phi(\eta - c) = \phi(\xi - c_0 - c) \geqq m$. Hence

$$\phi(\eta - c) \geqq \phi(\eta) = m,$$

for every complex number c, and η is in $\mathfrak{K}$ and not in $\mathfrak{C}$ since η in $\mathfrak{C}$ implies that $\xi = \eta + c_0$ is in $\mathfrak{C}$. We prove the

LEMMA. *Let* λ *in* $\mathfrak{K}$ *have the property that* $\phi(\lambda - \mathrm{c}) \geqq \phi(\lambda)$ *for every complex number* c. *Then*

$$\phi(\lambda - \mathrm{c}) = \phi(\lambda) \tag{15}$$

for every complex c for which

$$\phi(\lambda) > \phi(c) .$$

We apply Lemma 9 and see that the complex nth roots of unity, 1, $\epsilon_2, \ldots, \epsilon_n$ are in $\mathfrak{C}$. This means that $x^n - 1 = (x-1)(x-\epsilon_2) \ldots (x - \epsilon_n)$ for any x of $\mathfrak{K}$ and if we replace x by λc^{-1} and multiply by c^n we have

$$\frac{\lambda^n - c^n}{\lambda - c} = (\lambda - c\epsilon_2) \ldots (\lambda - c\epsilon_n) .$$

Since $\phi(\lambda - c\epsilon_i) \geqq \phi(\lambda)$ for every complex number c we have

$$\frac{\phi(\lambda^n - c^n)}{\phi(\lambda - c)} \geqq [\phi(\lambda)]^{n-1} .$$

But then $\phi(\lambda - c)[\phi(\lambda)]^{n-1} \leqq \phi(\lambda^n - c^n) \leqq [\phi(\lambda)]^n + [\phi(c)]^n$. By division by $[\phi(\lambda)]^n$ we obtain

$$\phi\left(\frac{\lambda - c}{\lambda}\right) \leqq 1 + \left[\phi\left(\frac{c}{\lambda}\right)\right]^n .$$

Since $\rho = \phi(c\lambda^{-1}) < 1$ by hypothesis, we have $\lim_{n \to \infty} \rho^n = 0$ and $\phi(\lambda - c) \leqq \phi(\lambda)$. This, combined with $\phi(\lambda - c) \geqq \phi(\lambda)$, implies (15) and our lemma.

We apply our lemma to

$$c = [2^{-1}\phi(\eta)]^{1/\rho}$$

so that

$$\phi(c) = |c|^\rho = c^\rho = \tfrac{1}{2}\phi(\eta) < \phi(\eta) .$$

Then $\phi(\eta - c) = \phi(\eta)$. Also $\phi(\eta - 2c) \geqq \phi(\eta) = \phi(\eta - c)$ so that $\eta - c$ is a λ of our lemma with $\phi(c) < \phi(\lambda)$ and $\phi(\eta - 2c) = \phi(\eta)$. Continuing with $\lambda = \eta - (n-1)c$, we obtain

$$\phi(\eta - nc) = \phi(\eta)$$

for every integer n. Now

$$2\phi(\eta) = \phi(\eta) + \phi(\eta - nc) \geqq \phi(nc) = n^\rho\phi(c) .$$

But $\lim_{n \to \infty} n^\rho = \infty$, and this is a contradiction. Hence ξ is in $\mathfrak{K}$, $\mathfrak{K} = \mathfrak{C}$, $\mathfrak{F} \leqq \mathfrak{C}$. We have proved

Theorem 5. *Every field $\mathfrak{F}$ with an archimedean valuation ϕ contains the*

field $\mathfrak{R}$ of all rational numbers and is equivalent over $\mathfrak{R}$ to a subfield $\mathfrak{F}_0$ over $\mathfrak{R}$ of the complex number field under a correspondence

$$\mathrm{a} \longleftrightarrow \mathrm{a}_0 \qquad (\mathrm{a} \text{ in } \mathfrak{F},\ \mathrm{a}_0 \text{ in } \mathfrak{F}_0)\,, \tag{16}$$

such that for some positive real $\rho \leqq 1$,

$$\phi(\mathrm{a}) = |\mathrm{a}_0|^\rho\,. \tag{17}$$

Moreover, the derived field $\mathfrak{F}_\phi \geqq \mathfrak{F}$ of $\mathfrak{F}$ is analytically equivalent either to the field $\mathfrak{R}'$ of all real numbers or to the field $\mathfrak{C}$ of all complex numbers.

The valuations defined by the same $\mathfrak{F}_0$ and different values of ρ are evidently equivalent and we may replace (17) by

$$\phi(a) = |a_0| \tag{18}$$

with no loss of generality. Thus every archimedean valuation of $\mathfrak{F}$ in the sense of equivalence will be obtained if we find all subfields $\mathfrak{F}_0$ of the field $\mathfrak{C}$ of all complex numbers equivalent over $\mathfrak{R}$ to $\mathfrak{F}$ and employ (18). In particular let us apply this to a field

$$\mathfrak{F} = \mathfrak{R}(\xi)$$

which is algebraic of degree n over the field $\mathfrak{R}$ of all rational numbers. We call the quantities of $\mathfrak{F}$ *algebraic numbers* and say that $\mathfrak{F}$ is an *algebraic number field.*

The quantity ξ is a root of

$$f(x) = x^n + a_1x^{n-1} + \ldots + a_n = (x - \xi_1) \ldots (x - \xi_n) = 0$$

with rational a_i and complex roots $\xi_1, \ldots, \xi_n$. If $\mathfrak{F}_0$ is equivalent over $\mathfrak{R}$ to $\mathfrak{F}$, then $\mathfrak{F}_0 = \mathfrak{R}(\xi_0)$ where ξ_0 is a complex number and a root of $f(x) = 0$. Thus $\xi_0 = \xi_i$, and we have proved that each $\mathfrak{F}_0$ is

$$\mathfrak{F}_i = \mathfrak{R}(\xi_i)\,,$$

equivalent to $\mathfrak{F}$ under a correspondence

$$a(\xi) \longleftrightarrow a(\xi_i)$$

such that, in the sense of equivalence of valuations,

$$\phi_i[a(\xi)] = |a(\xi_i)|\,.$$

Evidently the derived field $\mathfrak{F}_\phi$ is analytically equivalent to the derived field of $\mathfrak{F}_i$. This field is $\mathfrak{R}'(\xi_i)$ by Theorem 11.12 and when ξ_i is real is the field $\mathfrak{R}'$ of all real numbers. When ξ_i is imaginary we have $\mathfrak{R}'(\xi) = \mathfrak{R}'(i) = \mathfrak{C}$, the field of all complex numbers. Thus $\mathfrak{F}_{\phi_i}$ is analytically equivalent to the field of all real numbers when ξ_i is real, the field of all complex numbers when ξ_i is imaginary.

EXERCISES

1. Let $\mathfrak{K}$ be a field which is algebraic of degree n over $\mathfrak{F}$ and $\mathfrak{K}_\phi \geqq \mathfrak{F}_\phi$. Then the degree of $\mathfrak{K}_\phi$ over $\mathfrak{F}_\phi$ is called the *ϕ-degree* of $\mathfrak{K}$. Prove by the use of our results on composites in Chapter VIII that if $\mathfrak{K}$ is normal over $\mathfrak{F}$ its ϕ-degrees are divisors of n.

2. Let ϕ be an archimedean valuation of $\mathfrak{F}$. What are the possible ϕ-degrees of $\mathfrak{K}$ over $\mathfrak{F}$?

6. Non-archimedean valuations of $\mathfrak{R}$. Consider a non-trivial non-archimedean valuation ϕ of the field $\mathfrak{R}$ of all rational numbers. Then

$$\phi(n) \leqq 1\,, \qquad \phi(a+b) \leqq \max\,[\phi(a), \phi(b)] \tag{19}$$

for every integer n and rational a, b. If $\phi(p) = 1$ for every prime p then also $\phi(a) = 1$ for every rational $a \neq 0$. This contradicts our hypothesis that ϕ is non-trivial. Hence there exists a prime p such that

$$\phi(p) = \rho\,, \qquad 0 < \rho < 1\,.$$

If q is a prime distinct from p then $\lambda q + \mu p = 1$ and $\phi(\lambda) \leqq 1$, $\phi(\mu) \leqq 1$, $1 = \phi(1)$ is at most the maximum of $\phi(\lambda) \cdot \phi(q) \leqq \phi(q)$, and $\phi(\mu) \cdot \phi(p) \leqq \rho < 1$. But then $1 \leqq \phi(q) \leqq 1$, $\phi(q) = 1$.

Every rational number has the form

$$a = p^\nu r\,, \qquad r = st^{-1} \tag{20}$$

where s and t are integers and st is prime to p. We call ν the *order* of a. It is an integer and is either positive, zero, or negative. If q is a prime factor of st we have $\phi(q) = 1$. Hence $\phi(s) = \phi(t) = \phi(st^{-1}) = 1$,

$$\phi(a) = \rho^\nu\,. \tag{21}$$

Conversely, let $\phi_p(a)$ be defined by (20) and (21) for $a \neq 0$ and $\phi_p(0) = 0$. Then $\phi_p(a)$ defines a non-trivial non-archimedean valuation of $\mathfrak{R}$. For if also $b = p^{\nu_1} r_1$ we have $\phi_p(ab) = \phi_p(p^{\nu_1+\nu} r_1 r) = \rho^{\nu_1+\nu} = \phi_p(a) \cdot \phi_p(b)$ since $r_1 r$ is a fraction whose numerator and denominator are both prime to p. We prove (19) by proving

Theorem 6. *Let* ϕ_p *be defined by* (20), (21) *and let* a *and* b *be rational numbers such that* $\phi_p(a) \neq \phi_p(b)$. *Then* $\phi_p(a \pm b)$ *is the maximum of* $\phi_p(a)$ *and* $\phi_p(b)$.

For we may let $\phi_p(a) = \rho^\nu > \phi_p(b) = \rho^{\nu_1}$. Since $\rho < 1$ this implies that $\nu_1 > \nu$, $a \pm b = p^\nu(r \pm r_1 p^{\nu_1 - \nu})$. The rational numbers $r = st^{-1}$, $r_1 = s_1 t_1^{-1}$ have numerators and denominators prime to p so that $(t_1 s \pm s_1 t p^{\nu_1 - \nu})(tt_1)^{-1} = r \pm r_1 p^{\nu_1 - \nu}$ has this property. Hence $\phi_p(a \pm b) = \rho^\nu = \phi_p(a)$.

When $\phi_p(a) \neq \phi_p(b)$ we have Theorem 12.6. If $\phi_p(a) = \phi_p(b) = \rho^\nu$ then $a + b = p^\nu(r + r_1) = p^{\nu + \nu_2} r_2$, where the denominator of r_2 is tt_1 prime to p, the numerator of $r + r_1$ is p^{ν_2} times that of r_2. Hence $\nu_2 \geqq 0$ and $\phi_p(a + b) = \rho^{\nu + \nu_2} \leqq \rho^\nu$ which is $\phi_p(a)$ and $\phi_p(b)$. We have shown that $\phi_p(a + b)$ is at most the maximum of $\phi_p(a)$ and $\phi_p(b)$. Thus ϕ_p is a non-archimedean, non-trivial valuation of $\mathfrak{R}$. This result is stated as

Theorem 7. *A function* ϕ *on* $\mathfrak{R}$ *to the field of all real numbers defines a non-trivial non-archimedean valuation of* $\mathfrak{R}$ *if and only if* $\phi = \phi_p$ *is a function defined by* (20), (21) *for* $0 < \rho < 1$ *and some prime* p.

Any two valuations defined by the same prime p are equivalent. For if $\phi_1(a) = \rho_1^\nu$ then $\phi_1(a) = \rho_1^{\log_\rho \phi(a)}$, $\phi_1(a) > \phi_1(b)$ if and only if $\phi(a) > \phi(b)$. We see in fact that the number ρ is not significant and that only the order $\nu(a)$ is important.

EXERCISE

Compute the valuations of the rational numbers $\frac{1}{4}$, 4, $\frac{20}{3}$, $\frac{9}{4}$, $\frac{15}{8}$, $\frac{1}{20}$ with respect to ϕ_2, ϕ_3, ϕ_5. Take $\rho = \frac{1}{2}$ in each case.

7. The p-adic number fields $\mathfrak{R}_p$. The derived field $\mathfrak{R}_{\phi_p}$ of $\mathfrak{R}$ according to the non-archimedean valuation ϕ_p will be designated by

$$\mathfrak{R}_p .$$

It is the same field for all values of ρ in (21) such that $0 < \rho < 1$. We call $\mathfrak{R}_p$ a *p-adic number field* and its elements *p-adic numbers*. Thus the rational numbers and the classes

$$A = [\alpha] , \qquad \alpha = \{a_n\}$$

with α a regular sequence not equivalent to a rational number, give the p-adic numbers. The field $\mathfrak{R}_p$ is complete with respect to the valuation

$$\phi(A) = \lim_{n \to \infty} \phi(a_n)$$

for any A of $\mathfrak{R}_p$. If α is a null sequence we have

$$\lim_{n \to \infty} \phi(a_n) = 0$$

and $\phi(A) = 0$, $A = 0$. Similarly, if $\alpha = a + \tau$ where a is rational and $\tau = \{t_n\}$ is a null sequence we have $A = [\alpha] = a$, $\phi(A) = \lim_{n\to\infty} \phi(a + t_n) = \phi(a)$. We now study arbitrary p-adic numbers $A = [\alpha]$, $\alpha = \{a_n\}$ regular.

Let $\alpha = \{a_n\}$ be not a null sequence. We apply (11.23) to obtain an integer n_1 and a $\delta > 0$ such that

$$(22) \qquad \phi(a_m) > \delta > 0 \qquad (m > n_1) .$$

Take $\epsilon = \frac{1}{2}\delta$ and apply the definition of a regular sequence. Then $\phi(a_m - a_q) < \frac{1}{2}\delta$ for $m > n_0$, $q > n_0$. Take $q > n_0 + n_1$ and apply (22) and Theorem 12.6. If $\phi(a_m) \neq \phi(a_q)$ then $\phi(a_m - a_q) > \delta$, a contradiction. Hence $\phi(a_m) = \rho^\nu$ for $m > n_0$, $a_m = p^\nu u_m$, $\phi(u_m) = 1$, $\alpha = p^\nu \mu$, $\mu = \{u_m\}$. A sequence $\alpha + \tau$ is equivalent to α if and only if $\mu + p^{-\nu}\tau = \mu + \tau_1$ is equivalent to μ. But if $\tau_1 = \{t_n\}$, $\lim_{n\to\infty} \phi(t_n) = 0$ we have $\mu + \tau_1 = \{u_n + t_n\}$, $\phi(u_m) = \lim_{m\to\infty} \phi(u_m + t_m) = 1$. However we have seen that $\phi(u_m + t_m) = \rho^{\nu_0}$ for $m > n_0$. Hence $\nu_0 = 0$, $\phi(u_m + t_m) = 1$ for $m > n_0$ and we have proved

Theorem 8. *A* p-*adic number* A *of* $\mathfrak{R}_p$ *has the valuation* $\phi(A) = 1$ *if and only if every representative regular sequence* $\{a_n\}$ *of* A *defines an integer* n_0 *such that* $\phi(a_n) = 1$ *for* $n > n_0$. *Every* p-*adic number has the form*

$$A = p^\nu u , \qquad \phi(A) = \rho^\nu , \qquad \phi(u) = 1$$

where u *has a representative sequence* $\{u_n\}$ *with* $\phi(u_n) = 1$ *for all* n.

The integer ν is uniquely determined by $A \neq 0$ and we call ν the *order* of A. For rational numbers it obviously is the order $\nu(a)$ already defined in $\mathfrak{R}$ and has the properties

$$\nu(AB) = \nu(A) + \nu(B) , \qquad \nu(A + B) \geqq \min [\nu(A), \nu(B)] .$$

We define the order of zero to be plus infinity and have defined $\nu(A)$ for every A of $\mathfrak{R}_p$.

The set of all numbers A of $\mathfrak{R}_p$ with $\nu(A) \geqq 0$ forms an integral domain $\mathfrak{J}_p$ in $\mathfrak{R}_p$. For we see that $\mathfrak{J}_p$ is closed under addition, subtraction, and multiplication and is a subring of $\mathfrak{R}_p$. Every quantity of $\mathfrak{R}_p$ has the form $a^{-1}b$ with b in $\mathfrak{J}_p$, and $a = 1$ or p^ν, $\nu > 0$, a in $\mathfrak{J}_p$. Since $\mathfrak{R}_p$ is a field $\mathfrak{J}_p$ has no zero-divisors and is an integral domain. We call its elements the *p-adic integers*.

The units of $\mathfrak{J}_p$ are the quantities u with $\nu(u) = 0$, that is $\phi(u) = 1$. For $\phi(u) = \rho^\nu = 1$ if and only if $\nu = 0$, $\phi(u^{-1}) = 1$, $\nu(u^{-1}) = 0$, u^{-1} is in $\mathfrak{J}_p$. Every quantity of $\mathfrak{J}_p$ is thus the product of a unit of $\mathfrak{J}_p$ by a positive or zero power of p. Let $p = qr$ with q and r in $\mathfrak{J}_p$ and have $q = p^{\nu_1}u_1$, $r = p^{\nu_2}u_2$, $p = p^{\nu_1+\nu_2}u_3$, so that $1 = p^{\nu_1+\nu_2-1}u_3$ has order zero. Then $\nu_1 + \nu_2 - 1 = 0$ and one of ν_1 and ν_2 is unity, the other is zero. Then one of q and r is a unit, the other is associated with p. We have proved that p is a prime or irreducible* element of $\mathfrak{J}_p$. Of course every associate of p is also a prime element of $\mathfrak{J}_p$.

The number p and its associates are the only prime elements of $\mathfrak{J}_p$ which are not units. For if q is a prime we have $q = up^\nu$, $\nu \geqq 0$ and q is a prime only when $\nu = 0$ or 1. Thus $\mathfrak{J}_p$ is a unique factorization domain.

The results above enable us to determine all ideals of the integral domain $\mathfrak{J}_p$. We may in fact prove that

$$(p)$$

is the only prime ideal of $\mathfrak{J}_p$. For let $\mathfrak{M}$ be an ideal of $\mathfrak{J}_p$ and let $p^\nu u$ be an element of $\mathfrak{M}$ whose order is the least order of all quantities of $\mathfrak{M}$. Since the order of every element of $\mathfrak{M}$ is $\nu_0 \geqq 0$ we have $\nu \geqq 0$ and $\nu_0 \geqq \nu$. But u^{-1} is in $\mathfrak{J}_p$, p^ν is in $\mathfrak{M}$, every element of $\mathfrak{M}$ has the form $p^{\nu_0}u = p^\nu a$ where a is in $\mathfrak{J}_p$, $p^\nu a$ is in (p^ν). Hence $\mathfrak{M} \leqq (p^\nu)$. Since $\mathfrak{M}$ contains p^ν it contains (p^ν). We have shown that every ideal $\mathfrak{M}$ of $\mathfrak{J}_p$ has the form

$$\mathfrak{M} = (p^\nu) = (p)^\nu \text{ or } \mathfrak{M} = (1)\,,$$

when $\nu = 0$. The ideal (p) is a divisorless ideal since any proper divisor of (p) has the form $(p)^\nu$ and contains (p) and must be (1), that is, $\nu = 0$. This proves that *the ideals of* $\mathfrak{J}_p$ *have a unique factorization and are the elements of an infinite cyclic group generated by* (p).

8. The series representation of *p*-adic numbers. The structure of p-adic numbers is completely determined by that of the units since every p-adic number has the form $p^\nu u$, u a unit. We let $\mu = \{u_n\}$ be a representative regular sequence of a unit u so that we may assume that $\phi(u_n) = 1$. The $u_n = s_n t_n^{-1}$, where s_n and t_n are integers prime to p. By a well-known elementary number-theoretic result the integral congruences

$$s_n \equiv t_n x \pmod{p^n}$$

have solutions $x = v_n$ with $0 < v_n < p^n$ and we may therefore write

$$s_n = t_n v_n + q_n p^n \qquad (0 < v_n < p^n)\,. \tag{23}$$

* Both the terms "prime" and "irreducible" are used in algebra. The term "prime" occurs most frequently in the literature on which the present chapter is based and we shall use it henceforth.

This gives

$$u_n = v_n + (t_n^{-1}q_n)p^n ,$$

so that since t_n is prime to p we have $\phi(t_n^{-1}q_n) = \rho^\nu \leqq 1$. But then

$$\phi(u_n - v_n) \leqq \rho^n .$$

For every $\epsilon > 0$ there exists an integer n_0 such that $\rho^n < \epsilon$ for $n > n_0$. Thus $\{v_n\}$ is equivalent to $\{u_n\}$. Also s_n is prime to p and (23) implies that v_n is prime to p, $\phi(v_n) = 1$. We have proved that every unit has a representative sequence of integers v_n prime to p and such that $0 < v_n < p^n$.

The integers v_n have the form

$$v_n = a_{n0} + a_{n1}p + \ldots + a_{n\,n-1}p^{n-1} \quad (0 \leqq a_{ni} < p,\ a_{n0} \neq 0) ,$$

which we derived in (5). Also there exists an integer n_0 such that $\phi(v_n - v_m) < \rho$ for $m \geqq n_0$, $n \geqq n_0$. Thus

$$v_n - v_m \equiv 0 \pmod{p} \qquad (m \geqq n_0,\ n \geqq n_0) .$$

Take $m = n_0$ and see that $v_n - v_m \equiv 0 \pmod{p}$, that is

$$a_{n0} - a_{n_00} = 0 .$$

Put $w_0 = a_{n_00}$ and see that

$$u = w_0 + p^e u_1 \qquad (0 < w_0 < p) .$$

Continuing with u_1, we ultimately show that

$$u = w_0 + w_1 p + \ldots + w_{k-1}p^{k-1} + p^l u_k \qquad (l \geqq k) ,$$

where $0 < w_0 < p$, $0 \leqq w_i < p$ and u_k is a unit. The integer

$$\mu_k = w_0 + w_1 p + \ldots + w_{k-1}p^{k-1} \tag{24}$$

is called the *kth convergent of* u.

The quantities μ_k were chosen so that $v_n - \mu_k \equiv 0\ (p^k)$ for $n > n_0$. Now the form of μ_k implies that $v_n - \mu_n = v_n - \mu_k - p^k y_k \equiv 0\ (p^k)$, when $n > k$, $n > n_0$. Thus $\phi(\mu_n - v_n) < \rho^k$ for any k and $n > n_k$. This proves that u has $\{\mu_k\}$ as a representative sequence, that is

$$\lim_{k \to \infty} \mu_k = u . \tag{25}$$

We now let a be any quantity of $\mathfrak{R}_p$. Then $a = p^\nu u$ where u is as in (24), (25). This proves

Theorem 9. *Every* p-*adic number has the form*

$$a = \lim_{n \to \infty} a_n ,$$

where

$$a_n = p^\nu(b_0 + b_1 p + \ldots + b_{n-1}p^{n-1}) \qquad (0 \leqq b_i < p,\ b_0 \neq 0),$$

the b_i *are integers, and* ν *is the order of* a. *Hence* a *is the sum of the infinite series*

$$p^\nu b_0 + p^{\nu+1}b_1 + \ldots \tag{26}$$

and we call a_n *the* nth *convergent of* a.

The infinite sum above is of course convergent with respect to the non-archimedean valuation we are using but not with respect to ordinary absolute value. This former fact is obviously the interpretation of $a = \lim_{n \to \infty} a_n$. But conversely every series (26) converges to a p-adic number. For we have already seen that it is sufficient that $\lim_{n \to \infty} \phi(a_{n+1} - a_n) = \lim_{n \to \infty} \phi(b_n p^{n+\nu}) = \lim_{n \to \infty} \rho^{n+\nu} = 0$, and this is true since ρ is a real number less than unity.

9. Algebraic extensions of an integral domain. Let $\mathfrak{J}$ be an integral domain with $\mathfrak{F}$ as quotient field, $\mathfrak{K} = \mathfrak{F}(\xi)$ be algebraic of degree n over $\mathfrak{F}$. We define a set $\mathfrak{E}$ of all quantities α of $\mathfrak{K}$ with the property that there exists a monic polynomial $f(x)$ of $\mathfrak{J}[x]$ such that $f(\alpha) = 0$. The quantities of $\mathfrak{E}$ will be called the *integers* of $\mathfrak{K}$, and we shall think of $\mathfrak{E}$ as an algebraic extension of $\mathfrak{J}$.

This terminology is partially justified, as in the theory of algebraic numbers where $\mathfrak{J}$ is the set of all rational integers, by the proof that $\mathfrak{E}$ is an integral domain containing $\mathfrak{J}$ as subdomain. Theorem 7.5 states that $\mathfrak{E}$ contains

$$1,\ \alpha + \beta,\ \alpha\beta,\ a\alpha$$

for every α and β of $\mathfrak{E}$ and a of $\mathfrak{J}$. Hence $\mathfrak{E}$ is a subring of the field $\mathfrak{K}$ containing $\mathfrak{J}$. It possesses no divisors of zero since there are none in the field $\mathfrak{K}$, and this demonstrates our statement above that $\mathfrak{E}$ is an integral domain. We next prove the

LEMMA. *Every quantity* γ *of* $\mathfrak{K}$ *has the form*

$$\gamma = \alpha a^{-1} , \tag{27}$$

where a *is in* $\mathfrak{J}$ *and* α *is in* $\mathfrak{E}$.

For the minimum equation of γ has the form $g(y) \equiv y^m + b_1y^{m-1} + \ldots + b_m = 0$ where the b_i are in the quotient field $\mathfrak{F}$ of $\mathfrak{J}$. Then the b_i are the quotients of elements of $\mathfrak{J}$ and if we choose a to be the product of the denominators of the b_i we have

$$ab_i = c_i \text{ in } \mathfrak{J}\,.$$

But $\alpha = a\gamma$ is a root of $f(x) = f(ay) = a^m g(y) = x^m + c_1x^{m-1} + c_2ax^{m-2} + \ldots + c_ma^{m-1} = 0$ which has coefficients c_ia^{i-1} in $\mathfrak{J}$.

If ξ generates $\mathfrak{K} = \mathfrak{F}[\xi]$ over $\mathfrak{F}$, then $\xi = \xi_0a^{-1}$ and ξ_0 is an integer of $\mathfrak{K}$ and generates $\mathfrak{K}$ over $\mathfrak{F}$. The above proof of the lemma actually implies that we may assume that the minimum function of a suitably chosen generator ξ of $\mathfrak{K}$ over $\mathfrak{F}$ has the form

$$f(x) = x^n + a_1x^{n-1} + \ldots + a_n \qquad (a_i \text{ in } \mathfrak{J})\,.$$

Every polynomial in ξ with coefficients in $\mathfrak{J}$ is an integer of $\mathfrak{K}$ since $\mathfrak{E}$ is an integral domain containing ξ and $\mathfrak{J}$. But it is not true in general that there are no other integers.

The equation $f(x) = 0$ defines conjugate fields $\mathfrak{K}_i = \mathfrak{F}(\xi_i)$ where $\xi = \xi_1$. Every quantity $\alpha_i = \alpha(\xi_i)$ of $\mathfrak{K}_i$ satisfies the equation

$$G(y) = (y - \alpha_1)(y - \alpha_2) \ldots (y - \alpha_n) \tag{28}$$

which has coefficients which are in the field $\mathfrak{F}$. If $\mathfrak{F}(\alpha_1)$ is a subfield of degree m of $\mathfrak{K} = \mathfrak{K}_1 = \mathfrak{F}(\xi)$ of degree $n = mr$ over $\mathfrak{F}$ we have

$$G(y) = [g(y)]^r \tag{29}$$

where $g(y)$ is the minimum function of α by Theorem 7.14, and is a monic polynomial of $\mathfrak{F}[x]$. But we have not proved that either $g(y)$ or $G(y)$ have coefficients in $\mathfrak{J}$ when α is an integer of $\mathfrak{K}$. This is however true in the case of

Theorem 10. *Let $\mathfrak{J}$ be a unique factorization domain and α be an integer of $\mathfrak{K}$. Then the coefficients of the minimum function of α are in $\mathfrak{J}$.*

For if $h(y)$ is a monic polynomial of $\mathfrak{J}[x]$ and $h(\alpha) = 0$ we infer that $h(y)$ is divisible by $g(y)$. But Theorem 2.16 states that $g(y)$ has coefficients in $\mathfrak{J}$ and so evidently does $G(y)$ of (29).

The norm

$$N(\alpha) = N_{\mathfrak{K}|\mathfrak{F}}(\alpha) = \alpha_1\, \alpha_2 \ldots \alpha_n$$

is plus or minus the final coefficient of $G(y)$. When $\mathfrak{J}$ is a u.f. domain and α is an integer of $\mathfrak{K}$ the final coefficient c_m of $g(y)$ is in $\mathfrak{J}$ and so is $c_m^r = \pm N(\alpha)$.

10. Algebraic extensions of p-adic number fields. Let $\mathfrak{R}_p$ be the p-adic number field which is the derived field of the field of all rational numbers according to the non-archimedean valuation defined by the prime p. We then consider an algebraic extension $\mathfrak{K}$ of finite degree n over $\mathfrak{F} = \mathfrak{R}_p$ and have $\mathfrak{K} = \mathfrak{F}(\xi)$. The field $\mathfrak{R}_p$ is the quotient field of the unique factorization integral domain $\mathfrak{J}_p$ of all p-adic integers. Thus $\mathfrak{K}$ contains the integral domain $\mathfrak{E}$ defined in the last section and consisting of what we called the integers of $\mathfrak{K}$. Moreover we may take ξ in $\mathfrak{E}$ and Theorem 12.10 states that α is an integer of $\mathfrak{K}$ if and only if the minimum function of α has p-adic integer coefficients. It also states that α is an integer of $\mathfrak{K}$ if and only if the equation $G(y)$ of (28) has p-adic integer coefficients.

An integer α of $\mathfrak{K}$ has a p-adic integral norm. But we may also prove the rather surprising converse. To do so we consider an arbitrary polynomial with p-adic number coefficients. By multiplying by an integral power of p we may assume that our polynomial

$$f(x) = a_0x^n + \dots + a_n \tag{30}$$

has coefficients in the set $\mathfrak{J}_p$ of all p-adic integers of $\mathfrak{R}_p$ and, since $\mathfrak{J}_p$ is a unique factorization domain, that these coefficients have no non-unit common factor. Thus $f(x)$ is what we called a *primitive* polynomial in Chapter II.

If $g(x)$ is any polynomial of $\mathfrak{J}_p[x]$, we write $\bar{g}(x)$ for the polynomial whose coefficients are the first convergents of the coefficients of $g(x)$. Hence the coefficients of $\bar{g}(x)$ are in the field $\mathfrak{G}$ of all residue classes modulo p. We now prove the

LEMMA. *Let* $f(x) \equiv g_0(x)h_0(x) \pmod p$ *where* $\bar{g}_0(x)$, $\bar{h}_0(x)$ *are relatively prime polynomials of* $\mathfrak{G}[x]$. *Then* $f(x) = g(x)h(x)$ *where* $g(x)$ *and* $h(x)$ *are in* $\mathfrak{J}_p[x]$, *and* $\bar{g}(x) = \bar{g}_0(x)$, $\bar{h}(x) = \bar{h}_0(x)$.

For we may replace $g_0(x)$ and $h_0(x)$ by congruent polynomials modulo p and can assume that $g_0(x)$ has degree r, $h_0(x)$ has degree s, $r + s = n$, and the leading term of $f(x)$ is the product of the leading terms of $g_0(x)$ and $h_0(x)$. Write

$$\begin{cases} g_{k-1}(x) = g_0(x) + pu_1(x) + \dots + p^{k-1}u_{k-1}(x)\,, \\ h_{k-1}(x) = h_0(x) + pv_1(x) + \dots + p^{k-1}v_{k-1}(x)\,, \end{cases} \tag{31}$$

where the $u_i(x)$ and $v_i(x)$ are polynomials of degrees at most $r - 1$ and $s - 1$, respectively. Then $g_k(x)$ has degree r, $h_k(x)$ has degree s and the product of their leading terms is the leading term of $f(x)$. The sequences $\{g_k(x)\}$, $\{h_k(x)\}$ have convergent coefficients and will converge to the desired polynomials $g(x)$, $h(x)$ if we can prove that the u_k, v_k can be chosen so that

$$f(x) \equiv g_{k-1}(x)h_{k-1}(x) \pmod{p^k}.$$

This is true for $k = 1$ and we assume it true for the exponent k and write

$$f(x) - g_{k-1}(x)h_{k-1}(x) = p^k q(x)$$

where $q(x)$ evidently has degree at most $n - 1$. We take the polynomials $\bar{q}(x) \equiv q(x) \pmod{p}$, $\bar{g}_0(x)$, $\bar{h}_0(x)$ with coefficients in the finite field $\mathfrak{G}$, whose quantities are the residue classes modulo the given prime p. Since $\bar{h}_0(x)$ and $\bar{g}_0(x)$ are relatively prime there exist polynomials $a(x)$ and $b(x)$ in $\mathfrak{G}[x]$ such that $a(x)\bar{h}_0(x) + b(x)\bar{g}_0(x) = 1$. Multiplying by $\bar{q}(x)$ we have $c(x)\bar{h}_0(x) + d(x)\bar{g}_0(x) = \bar{q}(x)$. Adding a multiple of p by a polynomial of $\mathfrak{J}_p[x]$ we have

$$c_0(x)h_0(x) + d_0(x)g_0(x) = q_0(x)\,, \tag{32}$$

where $q_0(x) \equiv q(x) \pmod{p}$ and, since the polynomial added has degree at most the maximum of r and s, we are sure that the degree of $q_0(x)$ is at most $n - 1 = r + s - 1$. Then we may replace $d_0(x)$ by its remainder on division by $h_0(x)$ and have $d_0(x)$ of degree at most $s - 1$. This of course changes c_0 and for the new $c_0(x)$ we have $c_0(x)h_0(x) = q_0(x) - d_0(x)g_0(x)$ which has degree at most $r + s - 1$. Since the degree of $h_0(x)$ is s the degree of $c_0(x)$ is at most $r - 1$. We take $c_0(x) = u_k(x)$, $d_0(x) = v_k(x)$ and have $q(x) \equiv v_k(x)g_0(x) + u_k(x)h_0(x) \pmod{p}$,

$$p^k q(x) = f(x) - g_{k-1}(x)h_{k-1}(x) \equiv p^k[v_k(x)g_0(x) + u_k(x)h_0(x)] \pmod{p^{k+1}}\,. \tag{33}$$

Then if

$$g_k(x) = g_{k-1}(x) + p^k u_k(x)\,, \qquad h_k(x) = h_{k-1}(x) + p^k v_k(x) \tag{34}$$

we infer that since p^{2k} is divisible by p^{k+1},

$$\begin{aligned} f(x) - g_k(x)h_k(x) &= f(x) - g_{k-1}(x)h_{k-1}(x) \\ &\qquad - p^k[u_k(x)h_{k-1}(x) + v_k(x)g_{k-1}(x)] \\ &\equiv 0 \pmod{p^{k+1}}\,. \end{aligned} \tag{35}$$

The lemma above is applied as follows. We let $g(x) = x^n + a_1x^{n-1} + \ldots + a_n$ be the minimum function of a quantity α which is algebraic over

$\mathfrak{R}_p$ and suppose that a_n is in $\mathfrak{J}_p$. Then either all the a_i are in $\mathfrak{J}_p$ or there is a least power p^t such that $p^t g(x) = f(x) = p^t x^n + b_1 x^{n-1} + \ldots + p^t a_n$ has coefficients in $\mathfrak{J}_p$. The polynomial $f(x)$ is irreducible in $\mathfrak{R}_p$ and is primitive. Also $f(x) \equiv x^k(c_0 x^r + \ldots + c_r) \pmod{p}$ where the c_i are integers between 1 and $p - 1$, $k > 0$ and c_r is evidently the first convergent of the last b_i not divisible by p. Since x^k is prime to $c_0 x^r + \ldots + c_r$ we use the lemma and obtain a contradiction. This proves that the a_i are all integral and that α is an integer of $\mathfrak{K}$.

11. The valuation theory of $\mathfrak{K}$ over $\mathfrak{R}_p$. A quantity α of $\mathfrak{K}$ *over* $\mathfrak{R}_p$ *is now an integer of* $\mathfrak{K}$ *if and only if*

$$N_{\mathfrak{K}|\mathfrak{R}_p}(\alpha) = N(\alpha) \tag{36}$$

is a p-*adic integer*. The elements of $\mathfrak{R}_p$ have the form $p^\nu u$ where u is a unit of $\mathfrak{J}_p$ and ν is any ordinary integer, the order of $p^\nu u$. We let π be an integer of $\mathfrak{K}$ whose norm has the least positive order f. Thus $N(\pi) = p^f u_0$, $N(\alpha) = p^\nu u_1$ for any α of $\mathfrak{K}$. Write $\nu = \lambda f + r$, $0 \leqq r < f$, and obtain $N(\alpha\pi^{-\lambda}) = p^r u$ where u is a unit $u = u_1 u_0^{-\lambda}$ and $r < f$. Then our hypothesis on f implies that $r = 0$ and

$$\alpha = \pi^\lambda U\,, \qquad N(U) = u\,,$$

where u is a unit of $\mathfrak{R}_p$.

The units of the integral domain $\mathfrak{E}$ of all integers of $\mathfrak{K}$ are defined to be the quantities U for which U^{-1} is also an integer. Thus U is a unit if and only if $N(U)$ and $N(U^{-1}) = [N(U)]^{-1}$ are integers of $\mathfrak{J}_p$. It is evident that U is a unit of $\mathfrak{E}$ if and only if its norm is a unit of $\mathfrak{J}_p$. Now $\alpha = \pi^\lambda U$ with U a unit of $\mathfrak{E}$ as above and if also $\alpha = \pi^{\lambda_1} U_1$ we have $\pi^{\lambda - \lambda_1} = U_1 U^{-1}$ is a unit, $N(\pi^{\lambda-\lambda_1}) = p^{f(\lambda-\lambda_1)} u_2$ is a unit of $\mathfrak{J}_p$, $p^{f(\lambda-\lambda_1)}$ has order zero and $\lambda = \lambda_1$. We have proved

Theorem 11. *Every quantity* α *of an algebraic extension* $\mathfrak{K}$ *of finite degree over* $\mathfrak{R}_p$ *is uniquely expressible in the form*

$$\alpha = \pi^\lambda U \tag{37}$$

where U *is a unit of the set* $\mathfrak{E}$ *of all integers of* $\mathfrak{K}$, λ *is an ordinary integer called the order of* α. *Then* α *is in* $\mathfrak{E}$ *if and only if* $\lambda \geqq 0$.

The quantity π is a prime element of $\mathfrak{E}$ and it and its associates are the only prime elements of $\mathfrak{E}$. Hence $\mathfrak{E}$ is a unique factorization domain. The proofs are exactly as in the corresponding theorems for p of $\mathfrak{J}_p$ and we shall not repeat them. The reader may also verify that the only divisorless ideal of $\mathfrak{E}$ is (π), and that every ideal of $\mathfrak{E}$ is a power $(\pi)^\lambda$, $\lambda \geqq 0$. The order $\lambda(\alpha)$ of any α of $\mathfrak{E}$ also is seen to have the properties

$$(38) \qquad \lambda(\alpha\beta) = \lambda(\alpha) + \lambda(\beta) , \qquad \lambda(\alpha + \beta) \geqq \min [\lambda(\alpha), \lambda(\beta)] ,$$

while we repeat that $\lambda(\alpha) \geqq 0$ for α in $\mathfrak{E}$, $\lambda(\alpha) = 0$ for α a unit.

The integral domain $\mathfrak{E}$ contains $\mathfrak{J}_p$ as a subdomain and contains p. The order of p is some integer e and

$$(39) \qquad (p) = (\pi)^e .$$

If we define

$$(40) \qquad \phi(\alpha) = \rho^{\lambda/e} , \qquad \phi(0) = 0 ,$$

for any α of order λ of $\mathfrak{K}$ we have

$$(41) \qquad \phi(a) = \phi_p(a) = \rho^\nu$$

for any a of $\mathfrak{R}_p$. For let $a = p^\nu u$ with u a unit of $\mathfrak{R}_p$ so that the second equality in (41) holds. Then u is also a unit of $\mathfrak{E}$ and $a = \pi^{e\nu} U$ where U is a unit of $\mathfrak{E}$, a has order $\lambda = e\nu$, and $\rho^{\lambda/e} = \rho^\nu = \phi(a)$ by (40). We now obtain

Theorem 12. *The function* $\phi(\alpha)$ *defined by* (39), (40), *is a non-trivial non-archimedean valuation of* $\mathfrak{K}$ *preserving the valuation* (41) *of* $\mathfrak{R}_p$ *and is the only such valuation preserving* (41). *Moreover,* $\mathfrak{K}$ *is complete with respect to* ϕ.

For by (38) we have $\phi(\alpha\beta) = \phi(\alpha)\phi(\beta)$, $\phi(\alpha + \beta) \leqq \max. [\phi(\alpha), \phi(\beta)]$, ϕ is non-trivial since (41) holds and ϕ is non-trivial and non-archimedean in the subfield $\mathfrak{R}_p$. Let $\psi(\alpha)$ be another valuation of $\mathfrak{K}$ which is non-trivial, non-archimedean, and is such that (41) holds. Then we take α an integer of $\mathfrak{K}$ and thus a root of $x^m + b_1 x^{m-1} + \dots + b_m = 0$ with integral b_i in $\mathfrak{J}_p$. This equation may be used to express every power of α in the form

$$\alpha^t = c_{t0} + c_{t1}\alpha + \dots + c_{t\, m-1}\alpha^{m-1} ,$$

with c_{ti} in $\mathfrak{J}_p$ and $\psi(c_{ti}) = \phi(c_{ti}) = \phi_p(c_{ti}) \leqq 1$ for every integer of $\mathfrak{R}_p$. Hence

$$\psi(\alpha^t) \leqq M ,$$

where M is the real number $1 + \psi(\alpha) + \dots + \psi(\alpha^{m-1})$. If $\psi(\alpha) = \sigma > 1$ then for some t, $\psi(\alpha^t) = \sigma^t > M$, which is impossible. Hence $\psi(\alpha) \leqq 1$ for every integer α of $\mathfrak{K}$.

It is evident that $\psi(\alpha^{-1}) = [\psi(\alpha)]^{-1} \geqq 1$ for integral α and hence $\psi(U)$

$= 1$ for any unit U of $\mathfrak{E}$. If also $\psi(\pi) = 1$ then $\psi(p) = 1 = \phi(p) = \rho < 1$, a contradiction, so that $\psi(\pi) = \sigma < 1$, $\psi(\pi^e) = \sigma^e = \phi(p) = \rho$, $\sigma = \rho^{1/e}$. Then $\psi(\pi^\lambda U) = \rho^{\lambda/e} = \phi(\pi^\lambda U)$ and we have proved that $\psi(a) = \phi(a)$. Finally, $\mathfrak{K}$ is complete by Theorem 11.11.

The exponent e of (39) is connected with the degree n of $\mathfrak{K}$ over $\mathfrak{R}_p$ in an interesting manner. Write

$$\mathfrak{N}(\mathfrak{M}) \tag{42}$$

for the ideal generated by the norms of the quantities of the ideal $\mathfrak{M}$. Since $\mathfrak{M} = (\pi)^\lambda$ every quantity of $\mathfrak{M}$ has the form $a\pi^\lambda$, $N(a\pi^\lambda) = N(a) \cdot [N(\pi)]^\lambda = u^\lambda N(a) \cdot p^{f\lambda}$ where we have already seen that

$$N_{\mathfrak{K} | \mathfrak{R}_p}(\pi) = u \cdot p^f . \tag{43}$$

Hence $\mathfrak{N}(\mathfrak{M})$ contains only multiples of $p^{\lambda f}$ and contains $u^\lambda p^{\lambda f}$ for $a = 1$. Since u is a unit, $u^{-\lambda}$ is in $\mathfrak{J}_p$, $p^{\lambda f}$ is in $\mathfrak{N}(\mathfrak{M})$, and every quantity of $\mathfrak{N}(\mathfrak{M})$ is a multiple of its quantity $p^{\lambda f}$. It is thus true that

$$\mathfrak{N}[(\pi^\lambda)] = (p)^{f\lambda} , \tag{44}$$

and in particular that

$$\mathfrak{N}[(\pi)] = (p)^f . \tag{45}$$

Now $(p) = (\pi)^e$ so that

$$\mathfrak{N}_{\mathfrak{K} | \mathfrak{R}_p}[(p)] = (p)^n = (p)^{fe} , \tag{46}$$

and thus that

$$n = ef . \tag{47}$$

We call f the *degree* of (π) and e its *ramification order*.

Analogous results are obtainable for relative fields. We let

$$\mathfrak{K} > \mathfrak{H} > \mathfrak{R}_p \tag{48}$$

where $\mathfrak{K}$ has degree n over $\mathfrak{R}_p$, $\mathfrak{H}$ has degree m over $\mathfrak{R}_p$ so that

$$n = mq$$

and q is the degree of $\mathfrak{K}$ over $\mathfrak{H}$. We then define a prime P of $\mathfrak{H}$ and have

$$(p) = (P)^{e_1} , \qquad \mathfrak{N}_{\mathfrak{H} | \mathfrak{R}_p}[(P)] = (p)^{f_1} , \qquad e_1 f_1 = m . \tag{49}$$

Now

$$(50)\qquad (P) = (\pi)^{e_2}, \qquad \mathfrak{N}_{\mathfrak{K}|\mathfrak{H}}[(\pi)] = (P)^{f_2},$$

since $\mathfrak{N}_{\mathfrak{K}|\mathfrak{H}}[(\pi)]$ is evidently an ideal of $\mathfrak{H}$. However,

$$N_{\mathfrak{K}|\mathfrak{R}_p}(\pi) = N_{\mathfrak{H}|\mathfrak{R}_p}[N_{\mathfrak{K}|\mathfrak{H}}(\pi)] = N_{\mathfrak{H}|\mathfrak{R}_p}(uP^{f_2}) = u_0 p^{f_1 f_2},$$

where u and u_0 are units, so that

$$(51)\qquad (p) = (\pi)^{e_1 e_2}, \qquad e_1 e_2 = e, \qquad f_1 f_2 = f,$$

where e and f are the ramification order and degree, respectively, of π. We call e_2 and f_2 the (relative) ramification order and degree, respectively, of π with respect to P and may show that the degree of $\mathfrak{K}$ over $\mathfrak{H}$ is their product. For $n = ef = (e_1 f_1)(e_2 f_2) = m e_2 f_2$, $e_2 f_2 = q$ as desired.

12. Ideals in algebraic number fields. The non-archimedean valuations of an algebraic number field $\mathfrak{K}$ of degree n over the field $\mathfrak{R}$ of all rational numbers depend essentially upon a number of results from the theory of algebraic numbers. The theorems are easy to understand in view of the discussion of ideals we have already made but the proofs cannot of course be given here. This is not due to any particular difficulty but to their length. Our goal is to find the valuation functions ϕ of $\mathfrak{K}$ over $\mathfrak{R}$ and we shall of course assume the necessary properties of $\mathfrak{K}$ in terms of which ϕ must be described.

We let $\mathfrak{J}$ be the ring of all (algebraic) integers of $\mathfrak{K}$ and consider the ideals, that is, invariant subrings, of $\mathfrak{J}$. It is natural for us to assume the

Fundamental Theorem of Ideal Theory. *Every ideal is uniquely expressible, apart from the order of the factors, as a product of divisorless* ideals.*

We next have

LEMMA 1. *Let α and β be integers of $\mathfrak{K}$ and $(\alpha) = (\beta)\mathfrak{M}$ where $\mathfrak{M}$ is an ideal. Then $\mathfrak{M}$ is a principal ideal and $\alpha = \beta\gamma$, γ an integer of $\mathfrak{K}$, $(\gamma) = \mathfrak{M}$.*

For α is contained in $(\alpha) = (\beta)\mathfrak{M}$ and must have the form $\alpha = \beta\gamma$ with γ in $\mathfrak{M}$. Hence $(\alpha) = (\beta\gamma) = (\beta)(\gamma)$ by Section 11.2. The uniqueness in the Fundamental Theorem combined with $(\beta)(\gamma) = (\beta)\mathfrak{M}$ implies that $\mathfrak{M} = (\gamma)$.

This result is used as follows. We wish to know if an algebraic integer α is divisible by β. Factor (α) and (β) into divisorless ideals and see if $(\alpha) = (\beta)\mathfrak{M}$ where $\mathfrak{M}$ is the product of the divisorless ideal factors of (α) which

* An ideal $\mathfrak{P}$ is called a *prime* ideal if the difference ring $\mathfrak{A} - \mathfrak{P}$ is an integral domain. In the case of algebraic numbers this concept coincides with that of divisorless ideal where, as in Theorem 11.1, we have the stronger condition $\mathfrak{A} - \mathfrak{P}$ a field.

remain when we group certain of them to form (β). If this is possible, $\alpha = \beta\gamma$ by the lemma and α in $\mathfrak{J}$ is divisible by β in $\mathfrak{J}$.

There is evidently no loss of generality if we replace the principal ideals in our factorizations by the integers themselves. Thus we say that an integer α is divisible by an ideal $\mathfrak{M}$ and mean that (α) is divisible by $\mathfrak{M}$, that is α is contained in $\mathfrak{M}$.

When $\mathfrak{B}$ and $\mathfrak{C}$ are ideals of our particular $\mathfrak{J}$ and $\mathfrak{B}$ is divisible by $\mathfrak{C}$ then it is known that $\mathfrak{B} = \mathfrak{C}\mathfrak{D}$ for an ideal $\mathfrak{D}$ of $\mathfrak{J}$. We shall assume and use this result.

If $\mathfrak{P}$ is a divisorless ideal of $\mathfrak{K}$ the ideal $\mathfrak{P}^2$ is divisible by (contained in) $\mathfrak{P}$ but does not contain (divide) $\mathfrak{P}$. Hence there exists an integer π of $\mathfrak{P}$ which is divisible by $\mathfrak{P}$ but not by $\mathfrak{P}^2$. We are now able to write

$$\pi = \mathfrak{P}\mathfrak{Q}\,, \tag{52}$$

where $\mathfrak{Q}$ is an ideal prime to $\mathfrak{P}$. The norm $N_{\mathfrak{K}|\mathfrak{R}}(\pi)$ of the integer π is the product of π by another integer σ of $\mathfrak{K}$. It is a rational integer which may be factored into prime factors p_i and

$$\pi\sigma = p_1 \ldots p_r\,. \tag{53}$$

Factor the p_i into divisorless ideals and see that for some rational prime factor p of $N_{\mathfrak{K}|\mathfrak{R}}(\pi)$ we have

$$p = \mathfrak{P}\mathfrak{M}\,,$$

where $\mathfrak{M}$ is an ideal. Then the Fundamental Theorem states that

$$p = \mathfrak{P}^e\mathfrak{P}_0\,, \tag{54}$$

where $\mathfrak{P}_0$ is an ideal prime to $\mathfrak{P}$ and e is called the *ramification order* of p with respect to $\mathfrak{P}$. Thus *every divisorless ideal arises from the factorization of a rational prime.* We next prove

LEMMA 2. *Let* $\mathfrak{Q}$ *be the ideal of* (52). *Then there exists an ideal* $\mathfrak{Q}_0$ *such that* $\mathfrak{Q}\mathfrak{Q}_0 = \tau$ *where* τ *is an algebraic integer prime to* $\mathfrak{P}$.

For $\mathfrak{Q}$ is not divisible by $\mathfrak{P}$, is not contained in $\mathfrak{P}$ and there exists a τ in $\mathfrak{Q}$ and not in $\mathfrak{P}$. Hence $\tau = \mathfrak{Q}\mathfrak{Q}_0$ for an ideal $\mathfrak{Q}_0$ and τ is not in $\mathfrak{P}$ and thus not divisible by $\mathfrak{P}$. Notice that we repeatedly use τ instead of the principal ideal (τ). We shall usually do so as we have already said.

We finally let $\mathfrak{P}$ and π be given as above and prove

LEMMA 3. *Every non-zero quantity of* $\mathfrak{K}$ *is expressible in the form*

$$\alpha = \pi^\lambda\beta\,, \tag{55}$$

where λ is a rational integer uniquely determined by α and $\mathfrak{P}$, and β is a quotient of integers in $\mathfrak{K}$ each prime to $\mathfrak{P}$.

For write $\alpha = \alpha_1\alpha_2^{-1}$ where α_1 and α_2 are algebraic integers. Then $\alpha_1 = \mathfrak{P}^{\lambda_1}\mathfrak{M}_1$, $\alpha_2 = \mathfrak{P}^{\lambda_2}\mathfrak{M}_2$, for rational integers λ_1 and λ_2 and ideals $\mathfrak{M}_1$ and $\mathfrak{M}_2$ each prime to $\mathfrak{P}$. We define

$$\lambda = \lambda_1 - \lambda_2 ,$$

and may write $\alpha_1\mathfrak{Q}^{\lambda_1} = \pi^{\lambda_1}\mathfrak{M}_1$ where $\mathfrak{Q}$ is given by (52). Then $\alpha_1(\mathfrak{Q}\mathfrak{Q}_0)^{\lambda_1} = \alpha_1\tau^{\lambda_1} = (\mathfrak{M}_1\mathfrak{Q}_0^{\lambda_1})\pi^{\lambda_1}$ as in Lemma 2. By Lemma 1 the ideal $\mathfrak{M}_1\mathfrak{Q}_0^{\lambda_1}$ is a principal ideal and

$$\alpha_1 = \delta_1\pi^{\lambda_1}\tau^{-\lambda_1}$$

where δ_1 and τ are evidently prime to $\mathfrak{P}$. Similarly, $\alpha_2 = \delta_2\pi^{\lambda_2}\tau^{-\lambda_2}$ and

$$\alpha = \pi^{\lambda}\delta_1\delta_2^{-1}\tau^{\lambda_2-\lambda_1} = \pi^{\lambda}\beta$$

where the numerator and denominator of β are $\delta_1\tau^{\lambda_2}$ and $\delta_2\tau^{\lambda_1}$, respectively, and are prime to $\mathfrak{P}$.

The greatest common ideal divisor of two rational integers is their ordinary integral g.c.d. This gives

LEMMA 4. *Let* q *be an integer prime to* p. *Then the integer λ of* (55) *for α =* q *has the value zero.*

13. Fields of $\mathfrak{P}$-adic numbers. Let $\mathfrak{K} = \mathfrak{R}(\xi)$ be an algebraic field of degree n over the field $\mathfrak{R}$ of all rational numbers. If ϕ is a non-archimedean valuation function of $\mathfrak{K}$ we have proved that there exists a rational prime p with $\phi(p) = \rho < 1$. The derived field $\mathfrak{K}_\phi$ of $\mathfrak{K}$ was defined in the last chapter so that both $\mathfrak{R}_\phi = \mathfrak{R}_p$ and ξ are contained in $\mathfrak{K}_\phi$. But then Theorem 11.12 states that

$$\mathfrak{K}_\phi = \mathfrak{R}_p(\xi) .$$

We shall obtain all such valuation functions ϕ.

We first define a set of valuation functions. Let p be a given rational prime and $\mathfrak{P}$ be a prime ideal factor of p. Use (54), (55) and define

$$\sigma = \rho^{1/e} , \qquad \phi(0) = 0 , \qquad \phi(\alpha) = \sigma^{\lambda} , \tag{56}$$

for every $\alpha \neq 0$ of $\mathfrak{K}$. This is of course possible only because we proved the integer λ unique in Lemma 3. Then by Lemma 4

$$\phi(p) = \sigma^{e} = \rho , \qquad \phi(\alpha) = \rho^{\nu} \tag{57}$$

as in (20), (21). The function ϕ preserves the valuation (20), (21) of $\mathfrak{R}$ and it is easily verified that it provides a non-trivial non-archimedean valuation of $\mathfrak{K}$. We shall prove

Theorem 13. *The valuation* (56) *with* $\sigma = \rho^{1/e}$ *is the only valuation of* $\mathfrak{K}$ *preserving the valuation* (57) *of* $\mathfrak{R}$.

For let ϕ be a valuation and pass to $\mathfrak{K}_\phi$. We define a prime quantity π_0 of $\mathfrak{K}_\phi$ and have shown that every quantity of $\mathfrak{K}_\phi$ is uniquely expressible in the form

$$\alpha = \pi_0^\lambda U ,$$

where U is a unit of the domain of all integers of $\mathfrak{K}_\phi$ and λ is the order of α. Every integer of $\mathfrak{K}$ is obviously an integer of $\mathfrak{K}_\phi$ and must have non-negative order. We define $\mathfrak{P}$ to be the set of all integers α of $\mathfrak{K}$ with positive order.

LEMMA. *The set* $\mathfrak{P}$ *is a divisorless ideal of* $\mathfrak{K}$.

For if α is an integer of $\mathfrak{K}$ and β is in $\mathfrak{P}$, we have $\lambda(\alpha\beta) = \lambda(\alpha) + \lambda(\beta) \geqq \lambda(\beta) \geqq 1$ and $\alpha\beta$ is in $\mathfrak{K}$. Also $\lambda(\alpha - \beta)$ is at least the minimum of $\lambda(\alpha) \geqq 1$ and $\lambda(-\beta) \geqq 1$ for α and β in $\mathfrak{P}$ and $\alpha - \beta$ is in $\mathfrak{P}$. We have proved $\mathfrak{P}$ an ideal. If $\mathfrak{P}$ were not a divisorless ideal we could use the fundamental ideal theorem and write $\mathfrak{P} = \mathfrak{Q}\mathfrak{S}$ where neither $\mathfrak{Q}$ nor $\mathfrak{S}$ is $\mathfrak{P}$. Then both $\mathfrak{Q}$ and $\mathfrak{S}$ contain integers of order zero and if we designate these by η, ζ we have $\lambda(\eta\zeta) = \lambda(\eta) + \lambda(\zeta) = 0$ whereas $\eta\zeta$ is in $\mathfrak{P}$ and has positive λ.

We have of course used the ideal theory for algebraic integers in a fundamental way in the above proof.

Next let π be an integer of $\mathfrak{P}$ whose order y is the least order of any integer of $\mathfrak{P}$. The integers of $\mathfrak{P}^2$ have the form $\alpha = \beta_1\gamma_1 + \dots + \beta_r\gamma_r$ with β_i and γ_i in $\mathfrak{P}$ and $\lambda(\beta_i\gamma_i) = \lambda(\beta_i) + \lambda(\gamma_i) \geqq 2y$, $\lambda(\alpha) \geqq 2y$ for every α of $\mathfrak{P}^2$. It follows that π is in $\mathfrak{P}$ and not in $\mathfrak{P}^2$ and that π is thus divisible by $\mathfrak{P}$ and not by $\mathfrak{P}^2$. Thus π is an integer of Lemma 3, Section 12.12. As in that lemma we may write $\alpha = \pi^\lambda\beta$. Since both numerator and denominator of β are prime to $\mathfrak{P}$ they are integers not in $\mathfrak{P}$ and have order zero. Thus β is a unit of $\mathfrak{K}_\phi$ and the order of α is

$$\lambda(\alpha) = \lambda \cdot y .$$

But then

$$\phi(\alpha) = \rho^{1/e \cdot \lambda \cdot y} .$$

It remains only to prove that $y = 1$.

To make this proof we notice that the quantities of $\mathfrak{K}_\phi$ are classes of regular sequences $\{\alpha_q\}$ with α_q in $\mathfrak{K}$. Now when $\phi(\alpha_q) \neq \phi(\alpha_m)$ we have

$\lambda_q \neq \lambda_m$ and may take $\lambda_q > \lambda_m$, $a_m - a_q = \pi^{\lambda_m}(\beta_m - \pi^{\lambda_q - \lambda_m}\beta_q)$. The second factor is prime to $\mathfrak{P}$ and $\phi(a_m - a_q) = \rho^{1/e \cdot \lambda_m \cdot y}$. As in the proof of Theorem 12.8, the regularity of $\{a_q\}$ implies that $\phi(a_m) = \phi(a_g)$ for a fixed g and every $m \geqq g$. But then we have shown that the valuation of the quantity a of $\mathfrak{K}_\phi$ defined by our regular sequence is

$$\phi(a) = \phi(a_g) = \rho^{(1/e)\lambda_g y}.$$

In particular consider the prime element π_0 of $\mathfrak{K}_\phi$ and let $\{\gamma_q\}$ be a representative sequence,

$$\phi(\pi_0) = \phi(\gamma_g) = \rho^{(1/e)\Lambda_g y} = \rho^{1/e}.$$

Hence $\Lambda_g y = 1$ for ordinary integers Λ_g and y and $y > 0$ has the value unity. Notice that then we may take $\pi = \pi_0$, that is π is a prime element of $\mathfrak{K}_\phi$.

We have proved our theorem and have determined all valuation functions of algebraic number fields of finite degree over the rational number field.

EXERCISES

1. What are all possible valuations of an algebraic number field?

2. The elements of $\mathfrak{K}_\phi$ are called *$\mathfrak{P}$-adic numbers* and $\mathfrak{K}_\phi$ is usually designated by $\mathfrak{K}_{\mathfrak{P}}$ when ϕ is non-archimedean and therefore defined by a prime ideal $\mathfrak{P}$. Generalize the proof of Theorem 12.9 to obtain the series representation of any $\mathfrak{P}$-adic number as a power series in π.

14. The literature. In recent years extensive bibliographies on algebra have been compiled and it would certainly be useless for us to try to give a competitive bibliography in a short space. So we refer the reader to these collections. For the theory of matrices there is a very large collection of references in J. H. M. Wedderburn's *Lectures on Matrices* and the tract by C. C. MacDuffee in Volume II of the *Ergebnisse der Mathematik.* Both are in English. For references to the theory of algebras see the *Ergebnisse* tract (in German) of M. Deuring. This is in Volume IV. Our last two chapters are used principally in the class-field theory, and probably the best expositions of this subject are in the rather inaccessible mimeographed notes of H. Hasse (Marburg, 1932–33) and in the paper of C. Chevalley, *Journal of the Tokyo University Faculty of Science*, II (1929–34), 365–474. The material of the chapters themselves was based on papers by J. Kürschák, *Journal für Mathematik*, CXLII (1913), 211–53; A. Ostrowski, *Acta mathematica*, XLI (1918), 271–84; H. Hasse, *Mathematische Annalen*, CIV (1931), 495–534; and the paper of Chevalley above.

GLOSSARY

This glossary was written to make available a convenient set of memory-refreshing definitions of the terms and symbols most frequently used in modern algebra. We presuppose an earlier acquaintance of the reader with these terms, and give suggestive but not necessarily rigorous definitions. Our list is better incomplete than bulky, and can be extended by use of the Index.

A *group* is a set of elements closed with respect to an operation such that the associative law and the law of two-sided division hold.

The operation may be given any convenient notation and the terms *multiplicative* and *additive* group refer to said notation. Our definitions will be made for multiplicative groups. They are carried into additive definitions by replacing product by sum, quotient by difference, etc.

A group is called *abelian* if $ST = TS$ for all group elements S and T. Additive groups are usually assumed to be abelian and then frequently called *moduls*.

A *cyclic group* is the set $[S]$ of all powers of a group element S.

The *product* $\mathfrak{H}\mathfrak{K}$ of any two subsets of a group is the set of all elements HK for group elements H in $\mathfrak{H}$, K in $\mathfrak{K}$. Either set may consist of a single group element S and we write $S\mathfrak{H}$, $\mathfrak{H}S$ for such products.

A subgroup $\mathfrak{H}$ of $\mathfrak{G}$ is called *normal divisor* of $\mathfrak{G}$ if $\mathfrak{H}S = S\mathfrak{H}$ for every S of $\mathfrak{G}$. All subgroups of an abelian group are normal divisors of it.

Let $\mathfrak{H}$ be a normal divisor of $\mathfrak{G}$ and call all sets $\mathfrak{H}S$ for S in $\mathfrak{G}$ the *cosets* of $\mathfrak{H}$. They form a group (with respect to the set product operation) called the quotient group $\mathfrak{G}/\mathfrak{H}$.

If $\mathfrak{H}$ is a subgroup of an additive abelian group $\mathfrak{G}$ we call the quotient group the *difference group* $\mathfrak{G} - \mathfrak{H}$.

The *order* of a group is the number of elements in it and is either infinity or a finite integer.

If $\mathfrak{G}$ is a group of permutations on n letters we call $\mathfrak{G}$ a *permutation group of degree* n.

A permutation group $\mathfrak{G}$ on $x_1, \ldots, x_n$ is called a *transitive group* if there exists a permutation in $\mathfrak{G}$ carrying any given x_i into a given x_j.

A *ring* is an additive abelian group closed with respect to multiplication such that the associative law and the two-sided distributive law hold.

A ring is called *commutative* if $ab = ba$ for all ring elements a and b.

If the non-zero elements of a ring form a multiplicative group we call the ring a *division ring*.

An element c of a ring $\mathfrak{A}$ is called a *scalar* of $\mathfrak{A}$ if $ca = ac$ for every a of $\mathfrak{A}$.

The set of all scalars of a ring $\mathfrak{A}$ form a commutative subring called the *centrum* of $\mathfrak{A}$.

If $\mathfrak{B}$ is a subring of $\mathfrak{A}$ and x is a scalar of $\mathfrak{A}$ the set $\mathfrak{B}[x]$ is defined as the set of all polynomials in x with coefficients in $\mathfrak{B}$. We call x an *indeterminate* over $\mathfrak{B}$ if no quantity of $\mathfrak{B}[x]$ is zero unless the coefficients are all zero. Otherwise we call x *algebraic* over $\mathfrak{B}$.

An automorphism of a ring $\mathfrak{A}$ is a (1–1) correspondence $a \longleftrightarrow a'$ of $\mathfrak{A}$ and itself such that $(a + b)' = a' + b'$, $(ab)' = a'b'$.

A subring $\mathfrak{B}$ of $\mathfrak{A}$ is called an *ideal* of $\mathfrak{A}$ if ba and ab are in $\mathfrak{B}$ for every a of $\mathfrak{A}$ and b of $\mathfrak{B}$.

If $\mathfrak{B}$ is an ideal of $\mathfrak{A}$ the difference group $\mathfrak{A} - \mathfrak{B}$ consists of cosets $[a] = a + \mathfrak{B}$ and forms a ring with multiplication defined by $[a][b] = [ab]$. This ring is called the difference ring $\mathfrak{A} - \mathfrak{B}$.

An element 1 of a ring is called its *unity element* if $1a = a1 = a$ for every ring element a.

A non-zero ring element a is called a *divisor of zero* if there exists an element $b \neq 0$ in the same ring such that either $ab = 0$ or $ba = 0$.

An *integral domain* is a commutative ring with a unity element and no divisors of zero.

A *field* is a commutative division ring.

Thus a *field* is an additive abelian group whose non-zero elements form a multiplicative abelian group such that the distributive law holds.

The *characteristic* of a field is the order of its cyclic additive group [1].

We call a field *non-modular* if its characteristic is infinity. Otherwise the characteristic is a prime and we call the field *modular*.

A *linear set of order* n *over a field* $\mathfrak{F}$ is any set equivalent to the additive abelian group of all n-tuples $(a_1, \ldots, a_n)$ with a_i in $\mathfrak{F}$ such that

$$c(a_1, \ldots, a_n) + d(b_1, \ldots, b_n) = (ca_1 + db_1, \ldots, ca_n + db_n)$$

for all a_i, b_i, c, d in $\mathfrak{F}$.

A set of elements $u_1, \ldots, u_n$ of a linear set $\mathfrak{L}$ over $\mathfrak{F}$ are called *linearly independent* in $\mathfrak{F}$ if $a_1u_1 + \ldots + a_nu_n = 0$ for a_i in $\mathfrak{F}$ only when the a_i are all zero.

A linear set $\mathfrak{L}$ of order n over $\mathfrak{F}$ is said to have a *basis* $u_1, \ldots, u_n$ and we write $\mathfrak{L} = (u_1, \ldots, u_n)$ over $\mathfrak{F}$ if the u_i are linearly independent in $\mathfrak{F}$ and every element of $\mathfrak{L}$ is a *linear combination* $a_1u_1 + \ldots + a_nu_n$ with a_i in $\mathfrak{F}$.

The notation $\mathfrak{F}(x)$ means the set of all rational functions of x with coefficients in a field $\mathfrak{F}$. We have already defined $\mathfrak{F}[x]$.

A field $\mathfrak{K}$ is called an *extension* of $\mathfrak{F}$ and we write $\mathfrak{K} \geq \mathfrak{F}$ and call $\mathfrak{K}$ *a field over* $\mathfrak{F}$ if $\mathfrak{F}$ is a subfield of $\mathfrak{K}$.

If ξ in $\mathfrak{K} \geq \mathfrak{F}$ is algebraic over $\mathfrak{F}$ we call the monic equation of least degree with coefficients in $\mathfrak{F}$ satisfied by ξ its *minimum equation* over $\mathfrak{F}$. The corresponding polynomial is called the *minimum function* of ξ over $\mathfrak{F}$, and is irreducible in $\mathfrak{F}$.

A field $\mathfrak{K}$ is called *(algebraic) of degree* n *over* $\mathfrak{F}$ if $\mathfrak{K}$ is a linear set of order n over $\mathfrak{F}$.

A field $\mathfrak{F}(x)$ is called a *simple* (transcendental or algebraic) *extension of* $\mathfrak{F}$ (according as x is an indeterminate over $\mathfrak{F}$ or is algebraic over $\mathfrak{F}$).

A polynomial and the corresponding equation are called *separable* if the equation does not have multiple roots. Otherwise we call them *inseparable.*

A quantity ξ in a field $\mathfrak{K}$ of degree n over $\mathfrak{F}$ is called *separable or inseparable over* $\mathfrak{F}$ according as the minimum function of ξ over $\mathfrak{F}$ is separable or inseparable.

A field $\mathfrak{K}$ of finite degree over $\mathfrak{F}$ is called *separable over* $\mathfrak{F}$ if every quantity of $\mathfrak{K}$ is separable over $\mathfrak{F}$.

A field $\mathfrak{F}$ of characteristic $p < \infty$ is called *perfect* if every a of $\mathfrak{F}$ has the form $a = b^p$ with b in $\mathfrak{F}$. Then there exist no inseparable fields $\mathfrak{K}$ over $\mathfrak{F}$.

Any polynomial $f(x)$ of degree n with coefficients in $\mathfrak{F}$ factors in an existing field $\mathfrak{K} = \mathfrak{F}[\xi_1, \ldots, \xi_n]$ over $\mathfrak{F}$ into linear factors $x - \xi_i$. The field $\mathfrak{K}$ is essentially unique and is called the *root field* of $f(x)$. The fields $\mathfrak{F}[\xi_i]$ are called its *stem fields.* The root field of a separable equation is always a simple extension of $\mathfrak{F}$.

If $f(x)$ is irreducible its stem fields are equivalent over $\mathfrak{F}$ and called *conjugate fields* over $\mathfrak{F}$.

A field $\mathfrak{N}$ of finite degree over $\mathfrak{F}$ is called *normal* over $\mathfrak{F}$ if $\mathfrak{N}$ is separable and contains all its conjugate fields.

If the product of any two elements of a linear set of order n over $\mathfrak{F}$ be defined so that the result is a ring $\mathfrak{A}$ we call $\mathfrak{A}$ *an algebra of order* n *over* $\mathfrak{F}$. When it is a division ring we call it a *division algebra.*

A *commutative division algebra* of order n over $\mathfrak{F}$ is thus a field of degree n over $\mathfrak{F}$.

An *automorphism over* $\mathfrak{F}$ of an algebra $\mathfrak{A}$ of order n over $\mathfrak{F}$ is an automorphism of the ring $\mathfrak{A}$ such that $(\lambda a)' = \lambda a'$ for every a of $\mathfrak{A}$ and λ of $\mathfrak{F}$.

The set of all automorphisms over $\mathfrak{F}$ of a normal field $\mathfrak{N}$ over $\mathfrak{F}$ forms a group called the *Galois group* over $\mathfrak{F}$ of $\mathfrak{N}$. We call $\mathfrak{N}$ *a cyclic field* if the group is a cyclic group.

If $f(x)$ is a separable polynomial with coefficients in $\mathfrak{F}$ and the factorization $f(x) = (x - \xi_1) \ldots (x - \xi_n)$ in its root field $\mathfrak{N}$ over $\mathfrak{F}$ we may apply the automorphisms S of $\mathfrak{N}$ over $\mathfrak{F}$ to the set of ξ_i to obtain permutations $P_S\colon\ \xi_i \longleftrightarrow \xi_i^S$. The set of all such permutations is called the *Galois group* over $\mathfrak{F}$ of $f(x)$.

A field $\mathfrak{F}$ is said to have a *valuation* ϕ if there is a function $\phi(a)$ on $\mathfrak{F}$ to the real number field such that $\phi(0) = 0, \phi(a) > 0$ if $a \neq 0$ in $\mathfrak{F}$, $\phi(ab) = \phi(a)\phi(b), \phi(a + b) \leqq \phi(a) + \phi(b)$.

A field $\mathfrak{F}$ is called *complete* with respect to a valuation ϕ if every regular (with respect to ϕ) sequence of $\mathfrak{F}$ converges on an element of $\mathfrak{F}$.

A valuation ϕ is called *archimedean* if $\phi(n) > 1$ for some integer n.

A valuation ϕ of $\mathfrak{F}$ is *non-archimedean* if and only if $\phi(a + b)$ is at most the maximum of $\phi(a)$ and $\phi(b)$ for every a and b of $\mathfrak{F}$.

The derived field $\mathfrak{F}_\phi$ of a field $\mathfrak{F}$ with a valuation ϕ is the least complete extension of $\mathfrak{F}$ with respect to ϕ.

The *$\mathfrak{P}$-adic number fields* are the derived fields of *algebraic number fields* (fields of finite degree over the rational number field) with respect to their non-archimedean valuations. The quantities of such fields are called *$\mathfrak{P}$-adic numbers.*

An algebra over $\mathfrak{F}$ with a unity element is called *normal* over $\mathfrak{F}$ if $\mathfrak{F}$ is its centrum.

The classical example of a normal algebra over $\mathfrak{F}$ is the *total matric algebra* of all square matrices of a fixed order and elements in $\mathfrak{F}$.

An algebra $\mathfrak{A}$ is called a *normal simple* (or *quadrate*)* algebra over $\mathfrak{F}$ if there exists a field $\mathfrak{K}$ of finite degree over $\mathfrak{F}$ called *a splitting field* of $\mathfrak{A}$ such that the *scalar extension* $\mathfrak{A}_{\mathfrak{K}}$ is a total matric algebra over $\mathfrak{K}$.

We close with a reference to the symbols used to represent sets of elements. They are the following Gothic letters and we put the Latin equivalent with each one: $\mathfrak{A}$ *A*, $\mathfrak{B}$ *B*, $\mathfrak{C}$ *C*, $\mathfrak{D}$ *D*, $\mathfrak{E}$ *E*, $\mathfrak{F}$ *F*, $\mathfrak{G}$ *G*, $\mathfrak{H}$ *H*, $\mathfrak{J}$ *J*, $\mathfrak{K}$ *K*, $\mathfrak{L}$ *L*, $\mathfrak{M}$ *M*, $\mathfrak{N}$ *N*, $\mathfrak{O}$ *O*, $\mathfrak{P}$ *P*, $\mathfrak{Q}$ *Q*, $\mathfrak{R}$ *R*, $\mathfrak{S}$ *S*, $\mathfrak{T}$ *T*, $\mathfrak{U}$ *U*, $\mathfrak{V}$ *V*, $\mathfrak{W}$ *W*, $\mathfrak{X}$ *X*, $\mathfrak{Y}$ *Y*, $\mathfrak{Z}$ *Z*.

* *Cyclic algebras* are such algebras. See the Index for their definition as well as the definition of scalar extension.

INDEX

INDEX